Calculating the World

Magnus Vollset, Rune Hornnes, Gunnar Ellingsen

Calculating the World

The history of geophysics as seen from Bergen

ICO
FAGBOKFORLAGET

Foreword

Ten years ago, the (Norwegian) book *I vinden* was published in connection with the Geophysical Institute's 90th anniversary. Historical material about the Institute has been published in other books previously. For some of us, at least, it was therefore unclear whether new perspectives on our history could be written a relatively short time after 2007. However, during the work with the last book, which focused on institutional history, it became evident that the Institute has a science history that is not well known and definitely deserves to be publicized. This is what the present book, *Calculating the World*, sets out to do. Here, the main topics are concentrated on activities, theories, instruments and interaction between staff at the Geophysical Institute and other geophysical communities.

When the Geophysical Institute was founded, a pioneering era in geophysics was taking place internationally. Several breakthroughs were made, including some very important ones by the small, newly established institute in Bergen that was funded mainly by private donations from local citizens. It is quite remarkable that within this institute, solutions were discovered for scientific problems that have puzzled mankind for generations.

Initially, the Geophysical Institute was an independent body, and its leaders made a deliberate decision not to engage in teaching. Today, as a university department, we do conduct extensive teaching activities in meteorology, oceanography, climate dynamics, biogeochemistry and, recently, renewable energy (related to solar wind and water). Approximately 130 students are affiliated with the Institute today, distributed among our various bachelor's and master's programs. Teaching is vital for our development, and it is a great responsibility to be chosen by young talented people to give them an education. We believe their competence is important for society, and we will work to recruit even more students in the years to come.

Compared to the pioneering era, there are additional and different overall drivers behind the development of the academic community in geophysics today. Modern technology and infrastructure drive a demand for forecasting services of a much greater variety and complexity than before. This presents us with a serious challenge. Even more important is that our global environment and ecological systems are approaching stress levels that can be critical due to the climate challenge, resource shortages, ocean acidification, and environmental pollution, to name the most important factors. Knowledge and its applications in geophysics are central components in finding solutions to develop a more sustainable society.

With increased demands, it is satisfying to observe considerable growth in the Institute's activities over the last few decades. At present, we number approximately 120 employees from 20 nations working in various areas, and we have a productive and dynamic collaboration with the other partners in the Bjerknes Centre for Climate Research (UniRes, the Nansen Centre, and the Institute of Marine Research).

This is indeed worth celebrating, and I take this opportunity to congratulate staff and students at the Geophysical Institute with the anniversary and with this book, which you are a part of. At the same time, I would like to thank the University's Rector, Dag Rune Olsen, for the financial contribution that made it possible to make this book project a reality. Lastly, I am very grateful to the three authors, especially the main author, Magnus Vollset, for their valuable, efficient, and productive collaborative efforts.

Bergen, May 2018

Nils Gunnar Kvamstø
Head of the Geophysical Institute

Acknowledgments

The occasion for writing this book is the 100th anniversary of the Geophysical Institute in Bergen. The authors are grateful for the economic support the Institute has provided, and the exemplary access, support and patience offered by the head of the Institute, Nils Gunnar Kvamstø. Likewise, we wish to thank the members of the book committee, who have provided valuable feedback that has greatly improved the manuscript: Peter Mosby Haugan, Peder Roberts, Svein Atle Skålevåg, and Ellen Marie Viste. We also wish to acknowledge the Department of Archaeology, History, Cultural Studies and Religion (AHKR) for hosting the project, and the research group Health, Welfare and the History of Science for feedback during the process.

Throughout the project, we have been met with open arms by geophysicists who have shared their time and expertise, when giving interviews, answering questions over the phone and by email, commenting on drafts, and providing access to source material. We are grateful for your generous help and guidance: Lennart Bengtsson, Helge Drange, Anton Eliassen, Arne Foldvik, Gunnar Furnes, Tor Gammelsrød, Yngvar Gjessing, Sigbjørn Grønås, Eystein Jansen, Ola M. Johannessen, Thor Kvinge, Erland Källén, Steinar Myking, Karsten Storetvedt, Svein Sundby and Svein Østerhus. We would also like to thank those who have helped us to gain access to images: Marianne Økland Borge, Frank Cleveland, Ilker Fer, Arne Foldvik, Ola. M. Johannessen, Moss Kunstforening, Nils Gunnar Kvamstø, Algot Kristoffer Peterson, Lasse H. Pettersson, and Ola Søndenå.

Finally, we wish to thank the archivists, librarians and others who have helped us to gain access to sources and literature that would otherwise have been unavailable, in particular Ola Søndenå at the University of Bergen Library, who catalogued the archives from the Geophysical Institute while this project was taking place; Berit Bjørndal, who organized access to the troves of documents hidden in the Institute's attic; and Kjersti Dahle at the University's Division of Human Resources for providing access to newer source material. We are also grateful for valuable feedback from an anonymous reviewer.

Bergen, March 2018

Gunnar Ellingsen, Rune Hornnes, Magnus Vollset

Contents

1
Calculating the world

The North Sea, July 9th, 1922: the research vessel *Armauer Hansen* is almost back in Bergen, a city on the west coast of Norway, after two and half months at sea. The ship had visited Lisbon and Casablanca, Funchal on Madeira, the Azores, sailed through the English Channel, and had just left port in Belgium for the last leg across the North Sea when it was hit by a strong storm:

"This time, the situation was more serious. The sea was in uproar, and the vessel was rolling very hard. Nevertheless, the storm passed, the damage was again repaired, and the Armauer Hansen was able to continue by sail. The next day we spent several hours navigating planks and wrecks from ships that had been swept away by the storm."[1]

In his report from the oceanographic cruise, Belgian zoologist Désiré Damas stressed that the scientists had probably not been in any real danger, praising an experienced captain, a dedicated crew and the vessel's sturdy design.[2] Still, the cruise represents a typical activity for geophysics in Bergen: going into the field to make observations, often bringing foreign colleagues to wrestle with the forces of nature. For ten weeks, the thirteen men onboard the 76-foot vessel had lived in close quarters, collecting observations, making repairs and arranging daily lectures for each other. Bjørn Helland-Hansen, professor in oceanography and head of the Geophysical Institute in Bergen, took hydrographical stations, which meant measuring temperatures and collecting water samples at set depths using Nansen bottles and Richter's reversing thermometers. Once the bottles were hauled back onboard, the salt and oxygen content was analyzed by his assistant Olav Aabrek in the small onboard laboratory that also doubled as sleeping quarters. Damas made stops to collect zoological species from different depths using specially crafted nets and bottom scrapers. Ernst G. Calwagen, manager of the Meteorological Observatory in Bergen, observed the clouds and measured temperatures, wind and humidity at the ocean's surface and from the top of the ten-meter-high mast. When the weather permitted, Calwagen did similar observations up to an altitude of 1000 meters using kites borrowed from the Deutsche Seewarte. The report from the cruise included an unbroken temperature curve from an altitude of 1000 meters to 1200 meters below the surface, demonstrating how the atmosphere and the oceans were recognized as a connected whole.[3]

The observations were collected to study the motion of the ocean and the atmosphere with the ultimate goal of making predictions, such as forecasting the weather. During the cruise, Calwagen was responsible for an onboard weather service, guided by reports that came in over the newly installed ship radio. Twice daily the Deutsche Seewarte sent "special notices for the *Armauer Hansen*" at the end of its weather transmissions from Königs Wusterhausen, including supplementary observations from Norwegian stations and line ships. Calwagen used this to compile weather maps and make forecasts using the new forecasting methods being developed back in Bergen. In his report, Helland-Hansen highlighted the field weather service as an important scientific result: "The expedition clearly proved that with the use of modern technology, even a small vessel can be fully informed about the weather situation over large areas, gain a good basis for determining its development and predict the conditions that the vessel will meet."[4]

Another typical example of the strand of geophysics being developed in Bergen took place one year earlier, when Vilhelm Bjerknes hosted an international conference in the summer of 1921 on the investigation of the upper atmosphere. As a local newspaper remarked cheekily:

"Usually, the success of a visit to Bergen depends on the weather, but in this case, the weather is up to the gentlemen themselves to decide. They do not fear a week of rain. After all, meteorologists are not like other people: they prefer unstable weather, so that they can investigate the phenomena of the atmosphere. Sunshine and clear skies are always a disappointment for these learned gentlemen."[5]

Two days later, the same newspaper proudly continued: "In line with its best traditions, the city of Bergen greeted the gentlemen with heavy rainfall, which appeared to please them immensely."[6] According to Jacob Bjerknes's opening lecture, the rain that welcomed the prominent guests was the 34th cyclone family that had hit Bergen in 1921, each having produced three to five days of rain.[7]

While the humid weather had been a constant on the western coast of Norway for millennia, the conference itself was not. The journalists could not hide their patriotic pride in having attracted leading meteorologists from twelve countries all over the world, and at least five different newspapers published detailed presentations of the most renowned guests and speeches from the official receptions. Some even printed daily updates from the scientific presentations and from the negotiations in the International Commission for the Investigation of the Upper Air.[8] In addition to agreeing on a new schedule for internationally coordinated releases of weather balloons over Europe, an activity that had started at the turn of the century but had been interrupted by World War I, the commission agreed on standards for how the observations were to be conducted and presented, and on a plan for compiling and publishing the collective results. Finally, the organizers succeeded in promoting the insights into weather forecasting developed by the Bergen school of meteorology, a landmark in the history of meteorology. By 1930, more than a hundred international researchers had research stays at the Geophysical Institute lasting two weeks or more.[9] In addition to developing new methods for prediction – instruments for observations and methods for calculations – the geophysicists put a premium on spreading their insights to the rest of the scientific community. The 1920s also saw a number of new geophysical research institutes being established in different countries, and Bergen was a center for developing and learning research methods, a center for calculations, and a place where directors could come to learn how geophysical institutes could be organized.

The Bergen school of meteorology, as presented by Jacob Bjerknes on the opening day of the conference, focused on using weather maps to investigate the life histories of air masses: the movements of air with similar temperature and humidity. One of their main contributions was the development of the polar front theory, which asserted that the cyclones that repeatedly hit northern Europe started as waves on the "front" where cold polar air meets the warmer southern air masses. Cyclones develop in families of three to five, where the first and the third are generally the strongest, and each consecutive cyclone follows a slightly more southerly path.[10] These insights had clear and practical implications for weather forecasting. The Bergen school also outlined the physical structure and life cycle of cyclones, and developed the concepts of warm, cold, and occluded fronts, which are still in use.[11] A year later, Jacob Bjerknes and his colleague Halvor Solberg would summarize the findings in the classic paper "Life Cycle of Cyclones and the Polar Front Theory of Atmospheric Circulation."[12] The polar front theory came with a physical model for motions in the atmosphere expressed as a series of equations.[13] The Bergen school has often been presented as the moment when weather forecasting became scientific.[14]

The title of this book, "Calculating the World," summarizes the geophysical vision in Bergen: to gather observations from nature, expressed as numbers; to uncover the laws of physics relevant for how nature moves; and to use mathematics to understand mechanisms and make predictions for the future. However, as this book shows, exactly what and how to observe, what methods to use in the calculations, the purpose of the predictions, and even what scale or part of nature to focus on have changed over time. So have the tools, how the activities are financed and organized, the external conditions, and how the researchers in Bergen have related to each other and geophysicists elsewhere.

The third in the world

When the Geophysical Institute in Bergen opened in 1917, it was the third geophysical institute in the world. The first had been established by Emil Wiechert at Göttingen University in 1898, and focused on seismic methods for studying the interior of the earth alongside a smaller program on atmospheric electricity and northern lights.[15] The second institute was established by Vilhelm Bjerknes at the University of Leipzig in 1913, and focused exclusively on developing methods for scientific weather prediction.[16] While the emphasis in Göttingen was on "geo," understood as the physics of earth or land, the term "geophysical institute" in Leipzig was chosen to signify an emphasis on theoretical analysis rather than observations, and to signal that there were plans for future expansion. From the outset, the Geophysical Institute in Bergen focused on the parts of the planet that are in perpetual motion: the oceans, the atmosphere and, from 1928, the magnetic field. Only in the 1960s did movements in the planet's outer crust become part of the geophysical repertoire, and this period lasted for less than thirty years. Although an instrument for monitoring earthquakes had been installed at Bergen Museum in 1905, the same year the country gained its independence, seismology in Norway was seen as part of physical geography, and grouped with geology.[17] Later, solid-earth physics and marine seismology for use in offshore oil exploration would be categorized as part of the geosciences, not geophysics.

Unlike its two predecessors, the Geophysical Institute in Bergen was not part of a university, but grew out of Bergen Museum's Biological Station. Norway had opened its first and only university in the capital city of Christiania in 1813, when Norway was still part of Denmark and was ruled from Copenhagen. The museum in the center of Bergen was established in 1825. Between 1864 and 1894, under the leadership of physician Daniel Cornelius Danielssen, the Museum's ambitions changed from collecting artifacts and specimens to becoming a center for research. Starting at the turn of the 20th century, the Bergen Museum explicitly aimed at becoming the country's second university, and establishing a geophysical institute was intended as a step in this direction. The architectural drawings for the institute's building, inaugurated in 1928, were labeled "Bergen University."[18] For various reasons, this goal would not be achieved until 1948. The first building project for Bergen University was to expand the Geophysical Institute with two new wings. In the meantime, a limited number of students would receive supervision and attend lectures in Bergen, but had to go to the University in the capital for their exams. Instead of teaching, emphasis was on research. As historian of oceanography Eric Mills has put it, Bergen "became the center of instruction in mathematical oceanography, drawing students from Europe and North America until the Second World War."[19]

Previous histories of geophysics can be divided in three broad categories: the history of geophysical institutes, the history of a single geophysical discipline, and biographies of individual scientists. An example of the first approach is the book written for the 90th anniversary of the institute in Bergen, *I vinden* (*In the wind*), 2009.[20] This genre has generally put more emphasis on administration and institutional aspects, including teaching, with a more superficial glance at the scientific activities. The second approach usually puts scientific practices and the genesis of new insights at center stage, and uses the scientific

disciplines that became self-evident categories after the Second World
War as demarcation. While there are several studies on the history of
weather forecasting, meteorology in general, or oceanography, they
seldom discuss how the geophysical specialties related to each other.
Robert Marc Friedman's *Appropriating the Weather: Vilhelm Bjerknes
and the Construction of a Modern Meteorology* (1989) offers the authoritative account of the Bergen school of meteorology. For oceanography,
we have found Eric L. Mills's *The Fluid Envelope of our Planet: How
the Study of Ocean Currents Became a Science* (2009) to offer the most
comprehensive outlook, covering the period up to around 1960. Both
focus on the science involved: its theory, practice, and conflicts, and the
evolution of disciplinary insights. And both aim to contextualize the
disciplines and how the main insights were developed in specific places
at specific points in time. Finally, the third approach focuses on careers
and lives dedicated to geophysics, highlighting contributions to the
field and often commemorating anniversaries or someone's passing.[21]
While the scope is indeed more narrow, the biographical approach
shows how individual scientists could sometimes switch between
disciplines and move between institutions. When read together, the
biographies show how the scientific collective changed over time.

Building on previous studies and new sources, this book seeks
to give an integrated portrayal of the geophysical sciences pursued
in Bergen. This approach allows us to highlight geophysics' shared
origins in both theory and practice, especially how the ocean and
the atmosphere at one point were seen as two sides of the same coin.
The book shows how the specialties slowly drifted apart into separate disciplines with little or no contact, before reuniting under the
umbrella of climate research only in the past few decades. One part of
the explanation is in how the studies were organized, both locally and
internationally; another is found in how the researchers have related
to the field. Oceanographers in Bergen identified as field scientists,
going on cruises to collect observations that could be analyzed later
back at the office. Meteorologists, too, analyzed observations at the
office, but their main field experience was in developing infrastructure
where others did the observations on their behalf. During and after the
Second World War, meteorologist Carl Ludvig Godske turned his back

on the Bergen school of meteorology and tried to turn meteorology into a field science, with limited success.

The oceans, the atmosphere and the planet's magnetic field have never shown much respect for political borders. Likewise, geophysicists from Bergen have, from the very beginning, promoted and been involved in organizing international collaboration in the study of these phenomena. How they have related to colleagues and findings produced elsewhere has changed over time. Using the geophysical research community in Bergen as a lens, the book aims to show how geophysics has changed over time also in the rest of the world.

Rather than portraying the history of geophysics as seen from Bergen as a story of uninterrupted success, we have focused on what geophysicists have seen as important at different points in time. This contemporary view reveals how the quest to understand the physics of the oceans and the atmosphere has pursued what in hindsight can be seen as both dead ends and spinning threads that were picked up only much later. The authors believe this approach gives a more realistic picture of geophysics and how it has changed over time than selecting only the highlights and presenting this as the norm. We have, however, emphasized how geophysicists in Bergen have influenced or were influenced by science globally. This means that local institutional aspects, including the postwar rise of science administration and education, have been downplayed and only mentioned when having a direct impact on the scientific history. The 1960s is a case in point when, after more than two decades of stagnation, teaching became a reason for finally expanding the staff. This went hand in hand with a sharp increase in research funding from the NATO military alliance.

The chapters

This book is organized chronologically, and shows how early geophysics in Bergen was shaped in the encounter between studies of fisheries, polar exploration, dreams of climate prediction, and a nationalist struggle for independence. Local benefactors were vital in financing the research community that would eventually shape geophysics both nationally and abroad. Bergen was at the center of spectacular expeditions, hosting famous scientists, and pursuing both big ideas and minute details. It was a place to send observations to have them analyzed, and for decades, spending time in Bergen to learn techniques and methods in marine sciences and meteorology was almost considered compulsory.

The following chapter begins with the *Vøringen* expedition in the 1870s, which was funded by the Norwegian parliament (*Stortinget*), and organized mainly by the Bergen Museum. The steamship with seven gentlemen scientists and a crew of thirty-three crisscrossed the North Atlantic over three summers, and marked the beginnings of systematic research in geophysics centered in Bergen. The world's first professor of meteorology, Henrik Mohn, was geophysics personified. His attempts to calculate the motions of the atmosphere, the currents in the sea, and how the oceans and atmosphere were connected were groundbreaking. While Mohn failed to create a school, Vilhelm Bjerknes, Bjørn Helland-Hansen and Fridtjof Nansen and their colleagues succeeded about three decades later. The chapter shows how sustained research efforts in geophysics were born in a larger Scandinavian context, where interests in fisheries, agriculture, new international organizations, polar exploration and calls for Norwegian independence came together with new methodological insights. Together this gave birth to physical oceanography as a standardized and disciplined activity, with classic texts, tools, methods and standards. Physical oceanography was a cornerstone of the Bergen school of oceanography. After the opening of the Geophysical Institute in 1917, and the recruitment of Vilhelm Bjerknes, the community would give birth to a Norwegian Geophysical Association, an influential Geophysical Commission, and the journal *Geophysical Publications*.

Chapter three shows how the new institute sought a balance between scientific curiosity and practical usefulness, collaboration and disciplinary research, as well as the practical struggles of tools and infrastructure. We show how the Bergen school of meteorology was started by a group of young men in Vilhelm Bjerknes's attic, who organized what has been described as a continuous colloquium where weather maps were treated as puzzles to solve by using new methods of analysis. We show how Helland-Hansen tried to develop his institute in light of a geophysical world-view where everything from the depths of the oceans to rays from the sun were connected, and how Vilhelm Bjerknes argued that this was both immature and based on unreliable methods. The chapter also discusses the extraordinary seven-year *Maud* expedition (1918–1925), which brought the now famous geophysicist Harald Ulrik Sverdrup to Bergen. We also investigate failures, such as Helland-Hansen's attempt at establishing a factory, and the curious attempt at commemorating polar explorers by remaking a mountainside overlooking Bergen in the style of Mount Rushmore.

It was not until 1928 that the Geophysical Institute constructed its own building, an event that was celebrated in the journal *Nature*.[22] The fourth chapter shows how this in itself did little to facilitate collaboration between the different sections. In the mid-1930s, Bergen was a meteorological hub for the coordinated exploration of the upper atmosphere over Europe. It was also the headquarters for Helland-Hansen, who finally succeeded in organizing a synoptic study of the North Atlantic that would set an example for international collaboration in postwar oceanography. At the same time, the new division for geomagnetism and cosmic physics soon changed from geomagnetism to particle physics. Toolmakers served as a link between oceanographers and meteorologists, who increasingly grew apart. The chapter focuses on the colorful Odd Dahl, whose adventures included crashing the *Maud* expedition's last airplane in the polar wilderness, crossing the Amazon jungle in a canoe, doing "useless" science trying to climb Mount Everest for the Carnegie Institution in Washington, D.C., developing current meters and particle accelerators in Bergen, and eventually developing Norway's first nuclear reactor. The chapter will also show how leading scientists, including Jacob Bjerknes, Harald Ulrik Sverdrup, Jonas Fjeldstad and Jørgen Holmboe, left Bergen in the lead-up to the Second World War.

Geophysics is more than ideas and individuals; it is also a practice and a way of life. How geophysicists related to the field is the topic of chapter 5. Diaries and letters from the 1920s and 1930s, written by oceanographer brothers Olav and Håkon Mosby and by the institute's secretary, Aagot Borge, portray field life at sea as a variety of experiences. While the perspectives of the scientists and the secretary are different, they share a striking similarity in their attention to the social atmosphere. A friendly comment, a worried captain or a beautiful moon mattered to all. In the field, personalities show themselves from other sides. Meteorologists, however, worked in a different way, and brought home different experiences. The chapter gives a glimpse of how an international experiment with radiosondes and balloons brought out the field scientist in Jacob Bjerknes. It sheds light on some of the variety of challenges that geophysicists met in the field, challenges that formed them as scientists and as people.

When looking at geophysics from Bergen, the Second World War marks a noticeable shift in the international geophysical landscape. From April 9, 1940, to May 8, 1945, Norway was under German occupation. While oceanographers and meteorologists elsewhere proved their worth in the war effort, and entered into a liaison with the military that continued after the war, geophysics in Norway had been at a standstill. Despite the Institute being the cornerstone of a new University, quite literally, the stagnation would last until around 1960. How the oceanographers dealt with no longer being a leading scientific center, but a small institute on the outskirts of a rapidly growing field, is the topic for chapter six. Helland-Hansen's successor, Håkon Mosby, turned to the world: in the 1950s and 1960s, he became one of the most influential actors in organizing postwar oceanography internationally and setting the research agendas. Starting in 1960, he ran NATO's Subcommittee for Oceanographic Research out of Bergen, which financed the development of new instruments and facilitated international studies in areas relevant to both academic and military interests.

Among the meteorologists, Carl Ludvig Godske chose a radically different strategy. In chapter seven, we see how he, during a period of rapid expansion of a global weather forecasting infrastructure, computers and new methods for bringing the dream of calculating the weather ever closer, bade farewell to the Bergen school of meteorology. While Godske was the one to introduce the computer age to Bergen, the computer was initially used as an advanced punch card machine for statistics and applied mathematics. Instead of tomorrow's weather, focus was on how weather behaves in a landscape, educating the masses and promoting an appreciation of nature. Under Guro

Gjellestad, the section for geomagnetism became entangled in heated debates on plate tectonics and the geological history of the planet.

In the last three decades, geophysical research has been organized, conducted and funded in new ways. Interdisciplinary collaboration has become commonplace, quasi-independent research centers were set up outside the formal structures of the University, and new sources of funding became available. In chapter eight, we investigate how these changes came about, and how this shift occurred in many other countries during the same period. We discuss the relative importance of economic conditions, people's ideas, values and "culture" in this transition, and show how the transition taking place in Bergen was facilitated at a number of specific sites.

Starting in the late 1980s and early 1990s, Bergen geophysics has gradually become involved in interdisciplinary climate research. This has brought about changes in both the content of the scientific inquiries and the ways science is conducted. It has also granted Bergen geophysics a high standing in the international climate research community. In 2006, an international evaluation committee established that the Bjerknes Centre for Climate Research was about to become "one of the leading centres worldwide."[23] On the other hand, the turn to climate research has brought geophysicists in Bergen onto the stage of public and political controversy. In chapter nine, we investigate how meteorologists and oceanographers in Bergen became part of the emerging field of climate research, and what this has meant for research questions and strategies. Finally, we ask whether the entry into this new interdisciplinary field has come with a downside: Do the recurring climate disputes and the slow progress in climate policy formation indicate that people have lost faith in the capacity of geophysicists and other climate scientists to calculate the world?

This book is the result of a collaboration between three historians from the University of Bergen: Gunnar Ellingsen wrote chapter 5; chapters 8 and 9 were written by Rune Hornnes; while Magnus Vollset penned chapters 1–4, 6–7 and 10. This book was written primarily with geophysicists interested in the history of their science in mind, but it is our hope that general audiences will also be as fascinated by the geophysical project of calculating the world as we have become.

2
The first Bergen school

On the morning of June 1st, 1876, the 144-foot steamship *Vøringen* left the port of Bergen. Onboard were seven gentlemen scientists and a crew of thirty-three. The vessel had been hired to serve the Norwegian North-Atlantic Expedition, which for three consecutive summers investigated the ocean outside the Norwegian coast, delimited in the south by the Faroe Islands, in the west by Iceland, and in the north by Jan Mayen and Spitsbergen. The mission was to examine "the depth of the sea, its temperature, the chemical composition of its water, the currents prevailing there, both at the surface and in the depths, the nature and geological formation of the bottom, meteorological and magnetical [*sic*] phenomena, and more especially all forms of animal and vegetable life."[1] The expedition was explicitly inspired by the British Challenger expedition (1872–76), which had circumnavigated the earth but not visited the North Atlantic, the sea most vital to Norway.[2]

Norway depended on the ocean as a source of food and work, as a means of transportation and for making the climate habitable. In the 1870s, the fisheries provided 24.9 percent of the country's exports, and employed some 90,000 fishermen.[3] Despite a population of less than two million, Norway possessed the third-largest sailing fleet in the world.[4] Furthermore, the warm ocean currents were recognized as being responsible for making Scandinavia habitable despite its northern latitude. As meteorologist Henrik Mohn and marine biologist Georg Ossian Sars put it in their application for state funding for the expedition: "off our coasts extends a tract of ocean which is the origin and preserver of our existence as a civilized nation; and that expanse of sea being as regards its physical conditions well nigh unknown."[5]

This chapter will show how the expedition with the *Vøringen* was one of the events that made it possible for the coastal town of Bergen in western Norway to establish itself as an internationally recognized center of geophysical research. Shortly after the turn of the 20th century, Bergen became a center for marine research, and two decades later the town's Geophysical Institute gave birth to the renowned Bergen school of meteorology. To a large extent, these later accomplishments have overshadowed the earlier history of geophysics in Norway. In this chapter we will detail how the expedition to the North Atlantic in the 1870s led to the very first attempts at using physics and mathematics to calculate the movements of the ocean and atmosphere. Further, we will show how geophysics in Bergen was built on field expeditions, polar exploration, and national institutions set up to support fisheries research. We will also show how the research programs aimed at calculating the movements of the ocean and the atmosphere were inspired by, and built on, research elsewhere in northern Europe. Finally, the chapter will demonstrate the close links between the ocean and the atmosphere, and how Bjerknes's circulation theorem was at the core of the first Bergen school: the Bergen school of physical oceanography.[6]

A Norwegian *Challenger*

The very first official publication from the *Vøringen* expedition described the physical contours of the ocean, which during the expedition was baptized "the Norwegian Sea": beyond a shallow continental shelf stretching out from the coast, the depths exceeded the sounding gear's capacity of 2000 fathoms (approx. 3700 meters). The ocean was separated by a ridge between Svalbard and Jan Mayen, and the major currents consisted of "two principal tracts, an eastern with the Gulf Stream, as it is called, flowing north, and a western, with the Arctic current, flowing south, along the shores of East Greenland."[7] In the Gulf Stream, temperatures stayed above zero down to about 500 fathoms (900 meters); below the temperature sank to about -1.3° C. In the East Greenland current, the temperature reached zero at depths of only a few fathoms. One of the surprising findings was that the salt content of the two currents seemed more or less identical, suggesting that they were somehow part of the same system, linked beneath the polar ice.[8]

The man responsible for studying the physical conditions of the sea and the atmosphere, and one of the two formal leaders of the expedition, was Henrik Mohn. Mohn had been born and raised in a merchant family in Bergen, but in the 1850s he had moved to the capital Christiania, to study theology at the country's university, the Royal Frederik's University (renamed the University of Oslo in 1939). After building his own telescope, he soon switched from studying the divine to studying astronomy, mineralogy and physics. In 1861, Mohn was employed at the University's astronomical observatory, which also collected meteorological observations. After five years, in 1866, Mohn became the first director of the Norwegian Meteorological Institute, and the world's first professor of meteorology.[9] In 1870, he created a storm atlas with case studies of how the weather developed over time. For each day he drew two pressure charts: one chart depicting pressure variations, and one temperature variations. He also published separate charts showing water vapor content in the atmosphere.[10] His well-received monograph *On Wind and Weather: Fundamentals of Meteorology* (1872) was translated into German, French, Italian, Spanish, Russian, Polish and Finnish.

Mohn's first study from the *Vøringen* expedition examined how various meteorological elements changed as the planet spun around its axis. This was called the diurnal period, the 24-hour rhythm of the atmosphere, and was an approach developed by the Austrian meteorologist Julius von Hann. Hourly observations of wind, atmospheric pressure, temperature, humidity, clouds, precipitation, sea-surface temperatures, and ocean waves, taken from a single moving point at sea, did not lend themselves to a case-study approach. The majority of the 150-page publication consisted of detailed discussions of the instruments and methods used for each kind of observation, such as correcting the wind for being measured on a vessel in motion, followed by almost fifty pages of condensed tables detailing each finding.[11] The analysis was classical climatology, namely calculating the average values for only one meteorological phenomenon and presenting the results as curves in the hope of eventually uncovering patterns. There were no attempts to investigate how the different phenomena were related. The results were relatively meager: the temperatures at sea peaked at around two in the afternoon, and dropped to a minimum between two and three at night. Humidity followed a similar pattern, but about two hours delayed. Pressure had its minimum between three and six in the early morning, a maximum in the afternoon, and what seemed to be a second smaller wave in the evening.

It was in the ambitious analysis of what happened below the surface that Mohn started what was to characterize geophysics in Bergen for more than a century to follow: the quest to identify and calculate the ocean currents. To Mohn, the two main drivers behind currents were wind acting on the surface and the distribution of heat and salt in the sea itself. Beginning in the atmosphere, Mohn used the monthly average distribution of air pressure measured along the Norwegian coast, on Iceland and on Greenland as a proxy for wind. With the help of equations he had developed with mathematician Cato Guldberg some years prior, Mohn calculated the average wind direction and speed.[12] Because of the rotation of the planet, Mohn noted, the currents in the Northern Hemisphere would deviate to the right of the wind. This was at odds with the leading theory by the German geographer Karl Zöppritz, who argued that ocean currents followed the average direction of the wind and developed over geological timescales.

In Mohn's view, currents were also influenced by differences in temperatures and density. He therefore mapped the temperature distribution in different places and at different depths. The method used was to conduct "stations," which meant taking water samples and measuring temperatures at set depths. The stations were part of a total of 32 "sections," series of observations carried out along a virtual line. These crosscuts made it possible to analyze the horizontal and vertical distribution of both temperature and salinity. The observations confirmed that the Gulf Stream entered the Norwegian Sea through the Faroe-Shetland channel, and was compensated by a deeper and colder current going south past Greenland. Through analyzing salt content and water temperatures, Mohn calculated the specific gravity at different depths in the water column. This led him to conclude: "The distribution of the density of the sea-water would thus appear to indicate ascending and descending movements in the ocean."[13] But differentiating between the motions caused by wind and those caused by changes in density, Mohn believed, would be practically impossible.[14]

Mohn then turned to the question of how the ocean influences the atmosphere, an interest directly motivated by working on the world's first mathematical model of the dynamics of the atmosphere. In the two-volume *Studies on atmospheric movements* (1876, 1880), written in French, Mohn and Guldberg had presented the very first attempt at using equations from physics to analyze the motions of the atmosphere.[15] The hope was that identifying the laws governing the dynamics of the atmosphere would aid in predicting storms. In line with contemporary scholars, such as Scottish mathematician and physicist William Thompson (later Lord Kelvin), mathematician Theodor Reye in Germany, and mining engineer H. Peslin in France, the cyclone model relied heavily on thermodynamics, the study of how gases expand or contract with changes in temperature and pressure.[16] The model proposed that the area in front of a moving cyclone consisted of warm and humid air. The heat would make the air rise, and thereby create an area of low pressure. The storm center would then move into this low pressure area, causing wind and, as the humid air rose and got colder, rain. By knowing the values for each variable, it was, in theory, possible to calculate the speed and direction of storms.

Since humidity played such an important part in feeding the storms, Mohn designed his own instrument to measure evaporation from the ocean surface to the atmosphere, and called it an atmometer. The instrument consisted of an open water-proof iron box, which was filled with seawater and suspended from gimbals on the aft of the vessel, reproducing the conditions in the surrounding ocean. Inside the box, Mohn placed a buoy with a stem supporting an evaporation bowl, to be filled so that the water levels matched. By observing how the buoyancy of the dish changed as the water evaporated, Mohn hoped to arrive at a rate of evaporation in different conditions and at different latitudes, which in turn could be used to aid weather prediction. At sea, however, the design soon revealed major deficits: on clear days, the water in the instrument would heat up much more than the surrounding ocean, increasing the rate of evaporation to unrealistic levels. On the other hand, as the water in the dish evaporated, the saltiness increased, which meant that the rate of evaporation was unrealistically reduced. In rough seas, rain and water splashing in and out of the instrument made the readings wholly unreliable. Installing a thin metal roof and placing the instrument in a wooden box suspended from gimbals helped somewhat (see Figure 1), but the apparatus was still susceptible to both soot and vibrations from the vessel's steam engine. Finally, the lack of waves in the small evaporation dish meant that the rate of evaporation differed fundamentally from that of the open sea: "The evaporation of sea-water as measured with our apparatus cannot, therefore, represent the full

Figure 1

Mohn's improved atmometer, built to
measure evaporation from the ocean
surface, was used in both 1877 and 1878.
An evaporation dish (a) was suspended by
a buoy (b) floating in water gathered from
the ocean surface. To counteract heating
and vibrations, the apparatus was put in a
wooden box and suspended from gimbals
on the aft deck of the vessel. The thin iron
roof above was added to avoid splashing
and soot. (Mohn 1883: 138)

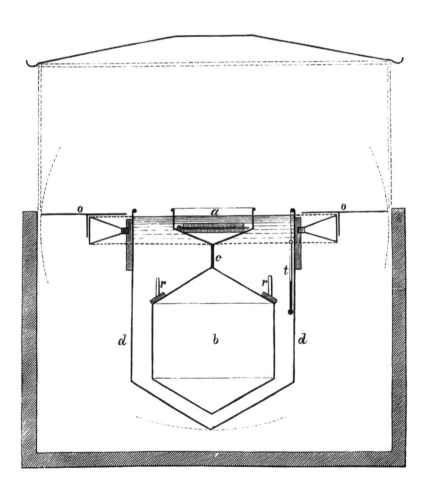

amount of evaporation at the sea-surface, but has, at most, only a relative value," Mohn concluded.[17] Although Mohn's improved atmometer ultimately failed, it illustrates both an early interest in the interaction between the sea and the air, and how making precise measurements of natural phenomena was recognized as a core challenge for geophysical field science as early as the late 1870s.

Accuracy was identified as key also when analyzing the water samples gathered during the expedition, and this introduced another set of challenges. The analyses were done by chemist Hercules Tornøe, who took part in the last two years of the expedition. Like the chemists on the *Challenger* expedition, he examined the content of air, carbonic acid, and salinity, as well as the specific gravity of the samples taken at the various stations. Although the potential implications of the findings received little attention in his report, it seems the hope was that the variations in chemical composition could be a key to determining and predicting both the movements of fish and the dynamics of the currents themselves. In accordance with contemporary science, emphasis was on procedure, the methods and equipment used, and presenting the results in long tables. Since the variations were minuscule, Tornøe argued, field science demanded a higher degree of accuracy than assumed by previous investigators. Chemical analysis could only be achieved in laboratory settings:

"The methods previously devised for determining the amount of salt in sea-water by which all observations with this object in view were taken on board, should unquestionably cease to be adopted, since they will not suffice, with the greatest care even, to attain the high degree of accuracy requisite for detecting such minute differences that are frequently found to occur."[18]

Over the following decades, Tornøe would continue to develop new methods for analyzing the chemical content of seawater, and in 1895 he presented the world's first instrument for determining salinity using electrical conductivity.[19] Tornøe's electrical salinity meter consisted of two glass cylinders connected at the bottom by a tube. Each cylinder contained an electrode. By running an alternating current through a water sample, and measuring the resistance between the electrodes, one arrived at the conductivity, from which one in turn could calculate the salinity. However, the temperature of the water sample had to be measured with an accuracy of 0.1° C, which could be problematic at sea. Furthermore, the instrument was prone to short circuiting. Even when it did work as intended, the instrument was far from ideal: since the resistance differed greatly with temperature, one had to use various tables and cumbersome calculations to arrive at the salinity, and the results were relatively inaccurate.

In 1896, Tornøe developed yet another instrument for in situ analysis of salinity, this time based on optical refraction.[20] This instrument consisted of a glass container with two chambers. One chamber was filled with fresh water, the other with the seawater sample. One could deduce salinity by shining a strong light through the instrument and measuring the difference in refraction between the two liquids. However, by the time the results and corresponding reference tables were published in 1900, the results were already regarded as too inaccurate for the instrument to be considered relevant. The refraction instrument had a brief afterlife determining the alcohol content of beer before eventually being abandoned.[21] Still, developing accurate instruments would remain a core practice for geophysicists in Bergen.

The motivation for analyzing the water content was twofold. First, the idea was that the chemical composition of the water determined the movement of different fish species. Identifying and predicting the changes in the water masses was, therefore, seen as key to improving the fisheries. Second, having identified density of the water as a product of salt and temperature, one could understand, and ideally predict, the currents that were responsible for the Norwegian climate. However, climate depended on a set of interrelated factors that were difficult to separate:

"Thus we come to the heat of the sun, the conformation of land and sea, and the rotation of the earth as principal factors in determining climate, atmospheric pressure, temperature, evaporation, precipitation, winds, and ocean currents. Any changes in these fundamental conditions will involve changes in the currents of the ocean, which may exert a great influence on climate."[22]

Despite being at the cutting edge of geophysical research, the *Vøringen* expedition did not spawn continued research efforts. On the contrary, for a long time Mohn's efforts were more or less forgotten. In 1926, geologist and geographer Werner Werenskiold pointed out that Mohn "had not received the recognition he should have."[23] Admitting that Mohn's oceanographic works were not as clear as they could have been and that some of the measurements and calculations were wrong, Mohn had still been the first to explain the dynamics of ocean currents, the relationship between currents, density and the earth's rotation, and a number of their implications. Several of Mohn's findings, Werenskiold argued, were later rediscovered and presented as new breakthroughs.[24]

Only recently have historians asked why Mohn's efforts were overlooked by his successors. Historian of oceanography Eric Mills has argued that Mohn was ahead of his time, and that in the 1880s and 1890s he had no ready audiences for his quantitative approach. Oceanography was at the time an "extensive" geographical science, exemplified by the British *Challenger* expedition (1872–76) circumnavigating the globe, rather than an "intensive" geophysical science aiming to understand the dynamics in a smaller region.[25] Historian Vera Schwach has pointed out that the *Vøringen* expedition was overshadowed by Norwegian polar exploration in the decades that followed.[26] The polar expedition put emphasis on discovery, adventure and daring sportsmanship, and gained much more public attention than the *Vøringen*, which had been a scientific expedition for and by the elite. Except for a stately dinner to celebrate the return of the scientists, organized by the Bergen municipality, the *Vøringen* expedition received limited public attention.[27] And as Mohn returned to the daily hassle of running a national meteorological institute, he had no direct scientific successors following his lead in studies of the oceans. Nor were there any oceanographic institutes ready to assume the mantle. When the next generation of geophysicists revisited Mohn's work in the first decade of the 20th century, they pointed out flaws stemming from inaccuracies in the measurements and mistakes in the relation between the wind direction at the ocean surface and the direction of the resulting current.[28]

Finally, the *Vøringen* expedition was financed in a way that for more than a decade would effectively block the development of a prolonged program for oceanographic field research in Norway. In the initial application, Sars and Mohn had recommended constructing a research vessel, estimated to cost between 160,000 and 190,000 NOK, with running costs of around 20,000 NOK for each of three seasons.[29] This proposal gained support from the Ministry of the Interior, the navy, and the directors of the Norwegian Geographical Survey. The budget was set to 246,500 NOK for constructing the vessel and financing the fieldwork. This was an investment equivalent to roughly 60 percent of the annual cost of running the country's only university. However, after a two-day debate in the Norwegian parliament, it was decided that the expense was to be divided in equal parts to be paid over three years. Although this did not reduce the actual costs, it meant that the expedition was forced to rent a vessel rather than building one, leaving Norwegian oceanography without a dedicated research vessel for more than two decades. Also, it is important to remember that the geophysical research was but a small part of the expedition: only three of the 28 volumes of reports published between 1880 and 1901 were on geophysics. In comparison, 22 volumes contained zoological studies describing species found during the expedition.[30] This is also how the expedition is remembered among Norwegian oceanographers, as an early beginning for physical oceanography but with a strong emphasis on studies of life in the oceans.[31]

Although the *Vøringen* expedition was a single event, Mohn was to define the research interests that would characterize geophysics in Bergen for more than 140 years to follow: the ocean currents, the movements of the atmosphere, the interactions between atmosphere and ocean, and the earth's magnetic field.[32] The research was approached by making observations expressed in numbers, using calibrated instruments and stringent methods. The numbers were in turn used to develop and test physical equations aimed at calculating the dynamics involved. Mills has summarized the expedition as follows: "Flawed though it may be in modern terms (...), Mohn's work on the Norwegian Sea was a *tour de force* of conceptualization, analysis and computation, aptly described as 'the first attempt at treating all known forces together in a single picture of ocean circulation.'"[33] Still, as we will see below, rather than being a disconnected antecedent, Mohn's work was an explicit inspiration to the next generation of geophysicists.

Into the Arctic unknown

While Mohn had introduced the geophysical vision of calculating the world, it was Fridtjof Nansen who brought geophysics to the masses, and pride to the nation, by combining scientific observations with daring expeditions, adventure, and discovery of the unknown.

After passing the entrance exams to the University in the capital, Kristiania, in 1880, Nansen chose zoology because of his love of nature, sports, hunting and the outdoors. Two years later, the 21-year old Nansen went on his first polar expedition with the sealing ship *Viking*. On Mohn's instructions, and using Mohn's instruments, Nansen measured ocean temperatures off Greenland. This was his first venture into physical oceanography.[34] After the expedition, Nansen took a position as junior conservator at Bergen Museum, which included working with the zoological material collected by the *Vøringen* expedition, as well as an extended research stay at the famous naturalist Anton Dohrn's Stazione Zoologica in Naples.[35] Four days after handing in his doctoral thesis on the structure of the central nervous system of hagfish, he departed to head an expedition across the Greenland ice sheet on skis.[36] The expedition was criticized by contemporaries as suicidal. It started with Nansen and five other men being dropped off in rowing boats 20 kilometers off the uninhabited east coast, with the closest civilization on the west coast some 600 kilometers away, across unknown lands. There was no turning back. Before reaching shore, the members of the expedition were caught by a current and forced to camp on a floating sheet of ice which carried them 380 kilometers south, after which they had to begin rowing back north along the coast before even making landfall. In his bestseller *The First Crossing of Greenland* (1890), Nansen explained that he believed in preparation, not luck, and that looking over one's shoulder would have been a waste of precious time.[37] In his obituary of Nansen, Swedish oceanographer Vagn Walfrid Ekman pointed out: "The Greenland expedition was by necessity primarily a powerful physical performance, but all chances for science were utilized. Prepare well, but never waste your time looking back."[38] After returning, Nansen married the singer Eva Sars, the youngest sister of Georg Ossian Sars, Mohn's collaborator in organizing the *Vøringen* expedition.

Nansen's next adventure into the Arctic unknown, the spectacular *Fram* expedition (1893–96), was motivated by science, adventure, and Henrik Mohn. The goals of the expedition were both to become the first to reach the geographical North Pole and to conduct scientific observations from this northern terra incognita. Planning began after Mohn, in 1884, held a lecture at the Norwegian Academy of Science and Letters on the naval exploration vessel *Jeanette*, which had sunk some 300 nautical miles north of the Siberian coast in 1881. Three years later, wreckage of the vessel was found near the southwestern corner of Greenland. This, Mohn explained, indicated that below the white spot on the map, an ocean current must flow from east to west underneath the polar ice. Nansen's idea was to construct a vessel that could withstand the ice, man it with a small and well-trained crew, stock it with food and equipment for five years, and use the current to drift with the ice over the North Pole. He recruited the shipwright Colin Archer to construct a 127-foot schooner with a fortified, wide rounded bottom and almost no keel, so that rather than being crushed, the ship would be pushed up by the ice. Financed by donations, state support and private funds from the sale of books about the Greenland expedition, and equipped with instruments and instructions on how to make oceanographic measurements from Mohn, the expedition set sail in the summer of 1893.

After 18 months in the ice, the vessel *Fram* was still unharmed, but Nansen calculated that the ice drift was so slow and unpredictable that it would take five years to reach the pole – if at all. By then, the expedition would have run dangerously short of food. Instead, the team was instructed to prepare to return once conditions allowed, while Nansen and expedition member Hjalmar Johansen would make a dash for the pole on skis, equipped with three dogsleds carrying food and canoes for the return over the open sea. Skiing northward on ice that was drifting south, the two reached 86°13.6'N on April 8, 1895, more than 300 kilometers further north than any man had ever set foot. There, with temperatures below -30° C, exhausted, and facing frozen ridges stretching into the horizon, they decided to turn south. In early August, the two reached the northern parts of Franz Josef Land, hunting polar bears, walruses, foxes, seals and birds for sustenance.

They built a stone hut and hunkered down, sharing a sleeping bag to keep warm. Only on New Year's Eve, after nine months alone on the ice, did the two finally change from formal to informal pronouns when addressing one another.[39] The following summer, after surviving 15 months alone in the Arctic, the two ran into an expedition led by English explorer Frederick Jackson at Franz Josef Land. Jackson had arranged the expedition after having been rejected as a crew member on the *Fram* because he was not Norwegian. After some time, Jackson took Nansen and Johansen back to northern Norway where they were reunited with the *Fram* and its crew.

Historian Narve Fulsås has described the return of the *Fram* in 1896 as the first modern media event in Norway, celebrating adventure, sportsmanship, science, and nationalism in a period when Norway was positioning itself as a polar nation and working to gain independence from Sweden.[40] Nansen's popular two-volume book *Farthest North* (1897) sold in huge numbers, and Nansen's lecture series in Europe and North America drew large crowds.[41] In comparison, the 28 volumes of reports after the *Vøringen* expedition published in Norwegian and English between 1880 and 1901 had been purely scientific and aimed at a specialized audience. After returning, the first independent research fund in Norway, the Nansen Fund, was established in his honor. Nansen was also awarded a professorship in oceanography at the University in Kristiania, where he edited a total of six volumes of scientific results after the *Fram* expedition, *The Norwegian North Polar Expedition 1893–96* (1900–06). In addition to observing that the *Fram* and the ice drifted at an angle 20–40° to the right of the wind, which Nansen argued was due to the earth's rotation and later led to the mathematical model known as the Ekman spiral, their most important observation was that the Arctic Ocean was over 3000 meters deep. Nansen would later express regret that his oceanographic observations from this remote area, like Mohn's observations from the Norwegian Sea, lacked accuracy. Unlike that of the *Vøringen*, the *Fram* expedition was celebrated as a national event, but neither of these expeditions would be recognized by later generations as founding moments for geophysical research: their observations did not have the necessary accuracy to be useful when new methods for calculations were introduced.

Going high and low with the circulation theorem

Instead of Henrik Mohn or Fridtjof Nansen, it was Vilhelm Bjerknes whom later generations came to highlight as a founding figure for geophysics in Norway. In 1880, Bjerknes had passed the introductory exam to the University in Kristiania with Nansen as his fellow student. When Nansen chose zoology, Bjerknes chose to specialize in physics in order to continue his father Carl Anton Bjerknes's studies on "action at a distance": how objects can influence each other without touching. While in Göttingen in 1856, mathematician C.A. Bjerknes had asked: "If two bodies move in a liquid, will they not then, through the liquid as intervening medium, mutually affect each other's movements? And will not an observer who sees the bodies, but not the liquid, believe that he is witnessing action at a distance?"[42]

The problem of action at a distance was especially relevant to the study of electromagnetism, and in the 1860s, Carl Anton Bjerknes had arrived at formulas that described how two bodies moving in a liquid mutually affect each other's movements. In 1881, when Vilhelm was 19, the two demonstrated an instrument at the first International Exposition of Electricity in Paris, which illustrated the effect and won a Diplome d'Honneur.[43] The experiment showed how two harmoniously pulsating balls submerged in a fluid acted upon one another as though they were electrically charged, attracting or repelling each other. The main purpose of the display was to serve as an analog for electromagnetism. Vilhelm Bjerknes continued his studies of electric waves at Heinrich Hertz's laboratory in Bonn in 1890–91, and developed equations for describing how electromagnetic waves penetrated and resonated with materials, including a method to determine wave lengths, and using them to transmit and receive electric oscillations.[44] This contributed to the development of wireless telegraphy. In 1893, Bjerknes was hired as a lecturer in mechanics at Stockholm's Högskola (Stockholm University College), and two years later he was appointed professor.

Vilhelm Bjerknes's entrance into geophysics began with a lecture in 1897 entitled "On a Fundamental Theorem of Hydrodynamics and Its Applications Particularly to the Mechanics of the Atmosphere and the World's Oceans."[45] There he introduced what was to become the theoretical basis for his research program, the circulation theorem. The theorem consisted of a set of basic equations with which real fluids could be calculated into the future, one step at a time. While Mohn's work had been one model for the movement of cyclones and one model for calculating ocean circulation, Bjerknes's approach was more fundamental: instead of presenting a complete theoretical package, Bjerknes's theory was more of an open-ended research program, which he invited colleagues to use as they saw fit.

A major drawback with earlier equations used to describe the circulation of fluids, in particular Hermann von Helmholtz's theorem of vorticity conservation from 1858 and William Thompson's (Lord Kelvin) theorem on the conservation of circulation from 1869, was that they applied to ideal fluids with no viscosity, density or friction.[46] This meant that circulation and vortex motions were eternal and could neither commence nor perish. Since what characterizes movement in the atmosphere and the ocean are vortices that come into being, exist for a while, and then vanish, the classic equations were relatively useless for studies outside of laboratory settings. In the 1897 lecture, Bjerknes argued that Helmholtz and Kelvin's equations were but special cases of two more general theorems which also encompassed temperature and pressure. Seven years later, Bjerknes presented his vision for using physics to calculate the world: if you know the state of the atmosphere or oceans in sufficient detail, and know the laws governing their motion, you can calculate the state of the atmosphere some time into the future.[47] Repeating the procedure faster than nature meant that you could see into the future.

To chemist and oceanographer Otto Pettersson, Vilhelm Bjerknes's colleague at Stockholm's Högskola, the circulation theorem arrived at the perfect moment. For more than a decade, Pettersson had been involved in studies of the hydrographic conditions in the Skagerrak and Kattegat straits.[48] The project, financed since the late 1870s by the Swedish Academy of Sciences and private donations, was tasked with finding out why the Bohuslän herring had returned after being gone since 1810. As in Norway, the economic importance of the fisheries translated into government willingness to pay for measures that were expected to benefit the industry. In 1870, three-quarters of the Swedish population depended directly on agriculture and fishing, and by the outbreak of war in 1914, historian Helen Rozwadowski has estimated that 50 percent of the population still depended on this livelihood.[49]

By 1890, the research had grown from extensive surveys to recruiting fishing boats for taking simultaneous water samples from different points in the Skagerrak-Kattegat area. The idea was that the measurements would give a synoptic view of the currents, and that this could be used to create a "weather map" of the sea. These maps could in turn help to identify correlations between hydrographic features of seawater, mainly temperature and salinity, and biological features, such as the appearance and disappearance of fish.[50] Bjerknes's circulation theorem was the tool Pettersson needed to turn the measurements into a dynamic image of ocean circulation.

The exploration of the ocean currents in the Skagerrak and Kattegat went hand in hand with international collaboration, and by 1894 Pettersson had recruited researchers in Denmark, Scotland, Germany and Norway. The goal was to replace the fishermen in fishing boats with trained assistants using research vessels and identical equipment. By the time Bjerknes had arrived at the circulation theorem, Pettersson and colleagues had started to arrange international meetings with the goal of institutionalizing the research collaboration. In 1902 this resulted in the establishment of the International Council for the Exploration of the Sea (ICES) with a secretariat in Copenhagen.[51] At both the first organizing meeting in Stockholm in 1898 and the next meeting in Gothenburg the following year, Vilhelm Bjerknes was invited to present his work.[52]

In Norway, too, the circulation theorem was first seen as a key to understanding what happened below the surface of the oceans. In 1898, as Fridtjof Nansen was working on analyzing the scientific results from the *Fram* expedition, he wrote to Bjerknes and described how his vessel had suddenly lost almost all momentum and refused to answer her helm. Tales of this mysterious "dead water" phenomenon being observed near coasts were not unheard of, always appearing with no forewarning and eerily forcing the ship to an unexplained halt, but the stories were often dismissed as simply figments of the imagination. Could Bjerknes have an explanation for the mysterious "dead water" phenomenon?

"In my reply to Prof. Nansen I remarked that in the case of a layer of fresh water resting on top of salt water, a ship will not only produce the ordinary visible waves at the boundary between the water and the air, but will also generate invisible waves in the salt-water fresh-water boundary below; I suggested that the great resistance experienced by the ship was due to the work done in generating these invisible waves."[53]

Bjerknes had his pupil, Vagn Walfrid Ekman, test the idea, and Ekman soon confirmed Bjerknes's suggestion both theoretically and in water tank experiments. When Nansen's bid to place the headquarters of ICES in Kristiania failed, and he instead became host to the quasi-independent "Central Laboratory" paid for by the Norwegian government, he recruited Ekman as his first assistant.[54] The laboratory worked to improve oceanographic instruments, produced standardized water used to calibrate chemical analyses of salinity, and developed new methods for analyzing results.

Prior to this, Ekman had also applied Bjerknes's circulation theorem to another phenomenon Nansen had experienced during the *Fram* expedition: wind makes the ice cover drift, but at a 20–40° angle to the right of the wind direction. Instead of merely noting, like Mohn had, that the angle was caused by the rotation of the earth, Ekman developed equations describing how the surface wind direction deviates, not just at the surface, but further into the deep.[55] Today this phenomenon is known as the "Ekman spiral." Ekman also developed a propelling current meter for direct observation of ocean currents, and derived an empirical formula for the mean compressibility of seawater as a function of pressure and temperature.[56]

In 1905, Vilhelm Bjerknes was invited to Columbia University in New York to give a series of lectures comparing hydrodynamic fields with electric or magnetic fields, a direct continuation of his father's research.[57] This resulted in an invitation to the newly inaugurated Carnegie Institution of Washington, where he gave a lecture on weather prediction as a problem of mechanics and physics and presented a program for making weather forecasting "an exact science."[58] The program had two parts: diagnosis and prognosis. In principle, diagnosing the three-dimensional state of the atmosphere at a certain time could be done through observation of pressure, density, temperature, humidity, and wind velocity. The second part, prognosis, was "simply" to use the laws of nature to calculate how the air masses would move a short time ahead. The results, in turn, created a new initial state, which again could be used to calculate one step further into the future. The lecture resulted in a grant from the Carnegie Institution, which was then renewed regularly until the Second World War.[59] The scholarship allowed him to hire assistants, many of whom became leading figures in Norwegian geophysics.[60] The laboratory study of action at a distance as an analogue for electromagnetism was thus abandoned in favor of studying movements in the real world. The idea of motion propagating layer to layer became a tool for describing, and eventually calculating, the dynamics of the atmosphere.

The attempts to use the circulation theorem on the atmosphere had begun immediately after Bjerknes's lecture in 1897, when Swedish student Johan Wilhelm Sandström volunteered to take on the task of working out tables and graphical methods for applying the circulation theorem to actual meteorological observations. For this, he used observations from higher altitudes, information that had not been available to Mohn and Guldberg. The upper air is where clouds form and move, where weather "begins." It is also a part of the atmosphere where topography and other local variations have less impact than near the ground.[61] Bjerknes later stated that it was Sandström who had convinced him to focus his research on how it could be applied to meteorology and oceanography.[62] When the funding from Bjerknes dried up, Sandström went on to work as an assistant to oceanographer Otto Pettersson, where he continued to use the circulation theorem to calculate ocean currents, illustrating how Bjerknes's approach was immediately used in the studies of the dynamics of the atmosphere and the oceans alike.[63] When Bjerknes returned from the United States, he hired Sandström as his first Carnegie assistant. Sandström also came along in 1907, when Bjerknes accepted a professorship at his alma mater, the University in Kristiania, and he was Bjerknes's coauthor for the first volume of the two-volume *Dynamic Meteorology and Hydrography* (1910–1911).[64]

The potential for using the circulation theorem was also recognized in Bergen, again in relation to the oceans rather than the atmosphere. In 1900, 22-year-old Bjørn Helland-Hansen was hired by the Biological

Station in Bergen to study physical oceanography, and on Nansen's suggestion he traveled to Stockholm to learn how the circulation theorem could benefit this work (Figure 2).[65] In 1900, Fridtjof Nansen had started practical investigations of the Norwegian Sea, and it was probably the presentations to what would become ICES that caused him to send his assistant to visit Vilhelm Bjerknes in Stockholm in 1901. Three years earlier, Helland-Hansen had had to cancel his medical studies after having lost two of his fingers to frostbite in a snowstorm during Norwegian physicist Kristian Birkeland's first Aurora Borealis Expedition from Tromsø.[66] With amputated fingers, Helland-Hansen was forced to change to the study of natural sciences, and starting in 1899, he studied with the Danish oceanographer Martin Knudsen in Copenhagen and helped analyze the findings from the Danish *Ingolf* expedition (1895–96). Knudsen was another of the architects behind ICES, and in 1901 he published the hydrographical tables that would establish a new and more accurate international standard for water density based on temperature, salinity and depth (pressure).

Based on what he learned in Stockholm, Helland-Hansen and Sandström developed a simplified set of equations, which could be used to analyze the dynamics of the ocean currents.[67] The field of research, at the time called hydrography and physical oceanography interchangeably, focused on the physical properties of the world's oceans, especially temperature, salinity, pressure, visibility, and oxygen content. These properties were interesting for two reasons: visibility and oxygen content determined minimum conditions for life, while temperature, salinity and pressure gave the relative weight of the water masses and could be used to calculate ocean currents. In 1912, Helland-Hansen pointed out, just as Mohn had done four decades earlier, that the oceans cover two-thirds of the surface of the planet, and the volume of the oceans is thirteen times the volume of dry land, but the lack of direct access to the depths meant that the physical properties were still relatively unknown.[68]

The core concept that Helland-Hansen and Sandström set out to establish was the "solenoid," a graphical representation of ocean dynamics based on static observations. These were based on hydrographical stations taken in sections, which was the same observation

Figure 2

Solenoids, as a function of pressure and
weight. (Sandström and Helland-Hansen
1903: 8)

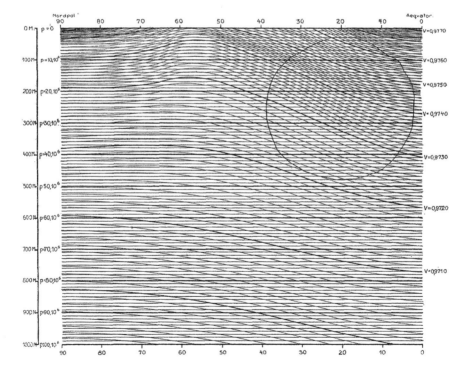

method Mohn had used. The chemical analysis used to calculate the precise weight of the water in each sample was similar to those of Tornøe. The next step, however, was novel: to plot the weight onto a grid, with pressure as the other variable. Graphically, each combination of pressure and density gave a specific shape, the solenoid. The shape of the solenoid showed the direction and intensity of the ocean currents (see Figure 2).[69] Compared to Mohn's calculations, the graphical representation of the solenoid was both simpler to produce and more immediately understandable.

The Bergen school of oceanography

When Helland-Hansen returned to Bergen, the institutional landscape for oceanographic research was relatively straightforward. The main scientific institution in town was the Bergen Museum.[70] In 1867, the museum had moved to a new building on Nygårdshøyden, a hill overlooking the center of town, and in the late 1890s, it was expanded with both a botanical garden and two new wings. Since 1892, the Bergen Museum had had its own biological station, and in October 1900 the Norwegian Fisheries Board was established as the continuation of a permanent commission to investigate and improve Norwegian fisheries that had been established in Christiania in 1864.[71]

Inspired by what he had learned in Stockholm, Helland-Hansen made the questions of how and where to make oceanographic observations and how to chemically analyze the salt content, as well as the method for calculating the solenoids, part of the international maritime research courses arranged by the Bergen Museum. The courses were important in making Bergen an internationally recognized center for oceanographic research and establishing the Bergen school of oceanography. Each course lasted between seven and ten weeks, and consisted of lectures and practical research covering a wide range of topics, including biology, zoology, plankton, fish populations (until 1906), physical oceanography, and bottom deposits of the fjords and ocean. In addition, there were weekly excursions with research vessels, and the observations and specimens collected were used in supervised laboratory work. In total, about 150 foreign and 15 Norwegian students attended the courses, with the number of participants each year varying between 9 and 23. The courses lasted until 1914, when they were cancelled because of the outbreak of war in Europe.[72] Several of those attending the courses went on to have careers in marine studies.

The courses also became a meeting place for polar explorers. In 1908 Roald Amundsen attended Helland-Hansen's course in physical oceanography as part of learning how to maneuver the vessel *Fram* to explore the Arctic Basin.[73] Nansen, whose professorship the same spring was changed from zoology to oceanography, looked forward to new and more precise measurements from the area. But when Amundsen, in 1909, heard that both Robert Peary and Frederick Cook claimed to have reached the North Pole, he instead headed south. Only after he had reached Madeira, west of Morocco, did he send a telegram to Nansen and explain to the crew that the expedition in reality was heading to Antarctica to beat Robert Falcon Scott to the South Pole. On December 14, 1911, Amundsen's expedition became the first to reach the pole. Scott's expedition reached the South Pole a month later, but perished on their way back. The tropical waters also damaged the vessel *Fram* beyond repair.

The publication that was most instrumental in establishing Helland-Hansen and Nansen at the forefront of physical oceanography was their influential 1909 monograph, *The Norwegian Sea*, based on cruises out of Bergen in 1900–1904 (Figure 3). In the book, the two oceanographers sought to establish a new beginning for their field. The *Vøringen* expedition and, especially, Mohn's mathematical model for calculating ocean circulation were recognized as heralding a new era in physical oceanography, but, they argued, the results from the expedition ought to be ignored.

"Owing to the imperfect methods of the time, the determinations of the specific gravity and the salinity of the sea-water were not sufficiently accurate, and by computing the densities from these determinations MOHN has arrived at misleading results as to the horizontal and vertical distribution of the density of the Norwegian Sea.

"MOHN'S discussion of his results and of the circulation of the Norwegian Sea is the first attempt to adopt a wholly mathematical method

Figure 3

The currents in the Norwegian Sea, as presented by Bjørn Helland-Hansen and Fridtjof Nansen in their classic *The Norwegian Sea* (1909). Their scheme of the general circulation "could hardly be bettered today." (Dickson and Østerhus 2007: 56).

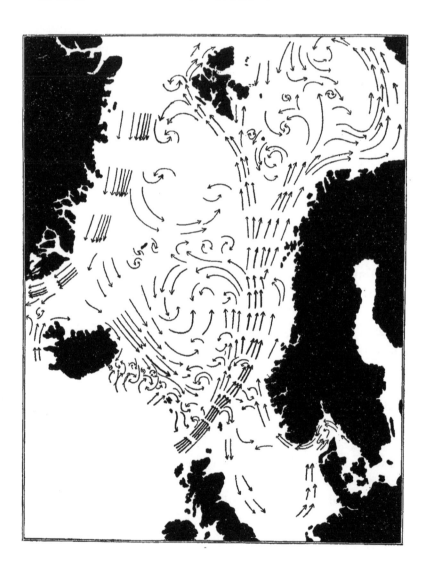

for the calculation of oceanic circulation; and it marks the beginning of a new era in Physical Oceanography. But as his discussions are based upon imperfect observation-material, it is not to be expected that the results would be very correct."[74]

The devil was in the details: the salinity in the Norwegian Sea only differed between 34.85‰ and 35.20‰. In order to investigate changes over time, or calculate the details in the currents, the chemical analysis needed for the salinity was an accuracy of 0.05‰. This standard was set by ICES in 1902, and had not been technically possible before Knudsen's work on analyzing the samples from the Danish *Ingolf* expedition (1895–96). It was "of no use," Nansen and Helland-Hansen argued, to make observations unless this accuracy was achieved, and that past measurements had to be abandoned was a logical implication, which applied both to the *Vøringen* expedition and Nansen's *Fram* expedition.

"The inaccuracy of these observations is especially fatal, as they are the only ones taken in a region where the difficulty of repeating them is particularly great. They will hardly afford material for future expeditions, equipped with modern and more perfect instruments and methods, to decide whether any changes in the physical conditions of the sea have taken place in that interesting region (...) One or two really accurate and trustworthy determinations of the salinity or the specific gravity of the bottom-water of the North Polar Basin would have been of far more value than the hundreds of observations actually made of this water during the *Fram* Expedition, at the expense of much valuable time. (...) The *quality* of the observations, especially in this deep part of the sea, is of much greater importance than their *quantity*."[75]

Increased precision relied on precision instruments. Of particular importance was the "Nansen bottle," named after its designer, Fridtjof Nansen. The device consisted of a metal cylinder that was lowered on a cable into the ocean. After it had reached the required depth, a weight was dropped down the cable. When the weight reached the bottle, the bottle tipped upside down, sealing the water sample from contamination. By fixing a series of bottles to the same cable, it was possible to take nearly simultaneous samples from different depths. The Nansen bottles were equipped with a reversing mercury thermometer, which made it possible to get an accurate temperature reading exactly as the sample was made. The Nansen bottle was the most commonly used oceano-graphic tool in the world until the late 1960s, when it slowly began to be replaced by the Niskin bottle, a water sampler where the reversing cylinder is replaced by a tube with spring-loaded caps on each end.

The combination of observations with increased precision and Bjerknes's circulation theorem made it possible to make completely new predictions. The measurements of bottom-water from the Polar Basin taken during Nansen's *Fram* expedition were a case in point: although the individual observations were too inaccurate to be used for studying ocean currents or fluctuations, the aggregated average of hundreds of bottom-water measurements gave a salinity of 35.10‰. In comparison, the bottom-water of the Norwegian Sea had an average salinity of 34.92‰. Helland-Hansen and Nansen reported: "From this fact important conclusions may be drawn with regard to the shape of the deep basin of the Norwegian Sea and that of the North Polar Basin, these basins being probably separated by a transverse ridge."[76] However, if the measurements from the North Polar Basin were on average 0.18‰ too high, the ridge did not exist.[77]

Finally, in addition to mapping ridges on the ocean floor and mak-ing a break with the past, the new accuracy seems to have given the oceanographers a strong belief that a new age was just around the cor-ner: an era where their science could offer a window to the future. "We think that these discoveries give us the right to hope that by continued investigations it will be possible to predict the character of climate, fisheries, and harvests, months or even years in advance."[78] In Hel-land-Hansen's chapter on physical oceanography in *The Depths of the Ocean* (1912), the second classic work in oceanography that came out of Bergen before the First World War, written by Johan Hjort and John Murray, he stressed that "very much more work will have to be done before we shall be able to solve many of the interesting and important problems relating to the great ocean waters."[79] But the intent was the same: through precise observations and physical calculations, it would, eventually, be possible to calculate the dynamics of the world's oceans.

The research would not have been possible without institutional backing or access to a research vessel. The fieldwork was conducted on the research vessel *Michael Sars*, a 125-foot steamer named after

G.O. Sars's father. The vessel belonged to the Norwegian Fisheries Board, which was headed by Johan Hjort. Hjort had begun medical studies at the University in Kristiania, apparently to please his father, at same time as Fridtjof Nansen. On advice from Nansen, who told him to "follow his dreams," Hjort switched to zoology. In 1892, at the age of 23, Hjort earned a doctorate from the University of Munich. Before moving to Bergen he had been a research fellow with the Norwegian Fisheries Investigations and for three years he was the director of Norway's first biological station in Drøbak.

When the Norwegian Fisheries Board opened in 1900, the Biological Station in Bergen had already been operational for eight years. Already during the *Vøringen* expedition, Mohn had suggested establishing a permanent station for investigating the biology of the coastal waters. A decade later, Nansen had revived the plans when he was working as a junior curator at Bergen Museum. The idea was to establish an aquarium where one could study living creatures, in contrast to the Bergen Museum, which he criticized for mainly storing dead specimens.[80] As Nansen quit his job for the expedition across Greenland, and then moved to Kristiania, it was his successor, Jørgen Brunchorst, who moved the plans of a biological station forward.

Inspired by Anton Dohrn's Biological Station in Naples, and similar institutions in Britain, Brunchorst suggested establishing "a seal basin, public aquarium and biological station."[81] This was prompted by public discussions on establishing a local zoo, which Brunchorst argued would be both expensive and complicated. An aquarium with local species, seals and a research institution attached, however, would solve the craving for public entertainment, facilitate scientific studies, and probably be self-financing. Brunchorst's plea resulted in a number of donations from local elites, and after a year's work, the Biological Station opened in September 1892. The proposed seal park opened in 1898, and soon attracted an average of 10,000 visitors annually. The following year, it was merged with the aquarium and established as an independent institution, relying solely on donations and ticket sales, while the research at the attached Biological Station was made part of the Bergen Museum. Rather than simply a source for financing research, the aquarium and the seal park must be seen as part of a larger civilizing project where the small, educated elite sought to educate and entertain the population through public engagement. It was this combination that created the institutional framework for oceanographic research in Bergen.

Already from that first autumn, the Biological Station attracted international researchers for prolonged residencies. The organized research residencies were mimicking the biological station in Naples, where both Nansen and Hjort had done laboratory work for their doctorates. The first taxidermist and caretaker at the Biological Station in Bergen, Nils B. Glimme, was often referred to as "a Nordic Lo Bianco" after the famous taxidermist and zoologist Salvatore Lo Bianco in Naples.[82]

The initial research aimed at benefiting the fisheries, with experiments such as hatching eggs from salmon and trout, testing different bait, and studying the spawning of cod, as well as examining the biological development of herring, trout, and salmon. The station also arranged international oceanographic summer courses, which started as a collaboration between the Biological Station and Hjort's Fisheries Board. The Bergen Museum contributed most of the teaching, while Hjort organized access to research vessels. The Fisheries Board also had its own research department, and in 1903 the Biological Station was transferred from the Museum to the Fisheries Board in an attempt to consolidate the town's marine research. This arrangement lasted only three years. When the Fisheries Board was reorganized as a permanent Fisheries Directorate in 1906, the Biological Station returned to the Bergen Museum under the leadership of Bjørn Helland-Hansen, and Hjort pulled out of the oceanographic courses altogether. Instead, as biologist Egil Sakshaug and oceanographer Håkon Mosby have pointed out, Hjort established his own Bergen school of fisheries studies focusing on developing new fisheries, improving fishing gear, and eventually studying the dynamics of fish populations.[83]

Over time, the two institutions would grow apart. Increasingly, the research at the Fisheries Directorate focused on fishing equipment and studies of fish populations for the benefit of the fisheries, while the zoological, biological, physical and chemical oceanography remained at the Biological Station. In addition to having different objectives and catering to different audiences, the personal relationship between

Hjort and Helland-Hansen started to deteriorate. The conflict intensified after the publication of *The Norwegian Sea* in 1909, and almost led to Helland-Hansen moving to Kristiania.

At the end of 1909, Nansen suggested that the University in Kristiania should establish a professorate for Bjørn Helland-Hansen.[84] By then, Nansen was a national hero, initially for his polar expeditions, and then for using his fame to argue the case for full independence from Sweden. After the union was dissolved in 1905, Nansen was the one sent to Denmark to convince Prince Carl and Princess Maud (later King Haakon VII and Queen Maud) to accept the Norwegian crown. From 1906 to 1908 he headed Norway's first diplomatic mission in London, where he became a close friend of King Edward VII. After the publication of *The Norwegian Sea*, Helland-Hansen had been offered a position in Berlin, and Nansen argued that the only way to keep him in the country was to offer him a professorship. In Bergen, he had no academic community to speak of, and the monograph proved that the two worked well together.

Historian Harald Dag Jølle has suggested that Hjort took Nansen's claim that Bergen lacked a relevant academic community as a personal insult.[85] Furthermore, Hjort was opposed to Nansen's role as a public figure. By letting wider audiences know they were working on methods to predict fisheries and climate, for instance, Nansen was doing science a disservice by creating expectations they could not possibly fulfill. On top of this, Hjort was in the middle of a divorce which was finalized in 1912, and both Nansen and Helland-Hansen sympathized with his wife, Wanda von der Marwitz, and not with Johan Hjort.[86]

Nansen's proposal also rubbed the board of the Bergen Museum the wrong way. The museum was working on plans to expand to become Norway's second university, and apothecary Johan Lothe, who was both a member of the museum's board of directors and an elected Member of Parliament, argued that the prospect of losing their most excellent scientist would be a huge blow to this plan. Instead, he suggested Helland-Hansen should be given the same title and conditions at the museum that he would otherwise receive in Oslo. This would both be cheaper to the state and avoid crippling centralization. In addition, *The Norwegian Sea* had proved that collaboration at a distance was not a problem, and the railway between Bergen and Kristiania that had opened in 1909 would only make it easier to arrange meetings. Eventually the suggestion from Lothe was approved in Parliament in March 1911. Helland-Hansen's salary as professor included a free pass to the railway.

Helland-Hansen's professorship sparked consolidated efforts to improve the facilities for oceanographic research in Bergen. The Biological Station was dated, and its research hampered by local pollution.[87] This led to ideas of establishing an oceanographic institute in its own building with laboratories, auditoriums, aquariums, modern equipment and a research vessel.[88] Again, the call was made to the citizens of the town, appealing to local patriotism, pride and fear of losing out to international competitors: "What is at stake is for Bergen Museum to strengthen its position as a center for modern oceanography in Northern Europe; it really boils down to a question of to be or not to be, if we want to be on the map among equivalent institutions elsewhere."[89]

Within months, the call had received 90,000 NOK in donations, which was used to build the specialized research vessel *Armauer Hansen*.[90] Part of the reason for the successful fundraising was that the Bergen Museum and its scientists were immensely popular among the locals. In 1887, Brunchorst had revived Norway's oldest popular science journal, *Naturen*, and the following year he was appointed director of the museum. In 1901, the museum began hosting five public lectures every week that soon drew audiences of up to 20,000 per year.[91]

After a failed attempt at organizing an oceanographic research cluster in collaboration with the Fisheries Directorate, the board of the Bergen Museum decided that the zoological research at the Biological Station should be made part of the museum, while Bjørn Helland-Hansen would be given a special geophysical institute to be opened in 1917. That the oceanographers at the Geophysical Institute focused solely on the physical aspects of the oceans was a way of avoiding further conflicts with both the Fisheries Directorate and the Biological Station: marine biology or fisheries studies were never considered research interests for the Geophysical Institute.

The first thing Helland-Hansen did when he was told he would finally get his own institute was, with the aid of Nansen, to reach out to Vilhelm Bjerknes. Since 1912 Vilhelm Bjerknes had been head of the Geophysical Institute in Leipzig, where he focused on using the circulation theorem for analyzing upper-air observations and improving weather forecasting.[92] The outbreak of war in 1914 had hampered the work greatly: "One after another the doctoral students were conscripted, five to perish. (...) Then came the turn to the Institute's assistants. Eventually there were at work only two women and my two still very young Norwegian Carnegie assistants."[93]

When Helland-Hansen sent Nansen to ask Bjerknes to return to his native country to head a section on meteorology at his Geophysical Institute, Bjerknes was not too difficult to persuade. In Germany, his students were dying faster than he could train them, there were food shortages, his family and friends in Norway were worried, and no one knew how long the war would last. In neutral Norway, his family wanted him closer to home, and he was being positioned to establish a physical institute once the planned Bergen University came to fruition.

Drifting apart

The term "the Bergen school" is most commonly used to refer to the Bergen school of meteorology, where Vilhelm Bjerknes was the leader responsible for making weather forecasting modern.[94] In his recent triple biography of key figures in the history of meteorology, historian James R. Fleming presents Vilhelm Bjerknes as the person who "put meteorology on solid observational and theoretical foundations."[95] In *Weather by the Numbers* (2008), historian Kristine C. Harper draws a direct line from Vilhelm Bjerknes to numerical weather forecasting using computers.[96] Vilhelm Bjerknes has also been celebrated on numerous occasions, and his narrative of the origins of the Bergen school is still central to the identity of many Norwegian meteorologists. However, in *Appropriating the Weather* (1989), historian Robert Marc Friedman showed that this origins story was part of the package that the Bergen schools members promoted to their students.[97] Starting the narrative earlier, and ending later, historians Yngve Nilsen and Magnus Vollset have discussed the different strategies employed to spread the

Bergen school methods and narrative, in particular as they were given responsibility for mass education of meteorologists during the Second World War.[98] This chapter, however, has tried to show that the study of the dynamics of the atmosphere and the dynamics of the oceans were more closely connected than they have often been portrayed.

The first use of Vilhelm Bjerknes's circulation theorem in Bergen was in the study of ocean circulation, and the method was central to how, before the First World War, Bergen established an internationally recognized center for oceanographic research. It was this first Bergen school – the Bergen school of oceanography – that brought Bjerknes to Bergen to continue his work on turning weather forecasting into a science through calculations. This, in turn, built on research interests that had been established by Henrik Mohn during the *Vøringen* expedition in the 1870s. However, while Mohn had been the first to attempt to calculate the world through combining numerical observations with a mathematical model based on physics, he failed to attract followers. Bjerknes's more fundamental circulation theorem, on the other hand, was treated as an open invitation to further research.

Bjerknes's circulation theorem arrived at the precise moment when exploration of the sea was about to become institutionalized at an international level. Eight different countries signed up for ICES in 1902, and it is today recognized as the world's oldest intergovernmental science organization. Bjerknes was invited to present his work at the preparatory meetings, and in Norway it was in the study of the oceans that the theorem was first used.

While it was the Biological Station that in 1917 gave birth to the Geophysical Institute, this had not been the only oceanographic research institute in Bergen. But after a falling out around 1910 between the head of the Fisheries Directorate, Johan Hjort, and the head of the Biological Station, Bjørn Helland-Hansen, the two institutions kept their distance. Instead of cooperation, research interests at the Biological Station, and later at the Geophysical Institute, would specialize in pure physical oceanography to not risk having to collaborate. When Bjerknes set up the Bergen school of meteorology, research on the dynamics of the atmosphere and the ocean would soon drift apart.

3
Useful curiosity

The arguments for establishing the Geophysical Institute in Bergen in 1917 encompassed both "groundbreaking scientific works" in the past, and the prospect of future "scientific and practical importance." The institute should provide "the best possible conditions for work within each individual discipline and for collaboration between them."[1] The researchers thus had to strike a balance between scientific curiosity and producing practical benefits, and between interdisciplinary collaboration and research interests specific to each discipline: physical oceanography and meteorology. This chapter will investigate how the meteorologists and physical oceanographers in Bergen found their balance between interdisciplinary collaboration and disciplinary unilateralism, and between curiosity-driven research and the goal of giving something practical back to the community that had made the institute possible in the years until it got its own building in 1928.

This period saw the birth of the landmark Bergen school of meteorology, by far the most famous event in the history of geophysics in Bergen. Under the leadership of Vilhelm Bjerknes, a group of young forecasters developed new methods and concepts for practical weather forecasting that over time would spread globally. The Bergen school is today generally recognized as the cradle of modern, scientific weather forecasting. However, as this chapter will show, forecasting was merely the tip of the iceberg: under the leadership of Bjørn Helland-Hansen, the geophysicists in Bergen used a variety of methodological approaches, with studies ranging from climate change to how the geophysical world was connected on an interplanetary scale. Bergen also became a center for calculation, analyzing results from routine measurements, and dramatic expeditions into the Arctic unknown, and it attracted researchers from near and far.

While both meteorologists and physical oceanographers took the dream of calculating the world as their starting point, there were different opinions as to what this meant in practice. Paradoxically, the Bergen school of meteorology was part of what made the different geophysical disciplines start growing apart. Before getting to the Bergen school and the other scientific activities, the chapter will begin by outlining the small but disorganized institutional landscape for geophysics in Norway, and show how the institute was but one of several initiatives aimed at bringing the different geophysical disciplines together.

Institutionalizing Norwegian geophysics

When the Geophysical Institute in Bergen opened in 1917, a handful of institutions concerned with physical movements in nature already existed. In the capital of Kristiania (renamed Oslo in 1925), the Norwegian Meteorological Institute had been established in 1866, the University's Astronomical Observatory that had opened in 1834 had its own department for geomagnetism, and Fridtjof Nansen ran an oceanographic laboratory. In Bergen, the Fisheries Directorate had been established in 1900, but the study of physical conditions in the oceans had by 1917 mainly been taken over by the Biological Station established in 1892. The Bergen Meteorological Observatory had been inaugurated in 1905, and was responsible for upper air observations, climatological studies, and issuing storm warnings for the fisheries. North of the Arctic Circle, the Norwegian state had, in 1912, turned the private observatory for northern lights at Haldde near Alta into a full-time institution, and there were proposals to establish a geophysical institute in nearby Tromsø.[2] Finally, a small observatory for geomagnetism was opened at Dombås in central Norway in 1916. However, the institutions and the small geophysical research community recognized that they lacked overarching coordination.

In August 1917, nine leading geophysicists met at Vilhelm Bjerknes's mountain cabin at Geilo and established a framework for coordinating geophysical research in Norway: the Norwegian Geophysical Association and the permanent Geophysical Commission.[3] The founders were Vilhelm Bjerknes, physicist Olaf Devik, oceanographer Bjørn Helland-Hansen, head of the Norwegian Meteorological Institute Theodor Hesselberg, physicist Ole Andreas Krogness, meteorologist and later oceanographer Harald Ulrik Sverdrup, and physicist Sem Sæland. Bjerknes's son Jacob and his first Carnegie assistant, the Swede Johan Sandström, also attended the meeting, and the group decided to invite Fridtjof Nansen to be a member. After the US had entered the World

War, in a time of shortages and a naval blockade, Nansen had been dispatched on a diplomatic mission to Washington to negotiate food supplies. Finally, Bernt J. Birkeland, manager of the Meteorological Observatory in Bergen; Oskar E. Schiøtz, professor in solid earth physics; astrophysicist Carl Størmer; and aurora borealis researcher Lars Vegard were immediately invited to be members of the association. In 1917, these were all the full-time geophysicists in Norway.

The purpose of the association and commission, in the words of Helland-Hansen, was "to secure collaboration between the individual institutions and researchers in the country, so that there could be a more intense and concerted effort around larger goals than has so far been possible for geophysical research in this country."[4] The association arranged annual meetings, and starting in 1920 it published a scientific journal, *Geofysiske Publicationer* ["Geophysical Publications"]. The commission had one permanent member from each of the three main geophysical institutions in Norway: the Meteorological Institute in Kristiania, the Geophysical Institute in Bergen, and the planned Geophysical Institute in Tromsø, which was inaugurated in 1918. The commission was soon expanded with two members elected by the association.

While the association was a forum for social and scientific gatherings, the role of the commission was to set national research agendas through coordinating research and resources. The commission met twice a year to discuss accounts and budget proposals from the major geophysical institutes, commenting on and occasionally amending the budgets before submitting them to the government. It also had a say in recruitment to permanent positions.[5] Moreover, the commission administered a research fund for geophysical research that had been set up in memory of physicist and northern lights researcher Kristian Birkeland, who had died in June in Tokyo on his way to celebrate his 50th birthday in Norway. As historian Yngve Nilsen has argued, the commission entailed moving the center of gravity in Norwegian geophysics from the Meteorological Institute, established in 1866, to the Geophysical Commission.[6] It was the geophysical commission that decided that the country's oceanographic and geomagnetic research should be directed from Bergen.

The purpose of the institute in Bergen was to bring together physical oceanography, meteorology, geomagnetism, and cosmic physics. These research avenues were seen as interesting in themselves, and for the purposes of prediction, interconnected. When the initiator, Bjørn Helland-Hansen, presented his plans to the board of the Bergen Museum in 1916, his main arguments were strong research traditions, including in polar research; that geophysics offered benefits to society and commercial applications; and that geophysics was a relatively inexpensive avenue for the young nation to build scientific prestige on the international stage. Geography also mattered: Norway had a long coastline and the warm currents made it possible to live farther north than anywhere else on the planet, which gave permanent access to the zone for northern lights. Whether studying the ocean, fjords, glaciers, or the polar north, Bergen was the natural base. Meteorologically the territory was climatologically diverse, and was positioned at the crucial boundary between "arctic" and "equatorial" air masses.[7]

At its inauguration in the spring of 1917, the Geophysical Institute consisted only of Helland-Hansen's section for physical oceanography. In March 1917, the board of the Bergen Museum agreed to establish a professorship for Vilhelm Bjerknes, who was then working as a professor in Leipzig, and in the fall, the Institute was expanded with Bjerknes's section for dynamic meteorology. Decisions regarding the institute as a whole were made by Helland-Hansen, who remained the head of the institute until 1947. Otherwise, the professors had full autonomy. As will be detailed shortly, the research environment was expanded with an independent weather forecasting unit in 1920, while a section for geomagnetism was established when the institute erected its own building in 1928.

The Geophysical Institute started out in the oceanographic laboratory in Joachim Frieles Gate 1 behind the Bergen Museum, but after only a year, it was clear that the premises were too small. The oceanographers moved to the old biological station, which had been abandoned after pollution in the neighboring Puddefjord had made it impossible to keep biological specimens alive. At the same time, the meteorologists relocated to a villa in Allégaten 33 some 500 meters away. When the new biological station opened at Herdla in 1922, the old station was sold to the navy, and the oceanographers moved to an office building in Lars Hilles gate 17, some 400 meters north of the meteorologists. In 1924, the meteorologists moved to Kalfarveien 59, about a kilometer outside the city center.[8] That the two departments were located in different parts of town impeded collaboration.

The Bergen school of meteorology

The most famous episode in the history of geophysics in Bergen is, by far, the brief life of the Bergen school of meteorology. In the history of meteorology, the Bergen school under the leadership of Vilhelm Bjerknes is seen as a beginning, a revolutionary event in which weather forecasting became scientific.[9] It has been regarded as "the unification of meteorology,"[10] "the premise for modern meteorology,"[11] and as where "the construction of a modern meteorology" took place.[12] The Bergen school has been acclaimed as "the beginning of meteorology as an exact science,"[13] and as where weather forecasting became "a universal science."[14] The Bergen school has been presented as a story about the turn of scientific meteorology from the laboratory to the atmosphere,[15] and as where atmospheric science was invented.[16]

In reality, the many histories about the Bergen school of meteorology consist of two separate narratives. The first focuses on Vilhelm Bjerknes, a physicist's journey into the world of meteorology, and the dream of calculating the weather. As we saw in the previous chapter, this theoretical aspiration began decades before Bjerknes arrived in Bergen. The circulation theorem, a research program aimed at predicting the movement of gases and fluids in the real world through the use of equations from mechanics, hydrodynamics, and thermodynamics, was announced in 1897 when Bjerknes was a professor in Stockholm. The research program set out to treat the real world as an initial value problem: first one needed sufficient insight into the present (initial) state of the atmosphere, and then one needed sufficient insight into the laws governing their development. The laws made it possible to calculate a new initial state a short time into the future. If one could repeat the procedure faster than nature, one could see into the future.

Before Helland-Hansen and Nansen recruited him to Bergen in 1917, Bjerknes had dedicated his efforts to applying his research program to the atmosphere. Starting in 1905 he received annual grants from the Carnegie Institution in Washington, which allowed him to hire assistants. As a professor in Kristiania from 1907, he had prepared two volumes of his larger work *Dynamic Meteorology and Hydrography: Statics* (with Johan Wilhelm Sandström, 1910) and *Kinematics* (with Theodor Hesselberg and Olaf Devik, 1911). In 1913, he moved to Leipzig where he analyzed and published observations from upper-air investigations,[17] and began assembling a three-dimensional image of the atmosphere expressed as equations. Bjerknes continued developing the theoretical fundamentals for calculating the weather in Bergen, which we will return to later in this chapter. However, these efforts have been overshadowed by the second narrative of the Bergen school of meteorology, in which the theories inspired a new practical approach to weather forecasting.

The second narrative of the Bergen school of meteorology concerns the development of new concepts, models and practices in synoptic weather forecasting, which started in Bjerknes's attic in Allégaten 33 in Bergen in the summer of 1918. This practical enterprise took place as a continuous colloquium. It was inspired by, but in practice quite detached from, the theoretical work on calculating the world through numbers and equations. When investigating the history of geophysics as seen from Bergen chronologically, it is this practical turn that stands out in the years of the Bergen school of meteorology.

Vilhelm Bjerknes arrived in Bergen with his Carnegie assistants Jacob Bjerknes, his son, and Halvor Solberg in the fall of 1917. As early as the following summer they established a temporary weather forecasting service, and it was this service that would eventually become the Bergen school of meteorology: responsible for identifying the fronts on the weather maps, developing a model for the life cycle of cyclones, and making Bergen a mecca where weather forecasters from all over the world would go to learn new methods.

The forecasting was set up as a response to the Great War, which had led to a halt in weather telegrams from the British Isles, the Faroe Islands and Iceland.[18] Given Norway's position on the coast of the

Atlantic, and with most rough weather arriving from the west, these observations were crucial for weather forecasting. The lack of forecasts was especially critical for the fisheries and the farmers: despite Norway being a neutral party in the ongoing conflict, the war hindered imports and there were emerging food shortages. For the first four years in Bergen, Vilhelm Bjerknes's salary came from private donations, and providing timely and accurate weather forecasts was both a way for the Geophysical Institute to fulfill its goal of being useful and for Bjerknes personally to offer something tangible to his benefactors.[19]

After reaching out to the Norwegian Navy, which operated manned signaling stations all along the coast, Bjerknes arranged to have them telegraph updated weather observations thrice daily. The stations were equipped with protractors, which he claimed made it possible to determine the wind direction with an accuracy of only five degrees.[20] It did not take long before the observations gave new clues to the inner workings of the atmosphere: "[The observations] showed a striking parallelism from station to station, until suddenly a new wind direction would dominate. This place of convergence or divergence would move from observation time to observation time."[21] In Leipzig, Bjerknes's team had produced the beginnings of a three-dimensional model of cyclones, characterized by "surfaces of discontinuity" between air masses with different temperatures, a model that was further developed by his son in Bergen in what became the first official "Bergen school" publication.[22] The model suggested that a storm center would be preceded by a "steering line" to the south of the path of the cyclone. The shifts in wind direction indicated that these structures were hundreds of kilometers long. "We could thus extend our maps out on the ocean using the lines we strongly suspected belonged to the clockwork of the cyclones. This gave us hope that it would be possible to identify cyclones before they hit land."[23]

However, in order to triangulate the center of the cyclone and hopefully calculate its speed and direction, a denser net of observations was required. Bjerknes traveled to the capital and had his colleague from the Geophysical Commission, Professor Sem Sæland, who was an elected member of Parliament, set up a meeting with Prime Minister Gunnar Knudsen. After fifteen minutes, where Bjerknes demanded that the number of telegraphing weather observations stations be increased from 9 to 90, he received 100,000 NOK to establish a trial weather forecasting service for the summer of 1918: "The hope, as I expressed it, was that the observations would show us the true face of the weather, and I referred to the portraits in newspapers being made up by dots; ten dots do not give a physiognomy, but ten thousand can give the characteristic wrinkles and lines that make a face recognizable."[24]

The first summer, for three months, trial forecasts were made by Halvor Solberg and Jacob Bjerknes, who were given offices at the Meteorological Observatory in Bergen. The main observations came from 70 stations along the coast, from Kristiansand in the south to Namsos in the middle of the country, and the forecasts were distributed to 1300 telephone and telegraph stations along the same coastal strip.[25] The daily forecasts were recognized as remarkably successful, and the trial was allowed to continue the following summer. However, in 1919 the forecasts were made in the attic in Bjerknes's residence in Allégaten 33. The argument, put forward by Bjerknes's colleagues in the Geophysical Commission in a letter to the Ministry of Church Affairs and Education, was the beneficial exchange of insights between the practical weather forecasters and Bjerknes's own theoretical investigations into the physical laws governing the motions of the atmosphere.[26] Bjerknes himself never drew weather maps, but Tor Bergeron, one of the first newcomers to join the weather forecasters in Bergen and recognized as a founding member of the Bergen school, later recalled the professor frequently turning up in the map-room with eyes gleaming with expectation, "Are there any new discoveries tonight?"[27]

The birth of the Bergen school did not come without conflict. In addition to doing upper-air observations with kites and weather balloons and calculating climatological statistics, the state-run Bergen Meteorological Observatory had since 1906 issued storm warnings for the fisheries. Its leader, Bernt Johannes Birkeland, did not appreciate newcomers infringing on his turf. For more than a decade, Birkeland's storm warnings were issued based on local experience and changes in atmospheric pressure. In comparison, the Bergen school consisted of inexperienced youths who approached weather forecasting by identifying air masses, air with similar temperatures and humidity, and used a model of ideal cyclones founded in mechanics and thermody-namics.[28] The concept of air masses came from Bjerknes's theoretical work, as did the conviction that in order to understand the atmosphere, all parameters must be taken into account.

When Bjerknes suggested that weather forecasting should become a permanent feature with state support, making the Observatory's storm warnings obsolete, Birkeland was furious. Moving the fore-casting away from the Observatory, Birkeland pointed out, had merely been a practical solution for a practical issue: in late spring 1919, Bjerknes had invited eleven students for a three-month course on weather forecasting at his home in Allégaten, and when the summer forecasts had begun, it was impractical to try to fit everyone into the Observatory. As Birkeland put it:

"The offices at the Observatory were not suitable, and in order not to ruin the whole course, I agreed to transfer all the weather forecasting for the three months to professor Bjerknes's institute. This is the true reason for the transfer. Bjerknes has abused my great benevolence in an impermissible way, in that he, against all assumptions, has hung on to the weather forecasting at his place."[29]

Birkeland's objections were ignored. In July 1920, the forecasting unit was made a permanent and autonomous unit under the Meteoro-logical Office: Vervarslinga for Vestlandet ["the Weather Forecasting Office for Western Norway"] based in the attic of Allégaten 33, under the leadership of 22-year-old Jacob Bjerknes. His fellow forecasters – Halvor Solberg, Svein Rosseland (who, when it turned out he was not very talented at drawing weather maps, was soon sent to Niels Bohr in Copenhagen to study physics instead), and the Swedes Carl-Gustaf Rossby, Erik Björkdal and Tor Bergeron – were all in their twenties. The forecasters never returned to the Observatory, and collaboration did not resume until Birkeland in 1921 accepted a new position at the Meteorological Institute in Kristiania, and was replaced by another Swedish meteorologist, Ernst G. Calwagen.[30]

The three-month course in 1919, which was one of several, was but one strategy for spreading what in letters, reports and publica-tions increasingly were referred to as "the Bergen school methods." Bjerknes went on lecturing tours to Kristiania, Stockholm, Lund and

Copenhagen in order to recruit candidates; he arranged international conferences, and invited meteorologists for extended stays as resident scholars. By 1928, about 60 meteorologists from 20 countries had had shorter or longer stays at the weather forecasting unit.[31]

Bjerknes had two reasons for proselytizing: first, he feared the lack of trained forecasters would soon mean an end to the weather forecasting enterprise.[32] According to a report from Bjerknes written in the fall of 1919, only eight people in the world fully understood the "Bergen school methods," all of them former assistants or colleagues.[33] Older forecasters, like Birkeland, often had their own methods based on experience, and Bjerknes was not optimistic for the prospect of conversion. Instead, he sought students about to finish their educations in physics and mathematics, in order to recruit young candidates who would grasp the physical fundamentals of weather analysis. However, when new positions were advertised, long hours for low pay resulted in very few applicants. Second, by spreading the methods the hope was that this would result in more precise and timely weather observations from foreign colleagues. In the fall of 1919, after arranging first a course for Norwegian, Swedish and English meteorologists, and then a course for German, Austrian and Finnish meteorologists, Vilhelm Bjerknes summarized:

"If we by these measures have succeeded in winning new forces to our domestic forecasting is still uncertain. However, it is certain that there among the foreigners, especially the English meteorologists, has been won an understanding of the new methods which will give manifold returns by telegraphic weather observations in the future being more tailored to our needs."[34]

Over the following years, the Bergen school would arrange international courses almost every year, taking up the tradition from the oceanographic courses that the Bergen Museum had arranged before the Great War.

So, what were the "Bergen school methods"? In addition to taking air masses and not atmospheric pressure as the starting point, experience with practical weather forecasting led to new methods and a new language for analyzing weather maps. In 1919, the term "the Polar Front" was coined, a term directly inspired by the war. The Polar Front signified the area where warm air from the south clashed with polar air masses, a virtual line along which all major storm centers hitting Norway from the west developed. In 1921, the terms "warm front" and "cold front" replaced what was previously called steering and squall lines, signifying the three dimensional lines of discontinuity where warm air masses overtook cold air masses, and vice versa.[35] The Swedish forecaster Tor Bergeron soon added the term "occluded front" to signify what happened when the cold front had overtaken a warm front and the cyclone collapsed. A hundred years later, these concepts are still in use. In 1922, Jacob Bjerknes and Halvor Solberg described the life cycle of cyclones: their genesis, characteristics, movement, collapse, and how they often appear in "cyclone families" of three to five.[36]

In 1949, Japanese meteorologist Sakuhei Fujiwhara, who in 1921 spent five months in Bergen learning the new forecasting methods, vividly recalled a lively and hectic forecasting community, with activities from early morning to late night. At 7 a.m., the first telegrams from the night's weather observations were gathered, and during the following hours, telegrams from 200 domestic and international stations were decoded, quality controlled, and plotted on maps. The forecasters then analyzed the maps, identifying air masses and comparing them with the previous maps to get an overview of the weather and its development. They were working on a strict deadline: By 10:45, the forecasts had to be ready to be telephoned to Bergen Radio, which in turn would transmit the "Meteo Prognostique Norvège" to ships at sea and foreign countries. A more detailed analysis of the map to produce forecasts for land districts and the fisheries followed, supplied at 11:05 with observations from 31 US stations distributed via the Eiffel Tower, along with observations from the *Maud* expedition in the Northwest Passage. These provided the basis for a circumpolar map, which was used for internal long-term forecasts. After an hour's break for lunch, a new set of observations would arrive, and the routine would begin again. Finally, at 7 p.m. the third and last batch of observations for the day arrived, with a deadline for telephoning forecasts to Bergen Radio at 9:45 p.m.

"On top of this, throughout the day, the meteorologists have to answer questions about the weather, for instance if a sick man dare cross the North Sea, if a ship owner wonders if it is safe to bring an empty freight vessel back from England, or whether to order their ships out of the Baltic Sea to avoid being frozen in, or if they should dare to let them remain until they are fully laden."[37]

There were also daily seminars where the forecaster presented the weather map, and explained the analysis behind the forecast. The seminars also discussed the previous day's forecasts and to what extent the prognosis matched later observations. Fujiwhara points out that the discussions were vital to developing the new language for weather analysis: "Then the words *warm front*, and *cold front* were often used and gradually became fixed words. At about this time they noticed the phenomen [*sic*] of occlusion. It was called Seclusion for a while then it was decided to call it as Occlusion just shortly before I left."[38]

Erik Björkdal, another of the young Swedish meteorologists working in Bjerknes's attic, later recalled that the discussions and analysis of the weather maps often continued into the night, as an everlasting seminar:

"Every map was a new problem that needed to be solved, and everybody contributed to the analysis as best they could. Every time we succeeded in revealing fronts or uncovering a new secret about cyclones, there was great enthusiasm. Office hours were non-existent. When a particularly interesting case arose, we could sit and discuss far into the night, yes, it happened once that a keen soul did not leave the house for a whole week."[39]

Meteorologists were not the only staff who needed training; so did the newly recruited observers. Just as the forecasters both learned from and contributed to Vilhelm Bjerknes's theoretical work, the knowledge exchanges with the observers went both ways: from fishermen and lighthouse keepers, the forecasters learned, for instance, that storm centers were preceded by distinct and recognizable banks of clouds. This was soon made part of the continuously developing cyclone model. In return, illustrating Bjerknes's craving for observations to be quantified with precision, he had a small forecasting instrument produced and distributed to the observers along the coast. The instrument, a wooden board with a rectangular hole in the middle with numbers next to it, was to be held at arm's length. By aligning the bottom of the hole with the horizon, the numbers gave the distance to the clouds in kilometers. Not only did this result in more accurate observations for the forecasters, but by measuring how the distance to the incoming clouds changed over time, the simple tool made it possible to detect the speed of the approaching storm center with great accuracy, and predict what time the rough weather would hit the coastline. In a popular account from 1920, Vilhelm Bjerknes proudly announced, "The instrument is now being distributed to our coastal stations. Individual reports with the accurate positions of storms have already more than once proven useful."[40]

Good relations with observers, fishermen, farmers, the navy, the business and shipping community, and Norway's first aviators turned out to be of vital importance to the survival of the weather forecasting enterprise. In the early 1920s, Bjerknes had hoped to create a centralized institute for circumpolar weather forecasting for the Northern Hemisphere, to be based in Bergen. However, when this failed to gain support, the idea of a national "central institute" came up, possibly moving the Meteorological Institute from the capital to Bergen. When this too failed, plans were made to centralize national weather forecasting in the capital, closing the forecasting units that had been established in Bergen and Tromsø in 1920.[41] Both institutes had opened just before the economic boom ended in the fall of 1920, and the government was soon looking for ways to cut costs. A strong reaction from local allies, including businessman Einar Blaauw, who donated a villa to the forecasters in Kalfaret on the condition that the forecasting unit in Bergen would continue, and backing from the Geophysical Commission (which included director Hesselberg at the Meteorological Institute), were all that saved the forecasting unit from closure. In 1924, Vilhelm Bjerknes and his forecasters moved to the new building about a kilometer outside the town. While this saved the forecasting unit from closure, it meant that the meteorologists were located even farther away from Bergen's other geophysicists.

Over time, Vilhelm Bjerknes grew increasingly frustrated. When invited in 1917, he had been told that the Geophysical Institute was the first step towards establishing a university, and that he would soon become head of a planned physics department.[42] Bergen had been booming, and during the First World War the total taxable income in Bergen had increased by a factor of ten, while taxable wealth had increased by a factor of five.[43] By the summer of 1918, Bergen Museum had received 170,000 of an estimated 250,000 NOK in private donations for erecting a building for the Geophysical Institute, which was to be the first university building. Plans were drawn up, and the architect drawings for an institute building were all labeled "University of Bergen."[44] However, rather than expanding, Bjerknes soon had to work for his weather forecasting unit to merely survive. First, a desperate need for rebuilding after the great Bergen city fire of 1916, in which 380 buildings were destroyed and 2700 citizens lost their homes, meant that new projects were put on ice. Next, an initiative to establish the Geophysical Institute as an international research institution, charged with bridging the gap between the great powers that had been at war, was defeated in the Norwegian parliament by a 54 to 50 vote. Rather than prioritizing geophysics, the majority view was that that the limited public research funds should be divided equally between science and the humanities.[45] Locally, the end of the war also led to a sudden drop in freight rates, which resulted in a stagnation of the whole economy. The institute building was postponed yet again. Lastly, in the fall of 1920, the Norwegian economy crashed: unemployment among trade unionists rose from 2.3 to 17.6 percent, and remained high until the Second World War; foreclosures more than tripled; and several banks declared bankruptcy or were put under state administration.[46] Despite several attempts, the planned Physics Department was postponed for a seemingly indefinite period. Increasingly frustrated by the lack of fellow physicists, Bjerknes traveled more and more, and in 1927, he gratefully accepted a professorship at the University of Oslo.[47]

That the meteorologists and the oceanographers grew further apart was not just due to the physical distance and Bjerknes's unhappiness: more importantly, the forecasters focused solely on the atmosphere and not the oceans. Both the practical cyclone model and Bjerknes's theoretical work on the atmosphere took as their starting point that the weather came from above: storms started and moved in the atmosphere. Further, the forecasters had to keep up with a rapidly changing atmosphere. To this end they established and maintained a network of weather stations, worked within strict deadlines to analyze the simultaneous observations, and then distributed their predictions. In comparison, the oceanographers went on expeditions at sea to gather their observational data, collaborated with chemists to determine the salinity of each sample, and worked to make sense of data series taken at different locations and at different points in time. The analysis did not have the same urgency. Furthermore, the head of the Geophysical Institute, Bjørn Helland-Hansen, was distracted by his own attempt at making oceanography useful by establishing a factory to process salt.

The salt works

To Bjørn Helland-Hansen, the end of the war would be the nail in the coffin for his plans to use his knowledge of oceanography to help establish a factory for producing table salt.[48] Having closed its last national salt works in 1860, Norway imported about 300,000 tons of salt annually. During the war, salt prices skyrocketed, mainly as a consequence of rapidly rising freight rates. Access to cheap hydropower, combined with Helland-Hansen's research that showed that the water in the Norwegian fjords had sufficient salinity for salt production at depths of only 40 meters, made the prospect of resuming national salt production attractive. A joint stock company was established in the fall of 1918, with Helland-Hansen and shipping magnates Christian Michelsen and Olaf Ørvig the main signatories. The company was valued at 10 million NOK, 35 times the estimated cost of a building for the Geophysical Institute.

The factory was placed at Osterøy, twenty kilometers north of Bergen, near a 135-meter waterfall that was being developed into a hydropower plant. The company hired two engineers, a chemist, and a director, with the goal of producing 60–70,000 tons of table salt annually. Then the war ended and prices on imported sea salt quickly dropped back to normal levels. It also became clear that the estimated production capacity was overestimated by a factor of ten. In an effort to save the plant, focus shifted to using the factory to produce magnesium. After chlorine and sodium, which make up table salt, magnesium is the third most abundant element in seawater. By scaling the production to 5000 tons of table salt and 100 tons of magnesium annually, the factory would run at a profit and increase the global magnesium production by about 20 percent.

In addition to his expertise on currents and the chemical contents of seawater, Helland-Hansen had influential contacts and lent the project scientific credibility and visibility. In a meeting of the national committee for public-private collaboration in 1920, which was noted in several major national newspapers, he argued that magnesium alloys were as strong as steel at a quarter of the weight and that they would soon be used in engines, electrical wires and suspension bridges. The oceans were "an inexhaustible source" of raw materials, and access to cheap hydropower made Norway uniquely positioned to corner the international market for magnesium production.[49] An enthusiastic letter from Helland-Hansen to Nansen a year later, in which he asked for advice on international patenting of several magnesium alloys, indicates that this was more than just propaganda.[50] Vilhelm Bjerknes, however, expressed frustration at Helland-Hansen's being distracted by his "darn factory," which got in the way of geophysical research collaboration.[51]

Helland-Hansen probably saw the factory as a way to secure further research funding, and the commercial success of his former university lecturer, physicist Kristian Birkeland, was a likely inspiration: in 1907, Birkeland, together with engineer and businessman Sam Eyde, opened a factory at Nesodden outside Kristiania that produced artificial fertilizer by fixing nitrogen from the air using electricity (the Birkeland-Eyde process). Within ten years their company, *Norsk Hydro*, had opened two new factories near hydropower plants at Rjukan and Glomfjord that both generated large profits. To Bjerknes, however, science was a goal in itself: entrepreneurship, like establishing factories based on scientific findings, should be left to entrepreneurs.

Helland-Hansen's factory did produce "several tons" of magnesium and was, in the early 1920s, the only manufacturer in Europe, but there was no demand for the product. Magnesium for tracer grenades had disappeared with the armistice, the need for magnesium for flash photography had already been met, and magnesium alloys did not catch on until years later. Production was shut down after less than a year, and a delay in the construction of the hydropower plant was used as a pretext to cancel the contract for long-term electricity deliveries. For a few years, the factory produced electric cooktops before the buildings were repurposed as a fish hatchery and a furniture factory.

While it is unclear how much effort Helland-Hansen actually invested in the factory, it seems to have added to Bjerknes's frustrations that the Geophysical Institute was not turning into the haven for hydro- and thermodynamics that he had envisioned when recruited from Leipzig. Bjerknes also remained unconvinced by the next research collaboration that Helland-Hansen and Nansen had pursued after publishing the *The Norwegian Sea* (1909): an attempt to use their ocean- ographic insights for climate forecasting.

Connecting the dots in climate variations

In *The Norwegian Sea* (1909), Helland-Hansen and Nansen had set a goal of climate prediction, and this would be the topic for their next large research collaboration. But what started as an attempt to use ocean surface temperatures to make seasonal weather forecasts would soon lead to the assembly of an integrated account of how different geophysical phenomena were connected on an interplanetary scale, from the oceans to sunspots. The study reflects Helland-Hansen's vision for his institute, and a rationale for gathering research into physical oceanography, theoretical meteorology, practical forecasting, geomagnetism and cosmic rays under the same roof. However, rather than unifying geophysics, the study contained even more seeds for the different geophysical disciplines to grow apart.

It was well known that Scandinavia had a more temperate climate than other areas at the same latitude, and that this was linked to the warm ocean currents bringing heat from the Gulf of Mexico. It was therefore logical that annual variations in seasonal temperatures could be explained by changes in the ocean temperatures. The start- ing point for Helland-Hansen and Nansen's study was to find out if surface temperatures from the Atlantic could be used to make seasonal weather predictions, such as whether it would be a cold winter or a dry summer.[52] To answer this, they had their German colleague, Adolf Hermann Schröer, analyze the temperature records from line ship journals for ten spring days between 1898 and 1917, found in the archives of the Deutsche Seewarte in Hamburg. They assumed that the surface temperatures would be representative of the temperatures in the uppermost ocean currents, and that it would be possible to find similar variation in temperatures some months later. Could the ocean surface hold the key to seasonal forecasting?

In 1912, in an appendix to Roald Amundsen's account of his expedition to the South Pole, Helland-Hansen and Nansen had argued that the relationship between temperatures in the sea and temperatures in the atmosphere was relatively straightforward. Water had a much higher heat capacity than air, and it was therefore common sense that changes in ocean temperatures would influence the temperature in the atmosphere.

"Water contains more than 3,000 times as much warmth as the same volume of air at the same temperature. (...) In other words, if the surface water of a region of the sea is cooled 1° to a depth of 1 metre, the quantity of warmth thus taken from the sea is sufficient to warm the air of the same region 1° up to a height of much more than 3,000 metres, since at high altitudes the air is subjected to less pressure, and consequently a cubic metre there contains less air than at the sea-level."[53]

When the results from analyzing the observations from the line ships were in, the fluctuations in the ocean surface temperatures seemed to confirm a strong correlation between temperatures in the ocean and over land, but the devil was in the details: instead of ocean temperatures affecting temperatures over land, the effect appeared to go in the opposite direction. Changes in temperatures in Stockholm, for instance, could occur days before changes in surface temperatures along the Atlantic coast. The solution to the riddle was found in meteorological observations. Using pressure gradients, the same proxy for wind speed and wind direction based on the rate of atmospheric pressure change that Mohn had used more than thirty years earlier, Helland-Hansen and Nansen found that when the wind came from the Arctic north, the ocean surface temperatures dropped, and vice versa. The same was true for surface temperatures measured at lighthouses along the coast. In other words, the atmosphere moved heat more quickly than the ocean currents, and the surface measurements revealed more about the prevailing winds than the temperature of the currents.

The study could have concluded that ocean surface temperatures were a blind alley for climate prediction, but instead Helland-Hansen and Nansen decided to examine how changes in atmospheric pressure happened. They promptly pointed to the sun, and the 11-year sunspot cycles. Identifying similar cyclic variations on Earth had been a main task for climatologists for decades, but efforts had so far failed to produce clear patterns. The two researchers had an explanation:

"We found that the changes in solar activity obviously influence the pressure distribution in the atmosphere, and that this in turn determines the temperatures at the earth's surface and in the ocean. But it is clear that the temperature variations in different areas must depend on the position relative to prevailing wind directions and their fluctuations: Increased solar radiation will in some areas lead to increased airflow, leading to cooling. (...) Previous researchers have not been sufficiently aware of this, and have therefore failed to understand that the effect of increased solar activities must be opposite in different places. On top of this, many have studied too large areas as one, so that opposite effects cancel each other out or there appears to be no relationship what so ever."[54]

The explanation for why the same phenomenon, increased solar radiation, could cause opposite effects on Earth was linked to latitude and location. As a rule of thumb, Helland-Hansen and Nansen argued, in places where the average temperatures were already higher than what could be expected based solely on latitude, such as Scandinavia, the effects of increased solar radiation would be increased temperatures. In areas colder than expected, more sun would lead to more wind, causing a cooling effect. In other words, stronger solar rays would lead to more energy in the atmosphere and more extreme weather.

Between first publication in German in 1917 and the expanded English publication in 1920, financed by the Smithsonian Institution in Washington, Helland-Hansen and Nansen added a section to integrate the latest findings from the Bergen school of meteorology:

in Scandinavia, changes in solar activity could influence the average position of the polar front between the arctic cold air and temperate hot air masses:

"The average position of the 'Polar front' fluctuates over longer periods because of changes in the solar radiation. In other words, the area for rough weather can move and produce more climatological fluctuations; in some places, this would have one effect, elsewhere the opposite. Continuing the study of these matters will hopefully produce such results that one in the future can predict the characteristics of the weather, not only for the coming day, but for much longer – weeks and months."[55]

Lastly, in addition to wind and temperature, it seemed that the sun had an effect on the earth's magnetic field. In the study, Helland-Hansen and Nansen pointed out that the pressure gradient over Scandinavia fluctuated in periods of 25 to 28 days, and that a similar rhythm had been identified in studies of the fluctuations in the magnetic field. Furthermore, these periods were more or less identical to the life cycles of sunspots.

The idea that geomagnetism and periodic weather fluctuations were connected was based on research by Norwegian physicist Ole Andreas Krogness. In a lecture at the Bergen Museum in 1915, Krogness had argued that the same solar rays that produced northern lights were connected with surface temperatures, albeit in confusing and opposing ways: "[There are] clear and remarkable correlations between the curves for magnetic storminess and curves for air temperature. But the effect varies greatly between the seasons, yes they are even occasionally opposite."[56] Krogness had been physicist Kristian Birkeland's student, and when Birkeland's observatory at Haldde in Norway was made into a permanent Auroral Observatory in 1912, he had become its first full-time manager.[57] When the Geophysical Institute's new institute building was ready in 1928, Krogness was appointed professor of terrestrial magnetism and physical cosmology.

The reference to Krogness's research was typical for Helland-Hansen and Nansen's study; in order to assemble their integrated account of the geophysical world, they had to rely on studies done by others. Most of the information on fluctuations in solar energy reaching the atmosphere was based on studies from the Carnegie-funded Mount Wilson Observatory in California, while the idea of extending the study of pressure gradients over land came from US meteorologist Henry Helm Clayton's study of the relations between solar radiation and temperatures in 30 places on all continents. Both the Danish Meteorological Institute in Copenhagen and Deutsche Seewarte in Hamburg had, for years, been producing maps of monthly pressure averages for the North Atlantic, which they used to compare pressure gradients and temperatures. The magnetic observations came from Krogness. Today, Helland-Hansen and Nansen's study would be labeled climate research, as it was characterized by seeing the earth's climate as one integrated system and required expertise from different fields. But as their study demonstrates, at this time geophysical publications were sufficiently nontechnical to be used by specialists from other fields.

Although Helland-Hansen and Nansen's study was initially well received, there were several reasons for critical remarks.[58] Statements such as "on average 2/3 of the solar energy is absorbed by the atmosphere, and only 1/3 reaches the earth's surface," for instance, were in stark contrast to the precision the two researchers demanded when it came to physical oceanography.[59] Still, the principal critique was not about imprecisions but the fundamental research methods. Their approach consisted of calculating statistical averages along different time scales, comparing different curves, and looking for correlations. Helland-Hansen and Nansen used 7-day, 10-day, 24-day, 30-day, seasonal, and annual averages; Krogness had used 3-day, 14-day, 27-day, 90-day, 6-month, 8-month, 2-year, 4-year, 11-year, 180-year and 1800-year periods. The comparisons consisted of temperatures in the atmosphere and ocean, pressure gradients, wind direction and wind speed, sunspots, magnetic storminess, tides and number of stormy days.[60] Was this merely playing with numbers and time scales, a blind search for patterns?

Vilhelm Bjerknes was the first to argue that the search for correlations was backwards and methodologically inferior. Opening a lecture on practical weather forecasting at the 1918 Geophysical Meeting in Gothenburg, he pointed out:

"There may possibly be several approaches to solving the problem of satisfactory practical weather forecasting. Possibly, there may among these be convenient avenues, which lead past or outside the straightest necessity for fully understanding the phenomena one is to predict. For such shortcuts, I have no interest. My sole interest is in the avenue that leads straight through the inexorable demand for full insight."[61]

Bjerknes did not mention Helland-Hansen and Nansen's study explicitly, but it must have been clear to the listeners that he actively distanced himself from his colleagues' newly published study. To Bjerknes, it was only through identifying the actual mechanisms behind the phenomena, the laws of physics, that weather forecasting – and geophysics in general – could become "an exact science."[62] Bjerknes had no faith in patterns of the past being a key to predicting the future, and saw empirical pattern recognition as a dead end in the geophysical quest to calculate the world.

Bjerknes's two most important theoretical works published during his nine years in Bergen demonstrate what he considered the way forward for geophysics, namely developing physical equations. The first paper was a dense 88-page elaboration on the dynamics of vortices and waves, including a discussion of how the insights could be applied to motions in the atmosphere.[63] The second paper developed equations for how waves propagate, and a framework for subdividing the atmosphere and oceans into layers based on pressure.[64] Like Helland-Hansen and Nansen's climate study, Bjerknes stressed that much work remained, but what is most interesting was their differences in approach. Helland-Hansen and Nansens's study contained 66 maps of pressure systems, currents and wind systems, 94 extensive tables with mean temperatures and observation points, and 139 curves comparing correlations between different phenomena. The analysis, almost exclusively, consisted of comparing averages over different time scales. In contrast, Bjerknes's first study had 134 equations representing the relationships between different physical forces in the atmosphere, 37 diagrams with abstract representations of how the equations described physical phenomena, 13 definitions of the terms used, and six short conversion tables. The second paper contained one table and 127 physical equations. To Bjerknes it was physics – isolating mechanisms and expressing them as physical equations – that held the key to prediction, not correlations. Between

1923 and 1945, Vilhelm Bjerknes received 26 nominations from nine persons for a Nobel Prize in Physics.[65]

Helland-Hansen and Bjerknes clearly respected each other. They sat on the board of the Bergen Museum together, discussed the workings of the Geophysical Commission, and collaborated on administrative tasks. Still, despite working in the same institute in the same city, Helland-Hansen and Bjerknes never collaborated on any scientific studies. The difference in methodological approaches was part of the reason. In a letter to Nansen in 1924, Helland-Hansen dryly noted: "Bjerknes surely works a lot, but the results are after all quite eccentric."[66]

A center for collaboration

While Bjerknes focused on the physics behind the weather more or less in solitude, only checking in with his forecasters in the attic every once in a while, it was the will to collaborate and to investigate how different elements in the geophysical world were connected that permeated the research projects Helland-Hansen promoted at the Geophysical Institute in the 1920s. In 1921, when Fridtjof Nansen accepted the post as High Commissioner for Refugees for the League of Nations (for which he received the Nobel Peace Prize in 1922), Helland-Hansen had to seek other collaborators. In a partnership with the Swedish glaciologist Hans W:son Ahlmann, he pursued the question of climate change.[67] Since 1915, Ahlmann had been a geographer at the University of Stockholm, and over the years he would visit the Geophysical Institute in Bergen on at least 22 occasions.[68] The attraction included academic collaboration, access to research facilities and nearby glaciers, and his marriage to Erica Maria Harloff, daughter of a wealthy shipowner in Bergen.

The starting point for the collaboration between Ahlmann and Helland-Hansen was not seasonal or annual variations, but how climate changed over millennia. In a paper published in 1918, the two set out to explain the last ice age. The cause for the ice age, they argued, was a land rise of about 300 meters, which had changed the interactions between land, sea and air. The land rise meant that only about half as much of the warm Gulf Stream flowed through the Faroe-Shetland Channel. Instead of heating northern Europe, the currents would merely cause about 50 percent more hoarfrost on land: "Thus the Norwegian sea, instead of becoming a source of heat, would be a cold sink."[69] The end of the last ice age was the same process in reverse: The land masses had sunk and caused an increased influx of warm water to the Norwegian Sea, heating the atmosphere. Warm winds from the south had brought heat over land, and snow and hoarfrost were replaced by rain for larger parts of the year. "Immediately, the inland ice will start melting with great speed, just as a spring break."[70] As with the study Helland-Hansen had conducted with Nansen, the argument was mainly qualitative, logically describing the assumed connection between the phenomena, and not an attempt at precise quantification or presenting equations for the mechanisms involved.[71]

The largest research collaboration came about in 1922, when the Belgian zoologist Désiré Damas chartered the Institute's research vessel *Armauer Hansen* for a two-month cruise via Portugal to Madeira and back via the Azores.[72] The expedition onboard the 76-foot vessel gathered 13 researchers with different specialties, who labored side by side and arranged daily lectures for each other. Helland-Hansen took 39 hydrographical stations, water samples from different depths using Nansen bottles and Richter's reversing thermometer. The samples were analyzed for salt and oxygen content in the onboard laboratory. Damas made 28 stops to collect zoological species from different depths using specially crafted nets and bottom scrapers. Ernst G. Calwagen, manager of the Observatory in Bergen, took meteorological observations: in addition to measuring temperature, wind and humidity at the ocean surface, the bridge, and from the top of the ten-meter-high mast, he took observations up to 1000-meter altitude using kites borrowed from Deutsche Seewarte. The report from the cruise includes unbroken temperature curves from 1000-meter altitude to 1200 meters below the surface, demonstrating how the atmosphere and the oceans were recognized as a connected whole.[73] An extensive travel letter in five parts was published in newspapers in Bergen and the capital.[74] Still, Damas was the only one to publish any scientific work based on the expedition.[75] In a letter to Harald Ulrik Sverdrup written in 1926,

Helland-Hansen explained that the intention had been to use the observations to examine the interactions between the ocean and the atmosphere, but after Calwagen died in an aircraft accident in 1925, it seems the observations were never fully analyzed.[76]

Calwagen was also responsible for an on-board weather forecasting service. The winter prior to the cruise, the vessel had been upgraded with electricity and a radio, and for the duration of the cruise, Deutsche Seewarte sent "special notices for the *Armauer Hansen*" at the end of its weather transmissions from Königs Wusterhausen, including supplementary observations from Norwegian stations and line ships. Twice a day, Calwagen compiled weather maps and made weather forecasts onboard. In his report to the board of the Bergen Museum, Helland-Hansen highlighted the field weather service as an important scientific result: "The expedition clearly proved that with the use of modern technology, even a small vessel can be fully informed about the weather situation over large areas, gain a good basis for determining its development, and predict the conditions that the vessel will meet."[77] Starting on November 1, 1922, the weather forecasting unit in Bergen broadcasted wireless weather forecasts to ships at sea twice a day.

The cruises – both longer expeditions and annual visits to the same fjords and ocean areas looking for variations and testing new equipment – as well as conferences, courses, lectures and weekly colloquia where different researchers took turns presenting their work, attracted a host of students and scientists to Bergen. This was the oceanographers' way of continuing the international research courses that had been interrupted by the war, but now with more emphasis on facilitating research collaboration than on teaching. By 1930, in addition to the 60 research visits to the weather forecasting unit where emphasis soon shifted from developing to disseminating methods, 45 international researchers had extended research stays at the Geophysical Institute, bringing with them a great variety of geophysical research interests. The 1920s also saw a host of new geophysical research institutes being established in different countries, and Bergen was both a center for learning research methods and a place where directors could come to learn how geophysical institutes could be organized.

Seven years in the ice

Polar exploration and geophysical research continued to be intimately connected in the 1920s, and nobody personifies this better than Harald Ulrik Sverdrup. In the United States, Sverdrup is remembered as "the founder of the modern school of physical oceanography," with reference to being the head of Scripps Institution of Oceanography in La Jolla, California, from 1936 to 1948, and as the author of the section on physical oceanography in the monograph *The Oceans: Their Physics, Chemistry and General Biology*, written with Martin Johnson and Richard Fleming, a publication frequently referred to as "the Bible of oceanography."[78] Sverdrup is honored with the unit "Sverdrup," which denotes a flow of one million cubic meters per second, and is used to measure the rate of transport in ocean currents. Since 1964, the American Meteorological Society has awarded the Sverdrup Gold Medal for outstanding contributions to the scientific knowledge of interactions between the oceans and the atmosphere. In Norway, to the extent that Sverdrup is remembered at all,[79] it is for his career after returning from the United States after the Second World War, both as head of the Polar Institute (1948–1957) and as the architect of the "Sverdrup plan," which, starting in the late 1950s, reformed the Norwegian universities from elite to mass education institutions. In Bergen, however, Sverdrup is venerated as a meteorologist and polar researcher.

Sverdrup started his career as an assistant to Vilhelm Bjerknes in 1911, and joined him when he went to Leipzig in 1912. Five years later, Sverdrup defended his doctorate on the North Atlantic trade winds.[80] When Bjerknes was called to Bergen, Sverdrup was recruited by the Norwegian polar explorer Roald Amundsen for his *Maud* expedition. After being the first to reach the South Pole in December 1911, Amundsen had set his sights on being the first to conquer both poles. The plan was to sail along the coast north of Siberia as Nansen had done with the *Fram* expedition some decades earlier, but to go farther northeast before letting the ship freeze into the polar ice cap. The *Maud* expedition, named after the specially constructed vessel built on the same principle as the *Fram*, was then to drift across the North Pole. Underway, the ship would function as a floating laboratory for oceanography,

meteorology, terrestrial magnetism, atmospheric electricity, studies
of the northern lights, and other geophysical subjects. Sverdrup's task
was to head the scientific work.

The *Maud* expedition was riddled with mishaps. The winter of 1918
was colder and had more ice than expected, and the ship froze in the
polar ice without reaching its intended point of entry. There it would,
in practice, remain stuck for two years. Amundsen almost drowned
after falling through a hole in the ice, broke an arm, was mauled by a
polar bear, and suffered from carbon monoxide poisoning. After the
first winter, two expedition members, Peter Tessem and Paul Knudsen,
decided to trek back to civilization, but perished before reaching their
destination. With Amundsen tied to the ship, Sverdrup had the occasion to spend almost eight months with the Chukchi people, resulting
in a monograph that was widely read in Norway.[81] Only in 1920, after
the ice broke and the ship sailed to Nome, Alaska, to resupply and bring
on new crew, were the isolated expedition members informed about
the outcome of the Great War. In 1921, stocked with new equipment,
food for seven years and coffee for twelve, the ice finally opened – only
to reveal a defective propeller. The crew had to take the vessel to Seattle
for more repairs.[82] Sverdrup spent his shore leave as a research fellow
at the Carnegie Institution's Department of Terrestrial Magnetism in
Washington, D.C., analyzing the electric and magnetic records from
the expedition and borrowing more equipment before, in summer 1922,
once again returning with the *Maud* to the ice.[83]

The second part of the *Maud* expedition did not fare much better
than the first. One expedition member, Søren Syvertsen, died from
illness.[84] Having experienced how unreliable the plan of drifting over
the North Pole had turned out to be, Amundsen had decided to bring
an airplane to conquer the pole by air, but the first airplane crashed on
its way to Seattle. The second plane made it to Alaska before crashing
in its first test landing. A third airplane that had been brought onboard
in crates was assembled after the ship had become stuck in the ice yet
again, but crashed in its third landing. In February 1924, almost six
years after the departure of what was planned to be a three-to-four-
year expedition, Amundsen had run out of funds and ordered his crew
to return to civilization. Locked in the ice, the expedition managed to

Figure 4

The *Maud* was a floating laboratory. As illustrated by Odd Dahl, the instruments included a bottom scraper (1), a bottom sampler (2), current meters (3), a depth gauge (4), water samplers (5), a plankton net (6), a tide gauge (7), electrical thermometers (8), an aurora camera (9), kites for high-altitude meteorological measurements (10 and 24), instruments to measure heat flux from ice gauges (11), weather balloons (12), onboard instruments to measure solar and electrical activities (13), an anemometer and weather vane to measure the strength and direction of wind (14), electrical thermometers (15), an onboard weather station (16), a rain gauge (17), balloons for observing wind directions at high altitudes (18 and 23), instruments for astronomical positioning (19), instruments for measuring electricity in the atmosphere (20), instruments for continuous recording of magnetic intensity and declination (21 and 22), and a reconnaissance plane (25). Illustration: Dahl, Odd and Reidar Lunde. *Odd Dahl og Maud-ferden*. Chr. Schibsteds Forlag, Oslo. 1976: 24–26.

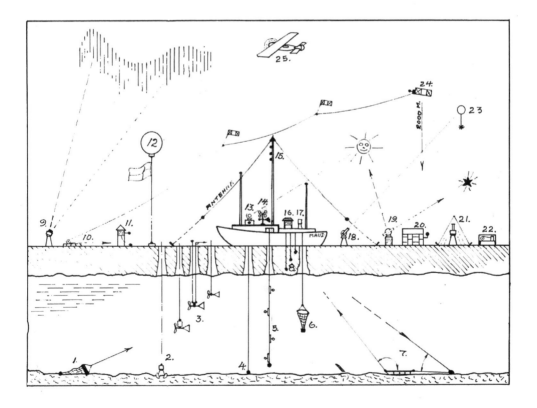

reach Seattle only in October 1925, 2,326 days after they had set sail. In the eyes of the public, the whole expedition was considered a failure. Sverdrup later summarized: "I hope that some of our contribution to the knowledge of the Arctic will be of lasting value, but if I am asked what I consider our greatest accomplishment, my answer is: 'That we parted friends for life.'"[85] He also stressed the value of "the closest possible contact with nature, a circumstance which to one who works in geophysics cannot be overestimated."[86] This virtue has been cherished by later geophysicists in Bergen.

Despite not reaching the North Pole, or even the deep Polar Ocean, the *Maud* expedition was considered a success by the Bergen geophysicists. Gathering unique and precise time series from the Arctic wilderness was considered more important than publicity stunts. Although planting a flag at the North Pole undoubtedly would have played nicely with donors and the general public, perhaps preventing Amundsen's bankruptcy, polar exploration and geophysical data gathering were beginning to drift apart. In the public mind, polar exploration consisted of daring adventure, conquest, and sportsmanship; to geophysicists it was a matter of preparation, risk management, routine, and endurance.

When the *Maud* went to Seattle for repairs, the first set of observations was sent to Bergen, and in 1922 the newly hired mathematician, Jonas Ekman Fjeldstad, and Nansen's assistant, Håkon Mosby, began analyzing the results.[87] After just a year, Fjeldstad showed that tidal waves propagated under the polar ice cap, which finally disproved the American oceanographer Rollin A. Harris's theory of a landmass under the North Pole.[88] Further calculations supported Nansen's theory that there existed a deep-sea ridge dividing the Arctic Basin in two.[89] In 1930, Fjeldstad defended his doctorate on tidal waves, and three years later he published *Interne Wellen* (1933), a general theory for waves between different strata of the oceans.[90] The theory was an expansion of Ekman's explanation for the "dead water" phenomenon that had been reported by Nansen in 1893. Among other things, Fjeldstad showed that continuous fluctuations could lead to a hierarchy of internal waves.[91]

The analysis of the *Maud* material was central to establishing Bergen as a geophysical center of calculations and analysis in the 1920s. Soon, hydrographical observations gathered by the Fisheries Administration and the Geophysical Institute in Tromsø, state-sponsored expeditions to the Arctic, and the privately funded *Norvegia* expedition to Antarctica (1927–31) were all analyzed at the Geophysical Institute.[92] Starting in 1922, weather observations from the *Maud* were transmitted via radio to Bergen every day and used in circumpolar weather maps used internally to explore the possibility of long-term forecasts.

The *Maud* expedition also connected Sverdrup to Bergen. After Vilhelm Bjerknes accepted a professorship at the University of Oslo, Helland-Hansen pleaded in several letters for Sverdrup to fill the open position as professor of theoretical meteorology at his institute. In the letters, Helland-Hansen highlighted that the Geophysical Institute was, under the Bergen Museum, dedicated to research, and not a university responsible for education. There were few administrative chores and few students to supervise, and Sverdrup would not have to lecture for more than about three hours per week: "Rest assured you will be given a very high degree of freedom exactly so that you can analyze your material."[93] Sverdrup accepted the call, and in 1926 he gave his an inaugural lecture on "Polar meteorology."

Sverdrup replacing Vilhelm Bjerknes as head of the department of meteorology gave new impetus for local collaboration, and Helland-Hansen and Sverdrup soon started arranging shared colloquia on meteorology and oceanography.[94] This seems to have been a conscious effort aimed at countering the drifting apart of geophysical disciplines. In November 1926, Helland-Hansen and Sverdrup both attended the first conference of the International Society for the Exploration of the Arctic by Means of the Airship, of which Nansen was president, in Berlin.[95] When Helland-Hansen, in 1927, traveled to Scotland, England, Germany, Denmark, Sweden, Oslo, and Trondheim, Sverdrup stepped in to head the annual oceanographic cruise to Sognefjorden. However, it was the continued analysis of the scientific results from the *Maud* expedition that would define Sverdrup's activities in Bergen. Over the years, Sverdrup received close to 100,000 NOK in funding for assistants and printing costs, which resulted in five extensive volumes of more than 2000 pages altogether.[96] In addition to groundbreaking observations and time series in meteorology, oceanography, astronomy, terrestrial magnetism, and biology, as well as studies of the northern lights, his main contributions to science were probably on how wind acts on the ocean and his systematic use of the curl of the vector current to simplify the geodynamic equations.

In 1930, Helland-Hansen arranged for Sverdrup to become the first research fellow at Chr. Michelsen Institute (CMI), an institution for pure research set up with funds from a legacy left by shipping magnate and former Prime Minister Christian Michelsen. Helland-Hansen remained director of the institute until 1955, and Chr. Michelsen Institute had its offices in the same building as the Geophysical Institute. The only difference was that the researchers had no administrative chores or teaching duties.[97] The offer was made after Sverdrup had received several attractive offers from the United States: in 1928 Sverdrup was offered a position as head of the Carnegie's Department of Terrestrial Magnetism, and in 1931, he was offered a well-funded research position at Woods Hole Oceanographic Institution in Massachusetts, which had opened a year prior, after initially declining the post as the institution's first director.[98]

Despite seven years in the ice, Sverdrup longed to go back to get the observations of the deep Polar Sea that he had failed to get during the *Maud* expedition. When the US adventurer Hubert Wilkins offered him the position as head scientist in the spectacular North Polar Submarine Expedition, an attempt to reach the North Pole using the rebuilt WW1 submarine Nautilus in 1930, he gladly accepted. Helland-Hansen and other colleagues tried to dissuade him up to the very departure: "I must admit I am quite skeptical, and from what I know about the expedition, I am not happy that you are participating. It might be that the risk is smaller, and the scientific yield higher than I believe, and that I might change my opinion when I learn more about the technical aspects."[99] Still, as head of a department, Sverdrup had full autonomy to choose what to pursue. Nor was Sverdrup the only one attending potentially dangerous expeditions: shortly after being hired as associate professor (amanuensis) in meteorology, Håkon Mosby had in 1927–28 attended the first of four privately funded *Norvegia* expeditions to Antarctic waters.[100]

The *Nautilus* expedition was a failure: The vessel lacked insulation and heaters, the fresh water system froze and the hull had leaks. When arriving at the polar cap, the stern planes that controlled diving depth were gone, and there were rumors of sabotage. The submarine did manage to become the first to operate under the polar ice cap by ramming itself under an ice floe, which led to further damage to the hull. Still, it resulted in two volumes of scientific results.[101]

In 1934, Sverdrup participated in a small expedition with Ahlmann and spent a summer on the Isachsen Plateau at Spitsbergen, collecting some 20,000 observations on heat exchanges between the atmosphere and the snow.[102] The resulting publication, like most of Sverdrup's work, shows how he managed to unite Helland-Hansen's insistence on collaboration and on investigating how different geophysical elements were connected, a strong empirical basis, and Bjerknes's ideals of developing physical equations for the precise mechanisms involved.

But then, as will be described in the next chapter, Sverdrup, too, left Bergen.

The curious case of the "Nansen monument"

The days of heroic polar exploration were coming to an end, and the death of Helland-Hansen's longtime collaborator, Fridtjof Nansen, in May 1930, was a turning point. In addition to being the recipient of the Nobel Peace Prize for his work with refugees displaced after the First World War, Nansen's greatest fame was as a polar explorer. While geophysicists and scientists in general were recognized as authorities and part of the elite, the Norwegian polar explorers were in a league of their own: Fridtjof Nansen's *Fram* expedition (1893–1896) and Roald Amundsen's planting of the Norwegian flag on the South Pole in 1911 were causes for national celebration. But by the 1920s, the areas of the world where no man had set foot were few and far between, and both the *Nautilus* expedition and the crossing of the North Pole with airships exemplifies how, in order to get attention and donations, the expeditions were exceedingly spectacular. Although polar exploration did not end, the deaths of Amundsen on a rescue operation in 1928 and

Nansen in 1930 marked an end to their heyday. Instead, the heroic polar explorers still lived on in the public imagination. This was especially evident in the curious case of the "Nansen monument," which was an attempt to create a Mount Rushmore-inspired memorial on the side of a mountain overlooking Bergen harbor.

The story of the Nansen monument began in the summer of 1934, when Helland-Hansen accepted an invitation from the German sculptor Ernst Müller-Blensdorf to assist in organizing the building of a polar monument celebrating mankind's brave expeditions into the inhospitable polar unknown.[103] The idea was formed shortly after the death of Nansen in 1930, but planning began in earnest only after the National Socialists came to power in Germany in 1933: Müller-Blensdorf lost his job as a lecturer at the Staatliche Kunstschule at Wuppertal, had his public contracts withdrawn, was declared a "degenerate," and had his studio destroyed, before he finally fled to Oslo.[104] There he teamed up with the Norwegian geologist and polar researcher Adolf Hoel, who was a lecturer at the University of Oslo as well as the first director of the Norwegian Scientific Exploration of Svalbard and the Arctic Regions. Müller-Blensdorf believed the monument belonged in Norway, in part because of its strategic position as a gateway to the Arctic, and in part to celebrate Norwegian efforts in polar exploration. Hoel was the one who suggested that the monument should be placed in Bergen, pointing out the rock types in the mountains surrounding the town and that Bergen was the harbor in Norway with the most cruise ships.

The monument was to be built in Sandviksfjellet, a 300-meter-high mountain just north of the entrance to the central harbor, and carved directly into the mountainside:

"The main figure, sitting on the top of the rock, shows Frithjof [*sic*] Nansen as explorer, gazing northwards. Giant groups, hewed out of the solid rock on both sides of the main figure, show the struggle of men against the powerful Nature and their victory over matter. Thus, out of the knowledge of the powers of Nature, the destination of man in the universe is recognized."[105]

Müller-Blensdorf estimated that 40 to 50 sculptors would be employed for twelve years, that the plan would cost four to five million

Norwegian kroner, and that this would make Bergen a world-famous tourist destination.[106] The idea was probably inspired by the Mount Rushmore memorial in South Dakota, which had begun construction in 1927. The famous faces of the four US presidents were completed between 1934 and 1939.

Helland-Hansen's involvement consisted mainly of assisting the sculptor in getting access to the grounds, which were on a military area controlled by the Norwegian Navy, and presenting the plans to the municipal government. He also agreed to join a working committee, which in addition to Hoel and Helland-Hansen consisted of polar explorer Hjalmar Riiser-Larsen and geophysicists Olaf Devik and H.U. Sverdrup from Chr. Michelsen Institute.[107] However, after reporting that the local politicians' enthusiasm was measured, and that an initiative from Müller-Blensdorf to establish a school for sculptors was unanimously voted down by the municipality, the Bergen geophysicists laid the plan to rest.

In Oslo, however, the plans continued to move forward at full speed: Müller-Blensdorf gathered letters and statements of support from a range of prominent figures in Europe, including Albert Einstein and Benito Mussolini, academics, polar explorers, sculptors, and leading members of various associations, as well as Fridtjof Nansen's son, Odd.[108] He also made at least two models of the monument, the largest in a 1:200 scale. In 1936, Hoel established an office for the "International Polar monument" in Oslo with a secretary and its own letterhead. The office authored promotional leaflets and contacted institutions such as the Norwegian Academy of Science and Letters, the Nobel Institute, the Bergen Museum and the Norwegian Foreign Ministry for support. According to the promotional material, the ambition was "to raise an International Monument for all Polar Exploration as a symbol of International understanding and collaboration"; to celebrate Western civilization, culture and technology; to praise the spirit of inquiry and exploration; and to be a symbol of strength, unity, intellectual freedom, understanding and international collaboration in a time of growing political unrest.[109] Rather than celebrating Nansen as a geophysicist or oceanographer, which would have fit well with choosing Bergen as its location, the planned monument illustrates how

Nansen's legacy as a heroic polar adventurer soon came to overshadow his contributions to science. This emphasis was more appealing to Hoel than to Helland-Hansen.

The first inkling that the monument was still on the agenda came in October 1936, when Helland-Hansen received a letter from his friend, Sem Sæland, rector at the University of Oslo, with one of Hoel's promotional letters attached. Sæland expressed surprise that Helland-Hansen had endorsed the enterprise and that he had agreed to be listed as a member of the organizing committee of a plan Sæland saw as utter nonsense: "(...) frankly, I find the whole plan ridiculous. The idea of such a gigantic monument in a mountain to me seems too German. The plan might produce jobs for some artists, but I have problems imagining why this business is worth begging for."[110]

The attachment to Sæland's letter showed that the "Nansen monument" had grown beyond reshaping the natural landscape in Bergen: the plans now included national contests where countries with a polar history should select what figure would best represent the country's polar accomplishments, and a scheme for an international Nansen day to collect "Nansen pennies" to finance its construction. This appears to have come as a complete surprise to the committee members in Bergen. Helland-Hansen promptly contacted Hoel and Müller-Blensdorf asking that no more promotion material be distributed before the people mentioned by name could at least agree on the wording.

The promise to involve the committee members before using their names in marketing was respected for about a year, when a promotional letter to be distributed by the Norwegian News Agency to all their subscribers was halted at the last second.[111] In a heated letter written in November 1937, Helland-Hansen argued that the idea of doing international fundraising for a monument was an insult to Nansen's memory: the fundraising would distract from Nansen Relief, a humanitarian organization established by Odd Nansen a year earlier to provide safe haven and assistance in Norway for Jews living in areas under Nazi control, as well as the Nansen International Office for Refugees that had been set up in Geneva in 1930, to continue his effort for refugees: "[These initiatives] deserve much more support, and it would be meaningless if the monument plans get in the way. (...) Now that the issue has been presented in this way, including your abuse of my name, I will have absolutely nothing more to do with it."[112] Also Olaf Devik withdrew, arguing that: "The best monument that can be erected to commemorate Fridtjof Nansen in this situation is humanitarian aid in his name."[113]

It is unclear to what extent the work on the monument continued after this, but it is likely that the final nail in the coffin was the outbreak of war and the German occupation of Norway from April 9, 1940, onward. This also put an end to the International Polar Exhibition that was planned to be held in Bergen in 1940. Yet again, Müller-Blensdorf was forced to flee, this time to Great Britain, where he was interned on the Isle of Man and then took up a teaching position in Somerset.[114] To Adolf Hoel, however, the occupation meant promotion. Since 1933, Hoel had been a member of the nationalist political party Nasjonal Samling (National Unity), initially second only to Vidkun Quisling. Hoel was made full professor at the University of Oslo, and from 1941 to 1945 he was appointed rector. After the war, Hoel lost his professorship and was sentenced to 18 months in prison as a collaborator. The plans for the Nansen monument were never resumed.

A geophysical capital

When established in 1917, the goal for the Geophysical Institute was to calculate the geophysical world for the benefit of both science and society. But there were different opinions on how to achieve this. The head of the institute, oceanographer Bjørn Helland-Hansen, championed methodological pragmatism, collaboration and research aimed at explaining how different geophysical phenomena were interconnected.[115] In comparison, Vilhelm Bjerknes was a purist: in order to make weather forecasting an "exact science" that could calculate tomorrow's weather, his aim was to develop precise physical equations for all mechanisms relevant to the dynamics of the atmosphere. Another reason for the lack of collaboration between oceanographers and meteorologists was purely practical: the weather forecasters needed up-to-date observations from the atmosphere and methods to quickly turn these into predictions. When trying to understand ocean currents or make seasonal forecasts, the researchers could incorporate more differentiated observations and had more time to do the analysis.

The one who best managed to bridge the gaps was Bjerknes's successor, Harald Ulrik Sverdrup. During the *Maud* expedition, he collected data from several geophysical disciplines, he investigated how different phenomena were connected, and he combined empiricism with developing stringent geophysical theory. It is therefore not surprising that Sverdrup's legacy is one of a meteorologist, an oceanographer, a field-scientist, and a science administrator.

The Geophysical Institute was to a large extent financed by local donations and, especially in the early years, both Helland-Hansen and Bjerknes were motivated by a desire to give something back to the community. Here Bjerknes was most successful, giving birth to a weather forecasting unit that would became world famous. Over the decades that followed, the Bergen school of meteorology would revolutionize the practice of weather forecasting around the world. Helland-Hansen's concurrent involvement in establishing a factory was not a success. Still, as this chapter has shown, the Bergen school was but one of several approaches that were pursued and developed at the institute in the 1920s.

In its first decade, the institute attracted geophysicists from all over the world. Some came simply to get data for their own research, others to learn and discuss research methods, and yet others to learn how to organize and run a geophysical institute at a time when many such institutes were being established elsewhere. Finally, Bergen became a hub for geophysical calculations. In addition to collecting its own observations, many institutions sent their data to Bergen to be analyzed. What ensued after the institute in 1928 finally got its own building and expanded with the planned section for geomagnetism is the subject of the next chapter.

4
Moving in and moving on

"The little northern town of Bergen, sea-port, fishing-haven, market town, has done more for science in the last two or three generations than many – not to say most – university towns."[1] These were the opening words of a lengthy review in the journal *Nature in* the summer of 1928. The occasion was that the esteemed geophysicists in Bergen on June 7, 1928, finally moved into "a new and splendid Geophysical Institute, (...) a handsome building, set in a fine avenue of old trees and built on a bluff commanding an extensive view over the fjord and the islands and out to sea." According to Scottish biologist D'Arcy Thompson, who wrote the account, the institute building meant a final recognition of decades of contributions to naturalism, meteorology, and oceanography. Most of all he praised the local population for having financed the building: "Many an opulent British town might learn the A B C of civic pride and patriotism from the town of Bergen."

This chapter will investigate the geophysical sciences based in the new building: What did the geophysicists work with, and in what ways did working under the same roof facilitate collaboration? After detailing the new building, which has remained the heart of the geophysical community in Bergen for ninety years, we will investigate Bjørn Helland-Hansen's great oceanographic project, Jacob Bjerknes's aerology, and how the new section for geomagnetism moved from attempts at seasonal weather forecasts to particle physics. We will also examine the instrument makers, and what happened to the many researchers who left Bergen in the 1930s.

Under the same roof

The Institute building ("Geofysen") had been planned for more than a decade. As early as 1919, more than half of the total 900,000 NOK in building costs had been collected, mainly through donations from local shipping magnates and the business elite.[2] Architect Egill Reimers's drawings were finished in the early 1920s. Still, postponements ensued due to the financial crash in 1920, finding a suitable plot, and uncertainty over whether the weather forecasting office in Bergen would survive to be a potential tenant. Only when the plan for merging the three Norwegian weather forecasting offices was abandoned in 1926, and the municipality offered a free plot on the premise that the institute would be a cornerstone of the planned Bergen University, did the building begin.

Geofysen was a brick building with a full basement, three floors, a central block with two more floors, and a tower with an observation platform on top. The lighting, colors, and columns in the 16-meter-tall lobby gave an impression of stretching from the depths below, to the cosmos above.[3] A Foucault's pendulum demonstrating the rotation of the earth hung from the ceiling. Above the entrance to the reception and meeting room on the first floor, a bronze plaque still reads "The citizens of Bergen erected this building." Stairs led up to a gallery with a lecture room with seats for 70–80 listeners or tables for 30 students, as well as further stairs to the third-floor library. Above this, narrower stairs led to two floors for the weather forecasting office, and two more floors in the tower for the radio and equipment.[4]

Each of the three main floors had wide doors leading to the wings. The first floor hosted chemical and physical laboratories, offices and workspaces for oceanographers, and offices for the Magnetic Bureau and the new section for geomagnetism. The second floor hosted the meteorological researchers, a private apartment for the head of the weather forecasting service, rooms for students, and guest offices. Starting in 1931, the researchers at Chr. Michelsen Institute moved in to some of the second floor offices.[5] The third floor had a pitched roof and hosted a large library, drawing rooms, photo laboratory, archives and a printing press. In addition, there was a full basement with workshops, a laboratory for hydrodynamic experiments, an apartment for the caretaker, the main boiler room, and storage rooms connected by elevator to the laboratories above. The basement also had access to a 140-meter-long unused railway tunnel some 15 meters below the building, which was occasionally used for experiments.

Every room was temperature-controlled, and five of the rooms were arranged for magnetic work, with iron and copper nets built in to the walls and window-frames to avoid electromagnetic interference. Specially designed window hatches made it possible to easily turn the laboratories into effective Faraday cages. The floors were coated with soundproofing materials, and all 32 offices were connected by telephones. In his review, D'Arcy Thompson praised the many practical details:

"Many small 'gadgets' strike one every here and there. The ceilings are all fitted with rows of screw-sockets, into which hooks or rods may be screwed for the suspension of cables, pipes, or apparatus of any kind. The smaller rooms have their walls covered with jute, on which charts may be pinned. The furniture, desks, tables, drawers, etc., is all standardized and interchangeable. I was struck by the beauty of the woodwork everywhere."[6]

"Geofysen" was designed to facilitate and expand geophysical research, and to put an end to different sections working in different parts of town. That it was featured in *Nature* was recognition of Bergen as a cradle of and center for oceanography and meteorology. But to what extent did working under the same roof facilitate collaboration?

Helland-Hansen's big project

When not busy inspecting institution buildings in England, Scotland, Germany, Denmark, and Sweden, going on lecture tours, collaborating on multidisciplinary climate studies, sitting on the boards of several research foundations, being head of the Geophysical Institute, leading the section for oceanography or, beginning in 1930, directing Chr. Michelsen Institute, Bjørn Helland-Hansen's main research interest was to study the dynamics of ocean currents. In 1916 he had introduced a new tool for identifying unique water masses, the T-S diagram.[7] The basic idea was that by drawing temperature (T) and salinity (S) as a curve on the same chart, it would be possible to objectively identify different water masses. While this was not explicitly brought up, the approach had similarities to that of the meteorologists identifying air masses on their weather maps. The T-S fingerprint could be used to separate currents from surrounding water masses, and to establish base values for investigating changes over time. Helland-Hansen's idea was simple in theory, but empirically demanding: in order to find the normal variations of different water masses, a large number of observations were needed. Producing and analyzing T-S diagrams would, over time, become a staple practice for physical oceanography, in particularly as a tool to detect anomalies and possible errors. In the early 1940s, Harald Ulrik Sverdrup commented:

"The T-S diagram has become one of the most valuable tools in physical oceanography. By means of this diagram characteristic features of the temperature-salinity distribution are conveniently represented and anomalies in the distribution are easily recognized. The diagram is also widely used for detecting possible errors in the determination of temperature or salinity."[8]

The introduction of the T-S diagram was intimately linked to Helland-Hansen's increasing worry that the demand for increased precision that he and Nansen had promoted in *The Norwegian Sea* (1909), discussed in the previous chapter, was not sufficient to solve the riddles of ocean circulation. The findings from the Faroe-Shetland channel, where the warm Gulf Stream enters the Norwegian Sea, were a case in point: instead of bringing clarity, precision brought confusion. In 1910, Helland-Hansen joined Johan Hjort and Sir John Murray's expedition that resulted in the landmark *The Depths of the Ocean* (1912) in order to get new measurements.

"During our previous investigations in the Norwegian Sea we discovered that the hydrographical conditions often varied very considerably within a short distance or in the course of a short period of time. The variations were not always of the same character. A number of eddies, both large and small, occurred apparently during the movements of the water-layers, and there were up and down movements in the boundary-layers – possibly big submarine waves or something of that sort – as well as distinct pulsations in certain currents."[9]

It was known that the oceanographic conditions in the Faroe-Shetland channel were affected by an underwater ridge – the Wyville Thomson Ridge, named after the chief scientist on the *Challenger* Expedition – some 500 meters below the surface. Helland-Hansen conducted 15 oceanographic stations in sections on both sides to find out what the ridge did to the currents, each station taken only 20 nautical miles apart. He also recruited the Scottish research steamer *Goldseeker* to do simultaneous measurements in order to get as accurate a snapshot of the situation as possible.[10] Finally, Helland-Hansen anchored the research vessel *Michael Sars* in the same spot for 24 hours, making hourly measurements of how temperatures at different depths changed over time.[11]

Not only did he find large differences on the two sides of the ridge, but the observations also changed over time. One feature consisted of huge underwater waves between layers of the sea, up to 35 meters in height, which seemed to follow the tides. This "discontinuity layer" was potentially important for calculating the dynamics of the currents, for understanding the biology of the sea, and possibly for fisheries: "The discontinuity-layer is often a boundary between two different worlds of living organisms, and it is a point of interest for the study of these to know if this boundary is moving up and down, for this would probably imply that the organisms themselves (possibly even shoals of fish) were also being moved up and down."[12]

Another more sobering finding was that previous studies had probably overestimated the importance of individual observations: rather than being representative of a stable situation in a large area, the many variations within a small area suggested that each individual observation reflected only an instant at a specific point in time and space. When reporting his results, Helland-Hansen stressed that despite the tight grid, his findings only applied to a single day, and that one needed more time series from different seasons, preferably taken over several years, before more general conclusions could be drawn. When Helland-Hansen got his own research vessel, the *Armauer Hansen*, in 1913, he immediately set sail for the Faroe-Shetland channel, and he returned there in 1914. Interrupted by war, he could not make the next trip until 1922. By then it was clear that calculating ocean currents based simply on density, a product of temperature and salinity, needed to be supplemented with T-S diagrams, time series and direct current measurements.

Making direct observations of ocean currents on the open sea was "much more difficult than might appear at first sight."[13] In addition to specialized equipment solid enough to withstand the pressure in the deep, the vessel had to be able to remain relatively stationary or it must be possible to measure its exact movements. Between 1923 and 1932, Helland-Hansen and Swedish oceanographer Vagn Walfrid Ekman conducted a series of cruises in order to test and improve on Ekman's repeating current meter.[14] The first generation was triggered from the surface by sending a running messenger down the sounding line, but organisms in the water sometimes stopped the messenger on the middle of the line. When this happened, the whole observation series was ruined. The final iterations included a new mechanism for starting and stopping the counter, and a clockwork where 47 small numbered lead balls were dropped into one of 36 chambers for every 100th revolution of the propeller. After hauling the instrument back up on deck, the distribution of the lead balls indicated both the speed and direction of the current over time.[15]

Regardless of improved tools, each instrument could not be used in more than one place at a time. In order to get a synoptic picture, Helland-Hansen needed collaborators, like he had had with the Scottish *Goldseeker*. In 1924, he reached out first to biologist Johannes Schmidt at the University of Copenhagen, and then to Captain C.F. Drechsel, general secretary of the International Council for the Exploration of the Sea (ICES) from 1908 to 1928.[16] In the letters, Helland-Hansen outlined a working hypothesis: in shallow water along the continental shelf and on the fishing banks, the currents were linked to the tides, and the observations therefore differed mainly by time of day. Farther out at sea, the hourly oscillations seem to be of less significance, but instead even a small change in position could produce quite different observations. This suggested that there existed previously unknown whirls and bends in the deep sea. In order to find out, he needed a tight grid of simultaneous observations: "Both direct observations and dynamic calculations show that the conditions are far more complex than previously thought. (...) Even a large and common current like the 'Gulf Stream' requires an especially tight station grid, not only in a few sections, but along a number of sections."[17]

Helland-Hansen was well aware that it was a similar research

program in the strait between Denmark and Sweden that had motivated the establishment of ICES in 1902, but Drechsel showed little
interest. ICES was reorganizing and extending its scope, and it still
retained a committee for physical oceanography, but it was fisheries
biology and population dynamics that crystalized as the main focus
of the organization. The organization did organize synoptic studies,
and physical oceanography was interesting insofar as it was relevant
to fisheries and fish migration.[18] Helland-Hansen's initiative ended
in nothing. Possibly, going directly to the general secretary and not
through Norway's representative rubbed Drechsel the wrong way.
Since its inauguration, Norway's representative to ICES had been
Johan Hjort, and by the 1920s, Helland-Hansen and Hjort were no
longer on speaking terms. In 1916, Hjort had resigned from his post
as fisheries director, and after five years abroad, he received a professorship in Oslo. In his letter to Drechsel, Helland-Hansen argued
that Hjort represented only himself:

"Norway's representative to the Council has no collaboration with
the fisheries director, who has been informed of meetings only after
they had taken place. A consequence of this lack of collaboration
is that Professor Hjort does not have the necessary insight into the
research conducted in Norway, and he cannot enter international
agreements without instructions from those responsible."[19]

The lack of support from ICES was part of the reason why the
Geophysical Commission decided to approach the International
Union for Geodesy and Geophysics (IUGG) in 1925, and to establish
national committees for meteorology, oceanography and geomagnetism.[20] IUGG was an umbrella organization established by the
International Research Council in the immediate aftermath of World
War 1. Initially, Norwegian geophysicists had been skeptical of the
organization because of its policy to exclude Germany, arguing that
Norway, as a neutral party in the First World War, had a special
responsibility to facilitate international collaboration between former enemies. Indications that the exclusion of Germany would be
revoked, coinciding with Germany's negotiations to join the League
of Nations at the Locarno Conference in 1925, added another reason
for the timing.

Despite repeating the mantra that "the research should be carried on by several cooperating vessels, as the observations from a single vessel would not satisfy the requirements of synchronism," it would be another decade before Helland-Hansen's plan for international collaboration finally bore fruit.[21] The breakthrough happened at the IUGG general assembly in Edinburgh in 1936, where Helland-Hansen was elected president of the International Association for Physical Oceanography (IAPO). Arguing that the theoretical investigations had reached much further than the gathering of empirical data, and that "observations scattered in the manner usual in oceanographic investigations may lead to quite misleading results with regard to the physical and dynamical conditions in the sea," the assembly supported his suggestion for an international survey of the Gulf Stream Area between America and Europe.[22] A committee was appointed, and after a test with three Norwegian vessels in the summer of 1937, the observation program that Helland-Hansen had promoted for close to three decades finally took place in the summer of 1938.[23]

Seven ships from six countries, working from the middle of May until the middle of July, took serial observations of temperature and salinity in sections with stations every 20 nautical miles in strategic parts of the North Atlantic. In total, 482 stations with nearly 7000 points of observation were collected, almost twice the number of observations taken in any previous season. That the operation was headed by Helland-Hansen onboard the research vessel *Armauer Hansen* was a source of local pride. Since acquiring the vessel in 1913, Helland-Hansen had used every chance to promote the advantages of small research vessels, both for low operating costs and for making stable oceanographic observations. Weighing only 57 gross tons, this was the smallest vessel in the expedition. In comparison, Germany had hired the ship *Altair*, which weighed 4000 gross tons.[24] At this point in time, however, the goal was not to spread methods or theoretical frameworks – this had already taken place – but to herald a new era of large-scale international collaboration in data collection.

However, when the preliminary results were presented in Washington in 1939, it was without German colleagues from the *Altair*. Midway across the Atlantic they had been called back: Germany had invaded Poland, and the Second World War had begun. Although a new expedition was planned for 1941 with an even denser network of observations, to be presented at the next IUGG meeting to be held in Norway in 1942, the war put a stop to Helland-Hansen's efforts. Still, his research program of conducting intensive surveys with simultaneous observations in dense networks, preferably organized as international collaborations, would set the tone also for postwar oceanography. Finally, Helland-Hansen succeeded in defining the research questions that physical oceanography would seek to answer for decades to come:

"What is the inner mechanism of the ocean? What about the stability of the currents? What is the effect of turbulence of different scale, from the irregular movements of individual water-particles to the interplay of huge bulks of water? What about the occurrence of eddies and the causes of their formation? To this may be added the importance of a systematic study of the interaction between the sea and the atmosphere, with regard to energy and mass."[25]

The last point, studying the interactions between sea and atmosphere, had been a research interest in Bergen since the *Vøringen* expedition, and it had the potential to involve both oceanographers and meteorologists. At the IUGG assembly in Edinburgh in 1936, Helland-Hansen helped facilitate a shared session between oceanographers and meteorologists aimed at air-sea interactions, and how concepts from one field could be used in the other.[26] In addition to the shared genesis in Vilhelm Bjerknes's circulation theorem, the interaction between sea and atmosphere had been a research interest for Ernst G. Calwagen at the Bergen Observatory, and for Sverdrup during the *Maud* expedition. With Helland-Hansen elected head of the IUGG's International Association for Physical Oceanography, and Sverdrup's successor Jacob Bjerknes as secretary for the International Association for Meteorology, the geophysicists from Bergen were perfectly positioned to facilitate international collaboration. But instead, it was the Swedish Bergen school missionary Carl-Gustaf Rossby at MIT, and his student, Raymond B. Montgomery, based at Woods Hole Oceanographic Institution, who in the 1930s were the main proponents for using methods from meteorology to study ocean currents and the interactions between air and sea.[27] Sverdrup was busy organizing Scripps Institution of Oceanography in La Jolla, California, Helland-Hansen was fully engaged in organizing his international oceanographic collaboration, and Jacob Bjerknes had, at the time, other interests.

Bjerknes reaches for the skies

Sverdrup's acceptance of the research position at Chr. Michelsen Institute in 1931 opened his professorship for Jacob Bjerknes. As a third-generation scientist, Jacob had virtually grown up with the circulation theorem: when he was born in 1897, his father Vilhelm Bjerknes was a professor in Stockholm, and when his father got a post at the University in Christiania in 1907, Jacob moved with him.[28] Jacob entered the University in 1914, but after two years, he and Halvor Solberg joined Jacob's father in Leipzig to become his assistants. In 1917, 19 years old, he published his first scientific paper, in which he outlined the principle of convergence and divergence lines.[29] Later the same year, father and son Bjerknes, and Halvor Solberg, moved to Bergen.

Based in part on his father's theoretical work on air masses, and in part on observations from the dense observation network, Jacob Bjerknes published the paper "On the Structure of Moving Cyclones" in 1919.[30] The paper outlined a new theory for the three-dimensional structure and movement of cyclones in the atmosphere, and became a cornerstone for the Bergen school of meteorology. The core of the model was that cyclones were asymmetric, with sloping surfaces separating cold air to the north and west of the cyclone center and warm air to the south and southeast. These surfaces were the convergence lines, later renamed fronts. Along the surfaces, warm air ascends, causing bands and clouds and precipitation, while cold air sinks and spreads along the ground. In the paper, Jacob Bjerknes outlined the distribution of vertical motion, clouds, and precipitation, the mechanics for its motion, and how cyclones play a role in the interchange of air between the polar and the equatorial zones.

As we saw in the previous chapter, combined with Halvor Solberg's description of the birth of a cyclone, and Tor Bergeron's description of its death, the Bergen school produced a four-dimensional model of the structure and life cycle of cyclones.

Jacob Bjerknes was appointed head of the Weather Forecasting Office for western Norway in 1920. Two years later, he was invited to Zurich, Switzerland, to consult with the Swiss Meteorological Institute. Using data from mountain peak observatories in the Alps, he could verify the existence of sloping frontal surfaces up to an altitude of 3000 meters.[31] For this work, he received a doctorate from the University of Oslo in 1924.[32] He also received funding from the Birkeland Research Fund to continue the research in Norway. In collaboration with his Icelandic student, Jón Eyþórsson, and the Norwegian Trekking Association, he established Norway's first meteorological mountain observatory. A small cabin, 6.3 × 4.5 square meters, which could also accommodate hikers, was built in Bergen, dismantled, and transported to the base of Fanaråken, a mountain in Jotunheimen. In the spring of 1926, the materials were transported to an altitude of 2062 meters by zipline and muscle power, and reassembled.[33] Until 1932, Fanaråken was manned by weather observers during the summer months, and from 1932 to 1978 it was in operation all year round.

Mountain observatories were not the only approach to making observations at higher altitudes: since the turn of the century, weather balloons had been sent up routinely throughout Europe, and analyzing these results had been the rationale for the Geophysical Institute in Leipzig that opened in 1913. In 1928, Jacob Bjerknes began collaborating with Jules Jaumotte and Erik Palmén on analyzing observations collected by weather balloons sent up during cyclones passing over Europe. Jaumotte was the director of the Belgian Meteorological Office, a pioneer in aerology who, after spending time in Bergen in 1921, had introduced the Bergen school methods to Belgian weather forecasting.[34] Palmén was a Finnish meteorologist working at the Institute of Marine Research in Helsinki, whose doctorate from 1926 used the Bergen school methods to study the direction and speed of cyclone movements.[35] Previous observations had shown that the altitude of the boundary between the troposphere (where the temperature declines with altitude) and stratosphere (where the temperature rises with altitude) could differ from seven to fourteen kilometers. In a popular lecture, Bjerknes compared the boundary to the surface of an ocean "with enormous swells, several thousand meters high, stretching for hundreds of kilometers."[36] The analysis of the cyclone data described, for the first time, that the waves in the upper atmosphere were connected to the cyclones below.[37]

In 1933, the first of a series of "swarm ascents" was carried out in collaboration between Jaumotte and Bjerknes, with weather balloons being sent up simultaneously from Ås, south of Oslo, and Uccle, just outside Brussels.[38] Bjerknes then organized synoptic weather balloon ascents in order to get a three-dimensional image of passing cyclones, a project in which a total of 18 European observatories participated. When a strong cyclone passed over southern Scandinavia in February 1935, 200 balloons were sent up over a period of three days, the largest synoptic upper-air experiment to date.[39] Based on these and other observations, Bjerknes's presentation at the 1936 IUGG assembly in Edinburgh showed how the rotation of the earth creates waves between wedges of high pressure and troughs of low pressure in the upper atmosphere.[40] This research was the empirical basis for Carl-Gustaf Rossby's study of long waves in the upper atmosphere on a planetary scale, which today are known as "Rossby waves."[41]

The weather balloons had clear limitations. Unlike the instruments the oceanographers used to study the currents in the deep sea, the weather balloons could not simply be hauled back on deck: they had to be found with the measurement instrument intact. Each instrument package was marked with instructions to return it unopened for a finder's fee, but many were either lost or damaged. In the synoptic study in 1935, 120 balloons were found, 60 percent of those launched. An overview from 1938 showed that of 285 balloon ascents from Norway, dating back to 1932, 196 had been recovered. Of these, 169 had intact observations.[42] In addition to only three out of five ascents resulting in useful data, it often took months between the balloons being launched and the instruments being returned. Although the weather balloons offered insights into the structure of the upper atmosphere (the balloons usually reached an altitude between 5 and 10 km), they were useless for practical day-to-day weather forecasting.

During the international aerological week in December 1937, Jacob Bjerknes was the first in Norway to use the radiosonde, which is a weather balloon equipped with a radio transmitter that makes it possible to obtain the observations instantaneously. The research into balloon-borne radio transmitters had begun at the Lindenberg Observatory in Germany in 1921, while the term radiosonde was coined by the French meteorologist Robert Bureau, who sent up his first radiosonde in January 1929.[43] In Bergen, it was the Finnish physicist and meteorologist Vilho Väisälä who introduced the new technology. As soon as he had left, the remaining instruments were handed over to instrument makers Odd Dahl, Matz Jenssen, and Helmer Dahl, who disassembled them in order to produce an improved and cheaper version, mimicking the how the same instrument makers collaborated with oceanographers to improve on their instruments.[44]

Sending up weather balloons had much in common with the oceanographers' sending equipment from the surface into the sea, but there were no efforts to link the atmosphere to the ocean as Calwagen had done in the early 1920s. Since it could take weeks or months for the upper-air observations to come in, Bjerknes's research, as Sverdrup's before him, was relatively disconnected from the daily needs of weather forecasters. The radiosondes were meant to improve this, this but their introduction was delayed by the war.

Bjerknes's successor as head of the weather forecasting office, Sverre Petterssen, also did research on the upper air, but with a different approach: instead of deeper insights into the physics and energy balance of cyclones, Petterssen's goal was to develop "objective" methods for weather map analysis, procedures that ensured that the forecasts would be the same regardless of which meteorologists did the weather map analysis, and to work to improve long-term forecasting. He focused on the kinematic characteristics of atmospheric pressure fields: *How* weather phenomena move, not *why* they move. The result was a set of practical equations for calculating how features on weather maps moved over time.[45] "The leading idea is: to develop methods for evaluating the instantaneous velocity and acceleration of the various pressure formations, such as: cyclones, anticyclones, troughs, wedges, fronts etc (...) to evaluate the displacement and variation in intensity during the forecasting period."[46]

By developing procedural methods, the forecasts would depend less on an individual forecaster and thus become "more scientific."[47] The equations also facilitated a division of labor: trained assistants could solve equations while the forecaster worked on the weather map. After the war, the approach was continued by Norwegian meteorologist Ragnar Fjørtoft, from 1955 head of the Norwegian Meteorological Institute, who developed a "graphical method" for integrating the equations, but it was little used in Norway.[48]

From geomagnetism to particle physics

In addition to finally working under the same roof, the inauguration of Geofysen meant that the long planned "Section C" for geomagnetism and cosmic physics could finally open. Since the Geophysical Commission, in 1917, had decided that the analysis of Norwegian geomagnetic measurements should be placed in Bergen, nothing much had happened until 1926 when the Rockefeller Foundation donated 75.000 US dollars for an observatory for northern lights and geomagnetism in northern Norway, on the condition that the state would cover operating costs for both the observatory and a Magnetic Bureau to analyze the observations. The Magnetic Bureau was placed in Bergen as part of the new Section C.

The first professor of Section C, Ole Andreas Krogness, was a long-term collaborator of institute-head Bjørn Helland-Hansen. From 1906 to 1912, Krogness had worked as an assistant to physicist Kristian Birkeland, analyzing observations on magnetic storms and telluric currents. Krogness was then appointed director at the world's first permanent northern lights observatory at Haldde near Alta in northern Norway, a 900-meter-high mountain peak where he lived with his family and made meteorological, magnetic and aurora borealis observations for six years. From 1915 Krogness collaborated with Olaf Devik in establishing a storm-warning service for northern Norway, and from 1918 to 1928, he was the first director at the Geophysical Institute in Tromsø, which was established on the initiative of Krogness, Devik and Helland-Hansen. Helland-Hansen and Nansen's integrated account of the geophysical world, first published in 1917, relied heavily on Krogness's studies of correlations between geomagnetic fluctuations and weather phenomena. Helland-Hansen and Krogness were two of the founders of the Geophysical Commission, and from their correspondence it is clear that they considered each other close personal friends.[49] In Bergen, Krogness worked on geomagnetism, established a magnetic station, and worked as a consultant and expert witness in court.[50]

In 1929, Krogness was joined by Karl Falch Wasserfall, who was hired to administer the Magnetic Bureau in Bergen. The Bureau's task was to analyze and publish magnetic observations from Norway. Wasserfall's research interest was geomagnetic mapping, looking for temporal patterns, and making seasonal weather forecasts. Before being recruited to Bergen, Wasserfall had worked for six years at the meteorological institute in Argentina, and from 1911 to 1914 he had been head of the magnetic observatory in Pilar outside Buenos Aires. After returning to Norway, he had analyzed the magnetic observations from Roald Amundsen's polar expedition on the *Gjøa* (1903–06) and the magnetic material left by the Norwegian pioneer in geomagnetic research, Christopher Hansteen.[51]

Both Krogness and Wasserfall worked within the research paradigm set out in Helland-Hansen and Nansen's integrated climate study from 1917, namely to study periodic fluctuations in the earth's magnetism, changes in sunspot activities, and the earth's temperatures, and to look for patterns that could be used in long-term weather forecasting. This approach had little in common with that of the weather forecasters, whose air-mass analysis was already making the Bergen school famous. As Wasserfall saw it, the dynamic "Bergen school meteorology" was useful in making accurate weather predictions for the next few days, but had little to offer when it came to long-term forecasting. Instead, he argued, "a promising basis for long-term weather prediction" was to compare periodic variations in the magnetic field and surface temperatures.[52] In the early 1930s, Wasserfall published a series of seasonal forecasts in the national newspaper *Aftenposten*, which predicted which "weather types" would dominate in the next three months.[53] Rather than arguing for practical applications, it seems Wasserfall simply believed the results were sound and that the best way of proving the seasonal forecasts correct was to make the predictions public. However, the work was conducted without collaborators and failed to gather widespread support: neither the forecasters nor the professors in theoretical meteorology engaged in Wasserfall's seasonal forecasts.

Krogness died of an illness in the spring of 1934, at age 48, and on his deathbed he expressed a wish that his friend, Olaf Devik, would continue his research.[54] Devik had been Krogness's closest colleague for decades. In 1911, after two years as an assistant to Vilhelm Bjerknes, Devik had joined Krogness as an assistant to Kristian Birkeland. After moving to Haldde with his family in 1915, Devik had headed the committee tasked with establishing a weather forecasting service for northern Norway. Three years later, when Krogness moved to Tromsø to become director of the Geophysical Institute, Devik joined him to head the weather forecasting unit.[55] Their roads had separated in 1922, when Devik moved to teach physics at the Norwegian Technical College in Trondheim, but the two had remained in touch. Ten years later, after defending a doctorate at the University of Oslo in 1932, on the causes for ice drift in rivers, Devik was recruited by Helland-Hansen to a position at Chr. Michelsen Institute. There Devik worked on developing methods for using sound and radio signals to make coastal navigation safer.

Instead of adhering to Krogness's deathbed wish, the board of the Bergen Museum decided to use the opportunity to change the research focus for Section C. Rather than geomagnetism and weather prediction, they wanted a stronger focus on cosmic physics. When Helland-Hansen informed Devik that he could not expect to be able to continue his work on radio navigation should he get the position, Devik withdrew his application. He did this not with a heavy heart: he was already part of the research environment in Bergen, and by withdrawing his application he would make it more likely that the community could be expanded with a new person.[56] While Wasserfall at the Magnetic Bureau continued his search for correlations between geomagnetism and weather phenomena, Krogness's successor, Bjørn Trumpy, would be guided by the firm belief that cosmic rays held the secrets to the structure of the atom.

Trumpy's interest in physics was awoken in December of 1916, when he as a teenager attended lectures on X-rays by Krogness and Devold, who at the time were advocating a Geophysical Institute in Tromsø.[57] Trumpy then studied chemistry at the Norwegian Technical College in Trondheim. After defending a doctorate in 1927 on spectral physics, he continued his research with physicists Max Born in Göttingen and Niels Bohr in Copenhagen on a scholarship from the Rockefeller Foundation. When Devik was recruited to Bergen in 1932, Trumpy took over Devik's teaching position in Trondheim. Trumpy also did research on molecular physics on the side: before moving to Bergen, he had published more than twenty papers in *Zeitschrift für Physik*. He would continue to publish in leading international journals throughout his career, and set a new standard for Section C.[58] During its lifetime, from 1928 to 1988, just over 50 percent of all publications from Section C were papers in international journals. This proportion was more than twice that of the meteorologists and oceanographers, whose publication practices after the Second World War would increasingly rely on producing reports.[59]

After being recruited to fill Krogness's post as professor in geomagnetism and cosmic physics in 1935, Trumpy pursued three separate but interrelated research projects: overseeing the Magnetic Bureau's mapping and surveillance of the magnetic field, investigating the nature of cosmic radiation, and experimenting with high voltage particle physics. None of the projects had any overlap with the research conducted by the oceanographers or meteorologists, and the research eventually gave birth to a physics department, where Trumpy became the first director.

In 1938, Trumpy and Wasserfall initiated collaboration the between Magnetic Bureau and the Norwegian Hydrographic Service, in order to produce a geomagnetic map of Norway.[60] The goal was to show the intensity, declination and inclination of the magnetic field, "especially for the sake of aviation, shipping and prospecting."[61] Knowledge about local deviations in the magnetic field could prevent accidents both at sea and in the air, especially when visibility was low, and magnetic abnormalities were believed to be caused by minerals that could be mined. The maps were finished some months after the Second World War reached Norway, and the country was occupied by Germany in April 1940. Instead of publishing the maps, Trumpy decided that they had low resolution, and that they had to be replaced with a new map with higher resolution. The number of observation points was increased from 119 to 1200, while the measurement schedule slowed noticeably. The summer of 1943, Captain E. Kjær, who had led the in-field observations, was arrested and sent to Germany, which further slowed the project. After the war, Trumpy explained: "The analysis was intentionally first taken up with full strength toward the end of the war, and was completed only after the armistice, so that the magnetic maps would not benefit the occupying forces."[62]

Trumpy's enthusiasm for geomagnetic mapmaking was overshadowed by his interest in cosmic radiation. In 1937, he built a cloud chamber – an early particle detector – to count and photograph the trails left by high-energy cosmic particles. By moving the instrument from the first floor offices to the abandoned railway tunnel 15 meters below the institute, he concluded that about 3/4 of the radiation was filtered by the intermediary rocks. By adjusting the angle, he concluded that the cosmic rays must originate in specific points in outer space, and that the radiation varied with time.[63] Trumpy's fascination, however, was not with the origins and frequency of cosmic rays, but their fundamental structure. He quickly rebuilt the detector, using lead plates to separate it into different chambers, and managed to produce secondary radiation that made it possible to differentiate between different components of the radiation. He interchanged the lead plates with aluminum plates to measure the mass and penetration of cosmic radiation, and to calculate the mass of mesotrons (now mesons), which are short-lived subatomic particles produced in the interaction between cosmic rays and the intervening materials.

His approach to the study of cosmic rays was a continuation of Trumpy's many publications on the Raman Effect, which is what most clearly illustrates his departure from Krogness's use of statistical curves to investigate northern lights and geomagnetic fluctuations.[64] The Raman Effect, which had won Indian physicist Chandrasekhara Venkata Raman the Nobel Prize in Physics in 1930, describes what happens when light interacts with the cores of atoms at a quantum level, and is investigated by measuring how filtered light reflects and scatters. The phenomenon explains why the sky is blue, and gave birth to the phrase "blue-sky research" for being curiosity-driven and initially having no obvious application. However, the effect made it possible to observe vibrations in the ultraviolet spectrum, and to examine nuclear cores of different elements by how they reacted to high-powered radiation. In addition to gaining new insights into nature's fundamental building blocks, the findings could in turn be used to identify unknown materials.

Trumpy's research also focused on using high-energy rays to make elements radioactive, and subsequently measure their half-life. To do so, he built a Van de Graaff accelerator at Bergen's Haukeland Hospital with a capacity of 1.7 million volts, which was in use in cancer treatment from 1942 to 1971. The accelerator was financed by a Norwegian Red Cross fundraising campaign, and built in collaboration with Trumpy, instrument-maker Odd Dahl, and head physician Sigvald Nicolay Bakke. Norway was a poor country, and the argument for the accelerator was that it was a cost-efficient avenue to radiation therapy: "One gram of radium costs more than one million NOK, but a high voltage facility could produce an effect comparable to at least a kilo of radium, for a far smaller sum."[65] The first year, what was probably the world's largest medical instrument treated an average of 30 patients per day. After hours, it was used for physics experiments. In 1942, Trumpy successfully irradiated beryllium, magnesium, aluminum, chromium, iron, nickel, copper, zinc, silver, cadmium, indium, gold, and lead, and the next year bismuth, strontium, antimony, selenium, and platinum, and measured their half-lives: "We assume that the atomic nuclei under high-energy radiation are brought to a higher energy state, and that they can exist in this excited state for some time before returning to their basic state."[66] In the experiments, Trumpy discovered that the same elements could be brought to different energy states, with different half-lives, called isomers. From his first twelve elements, Trumpy identified 35 different isomers. He also found methods for irradiating elements to metastable states using lower voltages by varying the wavelengths used. Although cosmic particles consist of high-energy radiation, the experiments had more in common with pure physics than geophysics.

Another ambition of the accelerator at Haukeland was to strengthen Bergen's position as a scientific (rather than geophysical) center, and to boost the plans for upgrading the Bergen Museum to a full-fledged university. In 1943, Trumpy was appointed head of the Bergen Museum, and in 1946, the foundation stone for the University of Bergen was laid by King Haakon VII. The first construction work for the University consisted of expanding the Geophysical Institute with two new wings, and when the University officially opened in 1948, Trumpy was elected as its first rector.

After Krogness's death and Wasserfall putting aside his seasonal weather forecasts, the research pursued at Trumpy's Section C was seen as too far removed to be relevant to the meteorologists and oceanographers. Even if reality was made up of nuclei and subatomic particles, they existed at a scale with no application to studying to weather systems and ocean currents. Trumpy did establish strong international publication traditions, but his research on the fundamental nature of matter and radiation had more to do with particle physics than with the geophysical vision of calculating the world.

The instrument maker

Geofysen had a workshop for carpenters and three workshops for instrument makers. To some extent, the instrument makers were the invisible glue between the different groups of researchers. In addition to maintenance, repairs and calibrations, the instrument makers worked to develop and improve instruments for meteorologists, oceanographers, and Section C alike. Trumpy's one-man particle physics research, for instance, would not have been possible without close collaboration with the most colorful instrument maker of them all: aviator, explorer, radio-pioneer, oceanographer, designer, and eventually builder of particle accelerators and nuclear reactors, Odd Dahl.

Dahl's first public appearance occurred in 1920, when the 22-year-old was invited by a local newspaper in Stavanger to demonstrate how he had established communication between a land-based station and a fishing boat at sea using a homemade radio. This got him in trouble with the Norwegian Telegraph Board, which had a monopoly on the airwaves. His next invention, turn signals for cars, was dismissed as overly complicated: sticking your arm out of the window is simple and it works, so why complicate matters? In his autobiography, he reflected: "From experiences like these, there are lessons to be learned. I learned the importance of timing: if you are too early, you cannot expect support for even the best idea. Later I learned something equally useful: if everyone agrees, you are too late."[67]

In 1921, Dahl graduated first in his class from the Norwegian Army Flight School at Kjeller, just as polar explorer Roald Amundsen was looking for a pilot to join his *Maud* expedition in an attempt to reach the North Pole with a flying machine. Dahl, whose only formal education before the pilot's license was middle school and some evening classes, was hired to pilot a small Curtiss biplane, man the expedition's wireless telegraph, be a film photographer, and step in as handyman. As we saw in the previous chapter, this very first attempt at using aircraft in the high Arctic was not a huge success: the biplane "Kristine" survived only three landings.[68] Stuck in the ice, Dahl instead served as an instrument maker and a technical assistant to H.U. Sverdrup, who was in charge of the scientific investigations.

Among the lasting inventions from the expedition was Sverdrup-Dahl's recording current meter, which made it possible to observe ocean currents in real time from the relative comfort of the vessel itself.[69] The operating principle was that the propeller would close an electrical circuit after a set number of rotations. The current's strength was measured by the interval between the pulses, while its direction could be measured by the strength of the pulses.[70] The instrument was also extremely robust: despite operating in subzero salt water, it was in continuous use for fourteen months without a failure.

When Sverdrup returned to the Carnegie Institution in Washington after the expedition, he brought Dahl with him. On Sverdrup's suggestion, Dahl was hired as a technical assistant to the Department of Terrestrial Magnetism with one condition: first he needed a vacation to get the polar cold out of his veins. During the next two years, Dahl and a friend crossed South America on foot and canoe, from Callo on the Peruvian coast to the mouth of the Amazon River. Officially, the goal was to pick up a postcard sent poste restante by the *Maud*'s crew to the US Embassy in Manaus, where the Rio Negro and Amazon River meet.[71] Two years after returning to the United States, Dahl spent a year trekking in Asia, making magnetic observations for Carnegie that, by his own admission, were quite useless. Part of the plan was an expedition to the summit of Mount Everest, but after being denied a visa to Nepal, this had to be canceled. When not inventing scientific instruments, Dahl was an adventurer.

At the Department of Terrestrial Magnetism in Washington, Dahl worked with the well-known physicists Gregory Breit, Merle A. Tuve, and Lawrence Hafstad, building instruments to examine how radio waves were reflected in the upper atmosphere and how heat was produced by microwaves, and building an accelerator for the department's new nuclear laboratory.[72] Both Tuve and Hafstad were of Norwegian descent, and together with Dahl, "the three musketeers" earned the American Association for the Advancement of Science Prize in 1931.

Dahl's return to Bergen occurred after Sverdrup "borrowed" him for a month to repair some of his oceanographic instruments that had been used on the *Maud* expedition, and starting in 1936 Dahl was hired as engineer and instrument maker for the Geophysical Institute and Chr. Michelsen Institute (CMI). Until 1956, CMI rented office and workshop space at Geofysen, with activities ranging from research in economics and geophysics to instrument making and vaccine development.[73] In addition to working on oceanographic tools, some of which were commercialized through the company Bergen Nautik, Dahl was involved in improving Vilho Väisälä's radiosondes for the meteorologists in the 1930s. Lastly, he collaborated closely with Trumpy, making the instruments necessary for studying nuclear physics.

Just as Sverdrup had been a door-opener for Dahl at Carnegie, Dahl's contacts in the United States, especially Tuve and Hafstad, would open doors for a host of Norwegian researchers over the years. In 1942, Tuve became the founding director of the Johns Hopkins University Applied Physics Laboratory, and starting in 1946 he was the director of Terrestrial Magnetism Research at Carnegie. Hafstad built the first nuclear fission reactor in the United States in 1939. When Tuve returned to Carnegie, Hafstad took over his laboratory and professorship at Johns Hopkins. Hafstad also served as technical chief of the Atomic Energy Commission in Washington, and later became vice president and head of research at General Motors Corporation.[74]

These contacts were crucial in 1946 when Dahl and Gunnar Randers, head of the Norwegian Defence Research Establishment, traveled to the United States to learn what it would take to build a nuclear reactor in Norway. The visit took place just a year after the Second World War, when the United States was still the only country that had harvested the "power of the atom" for weapons and reactors, when nuclear research was still classified as part of the secret Manhattan Project, and only months after the Truman Doctrine was announced. Still, the two Norwegian scientists were given the information they needed, including the average number of neutrons produced in uranium fission, which is fundamental to controlling a nuclear reactor:

"The conversations took place with ten men around a table, Randers, me and our 'opposition.' We presented our schematics for the reactor and our thoughts on how we had planned to proceed, and then we had to interpret their reactions to the best of our abilities. It was clearly a game, but since both parts played by the same rules we eventually managed to communicate quite precisely, without 'slips of the tongue.'"[75]

Open doors to deep military secrets were not only due to personal contacts, publications, or instrument construction; Norwegian scientists were held in high esteem in the United States in general and were seen as ambassadors for a thriving scientific community. The "three musketeers" had all identified as Norwegians, and in nuclear physics, the Norwegian scientist Rolf Widerøe's inventions of the resonance accelerator and the betatron were revered. In oceanography, Helland-Hansen had given a successful lecture tour in 1935 and was elected leader of the International Association for Physical Oceanography in 1936, and in 1938–39 he was Hitchcock professor to the University of California at Berkley.[76] Likewise, Sverdrup was highly respected as head of the Scripps Institution. In meteorology, Jacob Bjerknes was leader of the International Association for Meteorology starting in 1936; Vilhelm Bjerknes, Carl-Gustaf Rossby, Jacob Bjerknes, Jørgen Holmboe and Sverre Petterssen were all respected as scientific pioneers, and the narrative of modern weather forecasting originating in Bergen was beginning to take hold.

When Dahl and Trumpy visited the United States in 1947 to get an overview of the state of high-energy physics, the two were welcomed with open arms at the Carnegie Institution of Terrestrial Magnetism and the National Bureau of Standards in Washington, D.C., the Department of Nuclear Physics at the University of Chicago, the Betatron Laboratory at the University of Illinois, the Department of Physics at Yale, the Palmer Institute and Cosmic Ray Laboratory at Princeton, MIT, Harvard, and General Electric's Research Laboratory in Schenectady, New York.[77] The insights they gained were used to plan the building of a new van der Graaff accelerator capable of producing 2 million volts, and a betatron with a capacity of 50 million volts. Both were placed in a shielded building next to Geofysen. They also learned the last pieces of the puzzle for the nuclear reactor, which made Norway the sixth country in the world to "go nuclear" in 1951, after the USA, Great Britain, France, the Soviet Union, and Canada. The reactor was constructed by Odd Dahl at Kjeller, north of Oslo. The same year he was the first to be awarded an honorary doctorate from the University of Bergen. Dahl was also involved in building the solar tower observatory at Harestua outside Oslo, which opened in 1954; he initiated the construction of a 25-gigaelectronvolt

Proton-Synchrotron at CERN, and was involved in establishing the Andøya Rocket Range.[78]

Moving on and opening doors

While the early 1920s were characterized by the Bergen school attracting international guests, the 1930s were marked by members of the Bergen school leaving and taking with them a research culture they had learned and developed there. In the fall of 1935, Helland-Hansen went on a lecture tour in the United States where he visited Woods Hole Oceanographic Institution, Massachusetts; the University of Washington in Seattle; and Scripps Institution of Oceanography, La Jolla, California. At Scripps, he was asked by the director, Thomas Wayland Vaughan, who had stayed in Bergen with his colleague, Harry Richard Seiwell, in 1932–33, to evaluate the operations and give advice on a possible successor. Scripps had been founded in 1903 as the Marine Biological Association of San Diego, with a research program focused on marine biology, and nine years later it was made part of the University of California. The institution had its own campus with laboratories and a research vessel. However, in his report, Helland-Hansen noted that the institution lacked focus and suggested researching coastal circulation and its impact on biology as a way to give the researchers a common goal. This philosophy reflected his own focus in Bergen: rather than extensive sampling, he promoted intense studies of a limited area. Reflecting a view of the physics of the oceans as foundation upon which ocean life depends, physical oceanography should also dictate the direction for marine biology. Lastly, he recommended Harald Ulrik Sverdrup as the best man to head the ambitious research program.[79] Sverdrup accepted a three-year position, starting in 1936.[80]

When Sverdrup arrived at Scripps, he found an institution with no overarching research theme, no creditable teaching program, limited financial resources, a research vessel capable only of short coastal day cruises, and low-paid researchers living in shabby living quarters existing more or less isolated from the rest of the academic community. The closest university campus was in Los Angeles, some 170 kilometers to the north. In the first of a series of letters home to Bjørn Helland-Hansen, written about two-and-a-half months after accepting the position, Sverdrup reported that although he enjoyed his colleagues, "either the University makes the institute into a real oceanographic institution – or they turn it into a biological station – and in that case I return home."[81] However, shortly after he arrived, an explosive fire decimated the research vessel at anchor at the San Diego Yacht Club, seriously injuring two people. This provided a new beginning: over the following years, Sverdrup organized a new boat capable of longer cruises, fired some of the staff, and reallocated funds. He reorganized the teaching, and made field experience and basic knowledge in all fields of marine sciences a prerequisite for specialization. He turned the weekly staff meeting from administrative gatherings to research seminars with presentations and detailed minutes on who said what during the discussions.

Almost every month, Sverdrup sent letters to Helland-Hansen in Bergen, sharing frustrations and joys, occasionally asking for advice. Although Scripps under Sverdrup can in no way be seen as an extension of the Geophysical Institute in Bergen, many of the choices Sverdrup made were directly inspired by his experiences in Bergen. The weekly colloquia followed the blueprint of the colloquia he and Helland-Hansen had organized in Bergen. The move from extensive studies to an intensive field research program, where all specialties worked on problems in a defined area repeatedly throughout a year to look for seasonal changes, was fully in line with what Helland-Hansen had recommended.

In their correspondence, the two offered continuous insights into the oceanographic community behind the scenes on both sides of the Atlantic. A repeated issue was to find a person to take over when Sverdrup's tenure was over, and the two had a frank discussion about the qualifications of their colleagues. Sverdrup talked about his relations with institutions such as the US Navy and the Coast and Geodetic Survey, and named researchers at Scripps and other institutions, while Helland-Hansen provided updates on colleagues in Norway and Europe. This was typical for Helland-Hansen, who saw it as important to keep an eye on research activities in his small but growing field. In 1929, when he was about to step down from the board of the Scientific Research Fund of 1919, for instance, he had asked to continue getting copies of the applications: "It is very valuable to pay attention to those individuals and issues that our scientific activities entail, and in this regard, applications submitted to the large funds are very enlightening."[82]

Sverdrup at Scripps made Bergen less attractive for foreign guest researchers. In a letter from Thomas Wayland Vaughan congratulating Sverdrup on the position, he explained how some of his future colleagues had reacted to the announcement.

"Roger Revelle, who did come up to me yesterday, has planned to go to Europe for about a year, leaving this country about the first of September to go first to the meeting of the International Association for Physical Oceanography at Edinburgh, and then on to Bergen to work with Helland-Hansen for a while, said to me after the announcement, 'I don't see any need in my going to Europe!'"[83]

Revelle was one of the researchers Sverdrup and Helland-Hansen would later discuss as a potential future director of Scripps, a position he held from 1950 to 1964. After attending the 1936 IUGG assembly in Edinburgh, Revelle did spend eight months in Bergen with his family. He later commented: "Those were bad days in Europe, with Adolf Hitler screaming his bloody nonsense over the radio, the on-going tragedy of the Spanish civil war, and some fascists in every country. Bergen was nevertheless a wonderful place for us. We learned to ski and became very good friends with several of the scientists at the Geophysical Institute and their families. I didn't learn much about oceanography, but I did learn a good deal about people."[84]

When Revelle arrived, many of those who had made Bergen a world capital for geophysical research had already left or were about to move. In 1921, meteorologist Halvor Solberg left to study in Göttingen and Paris, and when he returned to Norway in 1927, it was to work with Vilhelm Bjerknes, who had left for a professorship at the University of Oslo in 1926. Tor Bergeron spent several years in Stockholm and Leipzig in the 1920s, before he made his final departure from Bergen in 1929. Sverre Petterssen, who had replaced Tor

Bergeron at the weather forecasting office and was made its leader when Jacob Bjerknes accepted Sverdrup's professorship in 1933, left for MIT in 1939. There he wrote two of the most widely used textbooks in weather forecasting: *Weather Analysis and Forecasting* (1940) and *Introduction to Meteorology* (1941). The former was translated into more than 20 languages. After the second textbook, Petterssen moved to Britain, where he became head of the division for upper-air forecasting at Dunstable, and was the meteorologist who got the D-Day invasion postponed.[85]

At MIT, Petterssen joined another of Vilhelm Bjerknes's assistants, meteorologist Jørgen Holmboe. Holmboe was hired when Sverre Petterssen was promoted, but already in 1936 he had been recruited as an assistant to Carl-Gustaf Rossby at MIT.[86] Rossby had only spent two crucial years in Bergen, 1919–1921, before moving on, but by the mid-1930s, Rossby was the main propagator of the Bergen school methods in the United States.[87] In 1939, mathematician Jonas Ekman Fjeldstad, who had mainly worked on internal waves, accepted a professorship in Oslo. Finally, the outbreak of the Second World War meant that Jacob Bjerknes, who was on sabbatical in the United States, got stuck and never returned. Instead, he and Holmboe established a Department of Meteorology at the University of California, which played a central part in educating US meteorologists for the war effort using the Bergen school methods.

The departure of leading geophysicists was not the only reason why the center of gravity for geophysics had begun shifting away from Bergen: starting in the 1920s, numerous new geophysical institutions were opening and looking for skilled researchers. In 1937, Thomas Wayland Vaughan catalogued 245 institutions doing oceanographic research.[88] In meteorology, more and more forecasting institutions were switching to Bergen school methods, and several universities opened departments for meteorological research. Even if new and modern facilities made Bergen attractive, there were simply more and more places where excellent geophysical research took place.

Working together?

It is difficult to assess to what extent the new building facilitated collaboration. Based on publication records, the geophysicists collaborated very infrequently, and if they did, it was mainly with researchers from other institutions. Out of 160 publications between 1928 and 1940, only twenty had multiple authors (12.5 percent). Of these, eleven were co-authored with researchers based elsewhere.[89] The in-house collaboration consisted of seven reports produced by the Magnetic Bureau and Section C, and two publications where the professor at Section C published with a student. Moving to the same building does not seem to have made much of a difference: in the decade preceding the opening of Geofysen, nine out of 49 publications were co-authored (18 percent), eight of which were written in collaboration with researchers from other institutions. The Bergen school of meteorology, which concurrently was described as an ongoing colloquium, resulted only in a single co-authored study on the formation of rain.[90]

Publication records reflect publication practices, and should not be taken at face value as a proxy for research collaboration. First, most of the studies coauthored with researchers elsewhere were based on collaborations taking place in Bergen or joint expeditions. Next, both the annual reports and correspondence indicate that the geophysicists saw themselves as members of the same research community, occasionally praising the spirit of collaboration.[91] The weekly interdisciplinary colloquia that Sverdrup and Helland-Hansen had started continued in the new building, and the proceedings indicate that the discussions on research results, ongoing projects, and publications were well attended by oceanographers, meteorologists, weather forecasters, and occasionally members of Chr. Michelsen Institute and foreign guests.[92] Although the records seem less complete than in the preceding decade, the number of annual guest researchers seems to have stayed stable throughout the 1930s. It seems unlikely that foreigners would visit to learn, and often make return visits, if they did not feel part of an inclusive research community.

International organizations were recognized as important avenues for collaboration, as exemplified by Helland-Hansen's initiatives toward ICES and IUGG to organize synoptic studies of ocean currents. The cold shoulder from ICES was part of the reason why the geophysicists in Bergen oriented themselves toward IUGG. Possibly, the organization's position as an umbrella over disciplinary associations contributed to the different trajectories of the different disciplines, despite their shared genesis. Thanks to IUGG, Helland-Hansen finally managed to organize his synoptic expedition. Although he also took part in facilitating a joint session on air-sea interaction between meteorologists and oceanographers, this was not followed up by research collaborations in Bergen. On the other hand, Bjerknes's weather balloon collaboration shows that larger collaborations could take place without an institutional superstructure. For both oceanographers and meteorologists, it was more important to get standardized observations from many locations than to try to connect the oceans and the atmosphere.

As head of the Geophysical Institute for three decades, Helland-Hansen put a premium on independent research. This was reflected in how he set up Chr. Michelsen Institute, which offered good salaries and complete independence for the members to pursue their own research interests. Likewise, the three sections of the institute were self-governing, and the professors had full freedom to pursue their own interests. The few documents we have come across that discuss hiring new employees emphasize the candidates being known figures with a strong publication record. This meant that research interests changed mainly when a professor was replaced, such as when Jacob Bjerknes replaced Sverdrup, or when Trumpy was recruited to Krogness's professorship. Despite Trumpy's move to pure physics, Helland-Hansen seems to have been committed to a vision where everyone worked on the same problem, to calculate the geophysical world, but attacked the problem from different angles. When he evaluated Scripps in 1935, for instance, his main concern was that Scripps lacked focus. In Bergen, all research consisted of gathering empirical observations, analysis, and developing physical insights that could eventually be used in prediction.

Of all the groups working at Geofysen, it was the instrument makers who had the most to do with the others sections. Odd Dahl and his colleagues in the basement assisted researchers in oceanography, meteorology, and geomagnetism alike, and did not shy away when Trumpy needed help constructing instruments for particle physics. But it seems that at the time, most instrument makers were seen as support staff, and not scientists. To the extent that instruments were named, it was after the researcher who came up with the idea, and not the instrument makers, who could spend years refining the design. Again, Odd Dahl stands out: awarding Dahl an honorary doctorate in 1951 was recognition that an instrument maker could also make significant contributions to science. However, he set the bar sky high by spending years doing field research, publishing a number of scientific papers, and constructing everything from current meters, thermosondes, and radiosondes to particle accelerators and nuclear reactors.

5
The ocean and atmosphere as field

"But in the sky the most lovely aurora lived and played in countless formations, draperies and hanks of beams and fluttering flakes up towards zenith. The totality of it was inexplicably mystical and captivating. And in our wake the mareel as a cheerful din of champagne – a radiant Milky Way in miniature."[1]

So wrote Aagot Borge, Bjørn Helland-Hansen's secretary from 1919 to 1952. In her diary from September 1947, Borge describes a trip home to Bergen with the research vessel *Armauer Hansen*. The starting point was Olden, the inner end of the long Nordfjord, where Borge and many of the oceanographers at the Geophysical Institute had spent the summer in their private cottages. Sailing out the fjord and down the coast, Borge was stunned by the different kinds of light in nature – in addition to the starry sky, aurora and phosphorescence of the sea, she was baffled by the intense sunlight, walls of rain with zig-zag lightning, and double rainbows. On board, life was pleasant as she played bridge with Bjørn, oceanographer Håkon Mosby and assistant Olav Aabrek. Gradually, however, during the night increasing wind and growing waves made life on board unpleasant. "It rolled and shook and rumbled and creaked, and I (...) got extra practice from the work of keeping myself within the berth."[2]

Aagot Borge gives an interesting glimpse into the field life of oceanographers. On board the *Armauer Hansen* scientists and others worked, ate, slept, discussed, and amused themselves with bridge. They became ill and took medicines, they returned to port because of bad weather, they vomited and went ashore, having to seek shelter in a closed summer house on an island until the vessel had been pulled off the rocks. Borge saw this field life from a certain distance, allowing her to discover the exotic aspects of it and describe it poetically. The scientists themselves also became poetic on occasion when moved by a beautiful sight, a dangerous situation, nostalgia, or other feelings or sensations.

This chapter will discuss how geophysicists in Bergen traveled into the field to seek answers to their questions before the Second World War. Work in the field has been a part of the geophysical practice, and perhaps an identity mark for geophysics in Bergen. In 1928, D'Arcy Thompson described it as one of the characteristics of the scientific staff in Bergen; they were "travelers and explorers as well as laboratory men."[3] Bergen's favorable position in the geophysical field had been one argument among others for placing the Geophysical Institute there and not in Oslo.[4]

Going into the field could be a journey into the unknown or a routine trip to a place that was visited annually. The field was a distant place or just outside the door. Field life was hardship in the Norwegian Sea or luxury in the Mediterranean, a detour from the normal or a return to the normal. The field also brought out different aspects of each person's personality.

Field life has changed throughout the 100 years covered in this book. In 1913, the field life of oceanographers could mean sitting for 24 hours to cover two tidal cycles in the *Armauer Hansen*'s lifeboat in the North Atlantic to take current measurements, eventually having to abort prematurely due to increasing wind and bad weather.[5] In the 1970s, hunting the same double tidal cycle in the Norwegian Sea could mean working 6-hour shifts on board the much more comfortable research vessel *Helland-Hansen*.[6] In 2013, field life could mean going to Antarctic waters on a two-month expedition time-shared with a range of other projects, disciplines, and purposes, leaving oceanographers with two days of fieldwork and the remaining time behind a computer.[7] The work intensity, comfort, danger, and "normality" of field life, despite their diversity, have changed over 100 years, but not linearly.

To meteorologists, field life had a different meaning. From early on, their fieldwork was about instructing and controlling other people's observations taken from lighthouses, farms, and towns in rural Norway. Meteorologists had built an infrastructure in the field, consisting of networks of stations supplying them with observations, allowing some meteorologists to complete whole careers without ever going into the field.

The field in oceanography and meteorology

The field has been contrasted to the laboratory or the museum as a venue of science with a different set of properties. While laboratory scientists or museum workers study elements of nature in depth and under controlled conditions, field scientists travel and get a briefer, but broader and more "real" impression of nature.[8] Geophysics in Bergen grew out of a museum tradition in which a similar dichotomy had developed to be clear and outspoken. Close companions of Bjørn Helland-Hansen, such as Fridtjof Nansen and Johan Hjort, had advocated for the advantages of practical field research and studying the scientific objects in nature rather than as "collected dead treasures" in museums.[9] From the start, Helland-Hansen had worked in a scientific community with strong aspirations to field life at the cost of scientific practices in museum collections.

However, the other professor at the institute, Vilhelm Bjerknes, did not have the same field experience. Bjerknes studied the atmosphere as a physicist and in a way that placed the meteorological practice in the map room. It was here, from the dots and lines on the maps, that new concepts, methods, and scientific problems had emerged from the attempts to solve the practical problem of forecasting tomorrow's weather. The field was important, but it was mostly inhabited by others than the meteorologists themselves.

The new disciplines of geophysics, dynamic meteorology, and oceanography that Vilhelm Bjerknes and Helland-Hansen had developed were founded on a set of natural laws taken to be universal. Picking places to practice exact science in the field meant balancing the complex, uncontrollable conditions in nature against the requirements for precision and control that characterized exact sciences. This must have been a familiar problem in Norwegian science. According to Robert Marc Friedman, field sciences were a significant part of Norwegian science, due to both tradition and economy.[10] Bjerknes and Helland-Hansen had good conditions for developing new disciplines on the basis of "physics in the field."

Robert E. Kohler has pointed out that while the field offers only unique places, the laboratory is characterized by a certain "placelessness": the claim that what happens in one laboratory could happen in any laboratory in the world is crucial and an essential property of laboratories.[11] When field-oriented geophysicists in Bergen picked their places for investigations and experiments, they went through a range of considerations. Both meteorologists and oceanographers evaluated the field according to the access it gave to natural "phenomena in their pure forms."[12] Bergen, with the surrounding coast and ocean to the west, was regarded as an excellent field in this respect.[13] Cyclones were continually arriving there like beads on a string, and ocean currents of different forms and origins were close at hand.[14]

In other words, Bergen as a field was attributed a kind of "place-lessness." The studies of cyclones and ocean currents were studies of types of natural phenomena that were considered to lead to knowledge about something more than the weather and ocean circulation along the west coast of Norway.

However, sometimes oceanographers chose a place in the field not for its "pure phenomena" and the opportunities they were believed to give to study something universal, but rather for its uniqueness. This uniqueness was in turn considered to give the place a key function in nature. The Faeroe-Shetland Strait, a key area in Helland-Hansen's understanding of the "inner mechanism of the ocean," was a unique, relatively small place considered to have a key role in the ocean. What happened here was regarded as the explanation, or at least a significant part of the explanation, for what happened in a large area of the North Sea and the Norwegian Sea. Studying the Faeroe-Shetland Strait led to broader insight into the ocean, not because what happened here was similar to processes elsewhere, but because the specific set of processes here affected large areas around it.

A glimpse out on the ocean – the *Armauer Hansen*

In an answer to a proposal by the secretary general of the International Commission for the Exploration of the Seas (ICES) Commander C.F. Drechsel, in 1924, Bjørn Helland-Hansen explained why a long expedition on the world's oceans of the kind that Drechsel had proposed would be of little use. The age of long expeditions with one vessel was over. His and Fridtjof Nansen's book *The Norwegian Sea* in 1909 had further advanced the study of the ocean, Helland-Hansen argued. One "should turn to a new principle: intensive investigations in every single area and not extensive over many areas by one vessel alone."[15]

Helland-Hansen described a transition in oceanography from broad mapping and collecting to detailed analysis. While Drechsel and the ICES had not yet begun this transition, Helland-Hansen had already been cultivating the "new principle" of specific, targeted cruises to smaller areas in the ocean for ten years. His most important instrument in this new scientific practice was the research vessel *Armauer Hansen*. Helland-Hansen had designed it himself.

Built at the Norwegian boatyard Lindstøl in Risør, it was completed in January 1913, almost five years before the Geophysical Institute was established. It was a vessel not at all typical of its time. In a period when the combustion engine was the prime emblem of the modern age at sea, the *Armauer Hansen* had sails. It was not made of iron or steel, as was the usual new custom building material in large ships, but of wood, from which ships in Norway had been built since the Bronze Age.

Helland-Hansen was proud of his vessel. His sources of inspiration were Fritdjof Nansen's *Fram* and Roald Amundsen's *Gjøa* – already legendary ships that had endured years in the harshest environments that any ship can be exposed to on this planet.[16] He lectured and wrote about the advantages of small research vessels. The *Armauer Hansen* was 23 meters long and 6 meters wide, and had berths for 12–13 people. Its engine had 40 horsepower.[17] Guests to the institute were interested in the ship, and some came to Bergen solely for the purpose of seeing it. Portuguese oceanographers had a similar ship built in Norway, the *Albacora*. Helland-Hansen also traveled abroad to tell his colleagues about it, and made sure that his colleagues saw it when it was in ports abroad.[18] During the International Union for Geophysics and Geodesy 1936 Congress in Edinburgh, the *Armauer Hansen* was there, and Helland-Hansen reported that many of the congress participants visited the ship to have a look.[19] It could be argued that this vessel was a symbol of a specific scientific method, discipline, and intellectual milieu. It embodied certain Norwegian national values, as well as some Bergen scientific values.[20]

The Armauer Hansen was a place of unpretentiousness, a place in which personal comfort had been sacrificed and cutting-edge ship-building technology had been eschewed to make it fit its very specific set of practical purposes. A small institute with a small budget needed a small ship with a small budget. Nevertheless, the *Armauer Hansen* was built for longer expeditions, such as to the Mediterranean. "In this little boat," D'Arcy Thompson explained, "the oceanographers in Bergen have repeatedly investigated the north-eastern Atlantic Ocean, all the way to Madeira and the Azores and far west of Rockall."[21] The sails significantly extended its range, and its seaworthiness, which was confirmed several times both by Bergen oceanographers and their guests, was a matter of a long and proud tradition of Norwegian ship-building, not of the latest technology.[22] Finally, the size was an attempt to answer the oceanographers' need to work smoothly on one point at a time on the sea surface, something that is arguably easier to do with a small, maneuverable vessel than with a big, heavy ship.

In this way, the *Armauer Hansen* was a place that invited a specific kind of heroism that primarily involved persistence. First, it required the patience to do stations day after day and week after week. "Stations and stations," Olav Mosby repeatedly notes in his diary from an expedition to Spitsbergen in 1927, hinting at the enervating effect of routine work on board. During the Second World War, this monotony must have been at its peak when up to 28 stations were taken from the *Armauer Hansen* daily from its anchor position in Olden Bay.

The monotony of work in the field could be accompanied by inconvenience: sore hands and backs, freezing, sweating, rolling, sea sickness, fog, and icebergs were only some of them. Olav and Håkon Mosby and the others aboard the *Armauer Hansen* south of Spitsbergen in 1927 did not bathe, and they slept in their clothes. It was a cold ship. Olav noticed this at the "Hotel Point" just off Longyearbyen when he slept over on another ship, *Ingeren*, that was to take him home to Bergen. Sweating immensely in his cabin, he turned off the radiator, opened every porthole, and pulled off all his clothes. He still sweated terribly. "The walls are tight here, compared to the drafty ones on Armauer," he reasoned. "That has to be the reason."[23]

Harsh weather conditions could turn boring routine work into something frustratingly difficult and sometimes dangerous. It was "absolutely impossible to walk on deck," Olav stated on July 23, 1927, due to the "disgusting" rolling and the slippery deck. Reading the temperatures on the water bottles was "pure torture, because one risks being thrown to hell."[24] However, danger was not a dominant part of field life. It was risked in small portions, but generally it was avoided through wise, sensible measures. Returning to harbor after just a few stations in the Isfjord on the southwest side of Spitsbergen on August 4 because of "untameable" weather was just such a measure.[25] Still, danger was in the back of Olav's mind when he saw the inlet of Isfjord for the first time after having covered more than 1600 kilometers crisscrossing the northern Norwegian Sea for nine days. His first association was with a huge, seething, jaw, and his line of thought ends with a simple fact: "This was where Iversen went down last year."[26] Luckily, Captain Iversen was still with them.

Oatmeal and meat stew were typical meals on board. Sometimes, when they had caught fish, the cook served the most delicious halibut. Alcohol was certainly on the menu on special occasions, which, judging from Olav Mosby's descriptions, seemed to occur much more frequently in port than at sea. Ashore on Spitsbergen, it seems to have been be a well-established ingredient in almost all types of social life, from baptisms to impulsive visits on board. At sea, alcohol was drunk in small amounts on birthdays or moments that were collectively felt as important. "Discovering" Bear Island after days with nothing but sea on the horizon was one such occasion.[27]

The field person

Sitting in the sun on a stone quay in Alversund north of Bergen, Aagot Borge writes in her diary about how the *Armauer Hansen* ran aground on its way from Olden to Bergen. It happened in the middle of the night, and everybody was rowed in the life boat to a nearby island. However, they were in shallow and calm waters, and the weather was good. In fact, Borge seems to have enjoyed the extraordinariness of the situation. She comments on the "miraculously beautiful, black night with a heaven full of big, clear stars" as another boat tried to pull them off the ground. To her, it brought a sense of adventure and allowed her to see other sides of her colleagues. From the stone quay she caught a sight of Bjørn Helland-Hansen which slightly distracted her writing:

"However, now Bjørn is shaving himself by the white table in front of the fairy tale house, and looks like the very summer. I would like to have stored that image in my mind."[28]

David Livingstone's point that people change according to where they are indicates that there must have been ship versions of many Bergen geophysicists. People on board a ship act differently because they respond to different surroundings. Physical and mental stress and exhaustion, extraordinary sensory impressions, and fear at sea bring out other reactions, other aspects of personalities, than does office laboratory life. Or, as Borge experienced, a positive sense of adventure and excitement made people act differently and see their colleagues differently.

At sea, other skills became more important than they would have been in an office or lecture room. These skills were possessed by people who represented other types of authority, accompanied by special types of knowledge and ideas. As we shall see, a clever, experienced captain with superstitious beliefs could significantly influence a cruise.

To Bjørn Helland-Hansen in the early 1920s, life at sea on the *Armauer* was a return to "normal."[29] "I feel how healthy it is to get out on the ocean with 'Armauer,'" he wrote to his mother in 1922. He ate and slept splendidly – life was "magnificent." Helland-Hansen's health was threatened not by the hardship at sea, he reasoned, but by stress and work overload on land. To him, field life at sea was a relief, good for body and soul.

But expeditions involved life ashore as well. Whether the expeditions went to the North Atlantic (1923 and 24), to the North Norwegian Sea and Spitsbergen (1927), to the Mediterranean (1930), or to the fjords, the *Armauer* always had errands in different harbors. Oil had to be refilled, messages and mail had to be collected and sent, people went ashore or came on board, people were visited, or they simply sought safe harbor because of bad weather. Life on land in foreign harbors was social, consisting of more or less formal dinners and parties, sightseeing, visits, and sometimes lecturing. In Lisbon, on their way to the Azores in 1931, Helland-Hansen dined with the Portuguese president and several ministers and gave a lecture to the Geographical Society together with his colleague and expedition companion, the zoologist Damas. They were "constantly" in the company of a cabinet minister or foreign minister. "But it is no doubt best to be at sea," Helland-Hansen concluded, longing for "getting to sea again and not having to be dragged around to be presented and join receptions and lunches and dinners."[30] The formal life ashore contrasted with the informal life at sea.

However, life ashore during expeditions was not necessarily entirely formal. In Spitsbergen, Olav Mosby describes the active social life in Barentsburg, consisting of relatively frequent gatherings among a relatively small group of people. It involved guided tours to the Dutch coal mining company there, a baptism and dinner for which Olav and his brother Håkon played the piano and the violin, respectively, short visits with strong drinks on board the *Armauer*, and parties into late nights ending in personal injury for Captain Thor Iversen. Olav did not complain about the formal life in Barentsburg. Nor did he praise the healthy life at sea.

Needless to say, shipboard life in the polar regions was radically different from shipboard life in the temperate and tropical zones. On his way southwards, west of the African continent, in October 1927, Håkon Mosby enjoyed stripping down to his underpants and a placing a handkerchief on his head.[31] Every morning started with a fresh shower from the seawater hose. Freezing in the icy northerly wind south of Spitsbergen, wearing their clothes day and night, was probably still fresh in his mind. Taking stations had turned from being difficult, unpleasant, hard work to being merely boring.

Sense impressions could dominate shipboard life in different ways. As Aagot Borge could not forget the lights in nature that she observed on her trip from Olden to Bergen in 1947, Helland-Hansen was fascinated by the light from stars, the rising moon, and the hillside city of Funchal, Madeira, in 1922.[32] Ice, fog, mountains, fjords, the midnight sun, the moon and cities by night made many beautiful sights for geophysicists at sea. "Too seldom is it so uniquely beautiful up here," Olav Mosby commented, referring to the bright sunshine and calm sea on the banks south of Isfjord on Spitsbergen.[33] Generally, beautiful sights occurred in nice weather, when nature otherwise showed her friendly side.

Odors sometimes made such an impression that they hindered or slowed work. Some types of seagulls smelled bad, as did rotting whales.[34] The laboratory offered scents that pleased or odors that displeased those who worked or slept there. The paraffin stove in the *Armauer*'s laboratory emitted an odor of "death and the devil," making telegraphist Olav Aabrek seasick.[35] Aabrek had his berth in the laboratory, in which work was done around the clock. The sounds in the laboratory, however, were crucial for Aabrek's ability to sleep. He claimed he could not sleep without the noise from the echo sounder. This became a problem in 1954, when the old echo sounder was replaced by a new, quieter model.[36]

Each person had a specific sets of skills. In 1927, Håkon Mosby discovered that he was a good shooter, and that he was good at sewing sails. Olav Aabrek lacked certain skills, Mosby grumpily noted, such as being able to repair the petroleum oven and operate the radio, and remembering to titrate in calm weather.[37]

Among the things occupying Olav Mosby's mind on board the *Armauer* were frequent dreams about horses. At first, Olav seems to have found this repetitive pattern of dreams merely peculiar. However, he slowly became concerned about how the news about these dreams would be received by the seamen on board. Whistling was not allowed, he noted, because seamen believed it would provoke a gale. Dreaming about horses, however, meant sinking.[38]

Initially, he did not tell anyone about it due to this superstition, but after having dreamt about horses and riding several times, he told the seaman Skoglund. The reaction was clear: "...he became clearly frightened; that brings misfortune." Skoglund was able to give a quite detailed analysis of the dreams. The fact that the horses galloped meant that terribly rough weather would come. "If one of them kicks it means death for the one whom the horse hits."[39]

Olav does not mention horses or dreams after this in his diary. The following day, however, he comments on the "lousy" mood of the captain, Thor Iversen, which seemed to be directed specifically toward Mosby himself. Iversen refused to let Mosby go ashore to see some polar foxes and bears that a hunter had brought to Barentsburg. He did not speak to Mosby at all after this, not even when Mosby spoke to him directly. Even though Mosby himself made no connection in his diary between Iversen's bad mood and his own relating of the dreams to Skoglund, to the reader their concurrence is striking. Four days later, Mosby had a farewell drink on board the *Armauer* and entered the steamer *Ingeren* that would take him to Bergen.

Thor Iversen held a significant position in Norwegian marine research and management. He facilitated cooperation and exchange of knowledge and experience between scientists and fishermen, and was one of few people that could fill this role of a "mediator" between practical and scientific life at sea.[40] Field life was essential in this role, and from Olav Mosby's descriptions we see another side of it. We see that individuals could influence life at sea, led by feelings and ideas peculiar to the field.

Observing westward and upward

In the summer of 1919, then 27-year-old Tor Bergeron spent a fortnight on top of Mount Lyderhorn outside Bergen. Near the coastline and more than 400 meters above sea level, Lyderhorn was traditionally an important navigation marker for sailors coming to Bergen from the south. Visibility was also exactly what Bergeron was after. However, rather than the view of a mountain, he wanted a view *from* it. Since his boyhood, he had observed clouds and looked for systems of clouds. As a new member of Vilhelm Bjerknes's group of young weather forecasters, he still considered clouds important.[41] From Lyderhorn he could, on clear days, see far into the air masses that approached the Norwegian coast from the west.

Vilhelm Bjerknes and his group of young researcher-forecasters were field-oriented in a different way than their seafaring fellow geophysicists, the oceanographers. Heroism in the Bergen school of meteorology never implied hardship in the field. Fieldwork was to a large extent left to other people, such as lighthouse keepers, military personnel, doctors, priests, or other reliable people who were favorably positioned to observe the atmosphere. This network of informants reading instruments, sending up balloons, and reporting to Bergen

was directly inspired by the dense network of meteorological observation stations in Germany, the "Feltwetterdienst."[42] It was a part of
the meteorological infrastructure.

To the Bergen meteorologists, fieldwork was concentrated into
shorter trips to inspect and maintain this infrastructure. Soon after
their arrival by train to Bergen in the summer of 1919, Vilhelm Bjerknes
introduced that year's new young researcher-forecasters to the Norwegian nature and weather through a cruise aboard the *Armauer Hansen*.[43]

Historian Yngve Nilsen distinguishes between the "pragmatic"
and the "dogmatic" versions of Vilhelm Bjerknes.[44] The "pragmatic"
version is the one who established a field weather service as a response
to societal needs in aviation, agriculture and fisheries. The "dogmatic"
version is the one who insisted on studying the atmosphere from a
physical point of view using his equations. The field weather service
that hosted the Bergen school and the field excursions that it involved
represented the "pragmatic" version.

Visibility, cloud cover and cloud forms were important parts of
the observations that the meteorologists received in their map room.
However, these weather elements were something they instructed their
observers to record and report, or they took the observations themselves from the tower on top of the new institute building in Bergen,
starting in 1928. Bergeron's stay at the top of Mount Lyderhorn during
the summer of 1919 had been an exceptional type of fieldwork for a
meteorologist. But meteorologists developed other ways of observing
the air masses above their heads.

Much ado about balloons

One of the limitations of the field weather service offered in Bergen
was found precisely in the meaning of the word "field." The precise
measurements on which the forecasts were made all came from the
earth's surface. The air masses above the ground were mainly studied
with the naked eye, in the way Bergeron had done on Mount Lyderhorn. When the Bergen meteorologists talked and thought about air
masses in three dimensions, their models enabled them to recognize

structures in the air masses from only one of its "sides," that is, along the surface of the earth. "The direct observational investigation of the upper atmosphere will in time show us to what extent this view corresponds to the truth," Vilhelm Bjerknes explained to the participants of a conference on aerology in Bergen in 1921.[45]

At the same conference, Bjerknes's Leipzig assistant, Olaf Devik, presented a paper on "A New Method of observing Balloons and its use in the daily Weather Service." Devik talked about how pilot balloons could be equipped with radio transmitters "in miniature," and how their positions as they ascended could be determined with the use of three movable antennas placed concentrically and perpendicularly to each other. He even suggested that the radio transmitter on the balloon could be modified to send information about the temperature.[46]

Here, Devik operated in a field of new technology and method. Ballooning was an old meteorological method for gaining data from the upper atmosphere, but there was a difference between using the balloon as a weather indicator and using it to carry instruments to unreachable places. Devik's suggestion involved both. For the first aspect, the balloon was the instrument, and observing its upward journey would provide interesting information about winds in different layers of the atmosphere. However, as low clouds could prevent the observers on the ground from seeing it, Devik suggested radio wave detection. And once the radio was there, it was only natural to consider how it could be used to report on the temperature on its way through the atmosphere, covering the second aspect.

Balloon ascents became Jacob Bjerknes's highest priority when he became professor of meteorology and leader of Section B at the Geophysical Institute in 1931. Throughout the 1930s, he participated in different international programs featuring balloon ascents. The balloons carried so-called meteorographs, which were self-registering instruments measuring temperature and pressure. Until 1937 these meteorographs were not equipped with a radio transmitter, but recorded the measurements on a meteorogram in the form of a thin line incised by a pen on a glass plate inside the instrument. As was discussed in chapter three, the released balloon was useful only if it was found and retrieved. To find a balloon with instruments that had risen to the stratosphere and been exposed to different winds on its way could be difficult. Jacob was focused on retrieval percentage, and at first he believed that the vast, uninhabited areas in Norway and Sweden would reduce it. To increase the probability of retrieval, Jacob Bjerknes considered carrying out such experiments in densely populated areas such as Belgium.[47] However, as the loss of instruments from releases in Norway turned out to be lower than expected, he changed the plan.

Balloons seemed to offer totally new insights if many of them were launched simultaneously. The people who had gathered in Bergen in 1921 represented the network that Bjerknes needed to realize such a plan. In 1934, the International Commission for the Investigation of the Upper Air promised that "countless balloons will be released from Sweden, Finland, Poland, Germany, Belgium, England, Spain, Italy, Austria and Hungary on the day that we telegraph that our ascents will take place." This would give the "most complete aerologic material ever collected for the study of a storm center."[48]

Early in 1935, Jacob waited for a suitable storm center to approach. As the Norwegian balloons would provide the northernmost observations, Jacob hoped for the center to hit south of Norway. In that way, his network of balloons, which spread out from Norway in the north to Italy and Spain in the south, would best be able to cover the whole cyclone. In the morning of February 15, 1934, as Bjerknes observed that a strong storm center would hit southern Scandinavia, he sent telegrams to 15 different observatories abroad. Around two hundred balloons were released across Europe during a three-day period. Forty of these were released in eastern Norway, from the meteorological station at Ås, south of Oslo.

Balloon observations constituted a special type of fieldwork. First, Jacob and the other meteorologists in Bergen did not handle the balloons; they had meteorological assistants at Ås perform the practical work. Second, the balloons were sent out on journeys to a place that was still unreachable for humans, the upper atmosphere. As the balloon rose, the air pressure around it would drop, and the balloon would consequently expand. At some altitude, the balloon would burst, and the instrument would fall to the ground. After this journey, the scientists would not know exactly where it had been. Third, the whole project depended upon whether or not anyone passed by and noticed the instrument. Fourth, if someone did, the experiment also depended upon whether or not the finder would make the effort to send it back to the Geophysical Institute in Bergen. To ensure this, a small notice was attached to the instrument announcing a finder's fee of five kroner (about 190 kroner today).

The recollection percentage from the February 1935 releases was around 70 percent, which made Jacob Bjerknes optimistic.[49] However, due to technical problems with the instruments, 90 percent of them were failures.[50] Bjerknes and his contact at Ås, Nils Russeltvedt, discussed the problem via letters in March. The pen that incised a thin line on the glass plate was a moving part. For some reason, in nine out of ten instruments, this pen had been lifted from the plate and thereby been prevented from leaving any marks on it. Russeltvedt was alarmed and distressed upon hearing about this from Bjerknes.

Russeltvedt described in detail the set of control routines used when releasing the balloons. As part of these routines, the pen would be examined on three occasions during the preparations for each specific release. The meteorograph was first mounted into a protective "basket" or cage and then hung into an instrument house. Then, they would turn on the pens so that initial values could be registered while they inflated the balloon. When the agreed time of release was approaching, the instrument would be attached to the inflated balloon, and the pens would be inspected. After having brought the whole arrangement to the open area for departure, the pens would be inspected again, just before the release.

Russeltvedt and Bjerknes's troubleshooting involved considerations of unfavorable relationships between the instrument's weight, the tension of the rubber strings in the instrument, and some intense oscillations during release. In the February releases, Russeltvedt had observed some incidents where the instrument cage had hit the ground just after release. During another release, the cage had experienced intense oscillations. "It looked plainly dangerous," Russeltvedt reported.[51] Bjerknes made the point that this was not a new condition, and hence was not tied to this new problem. He argued that the explanation of the 90 percent failures must be a systematic error, and suggested a modification to the instrument. At the end of his letter, Russeltvedt suggested that Bjerknes come out into the field: "I would very much like to get to the bottom of this and I look forward to your considerations about the matter. Surely, it would have been fine if you studied these things together with me on the spot."[52] The problem was eventually found. They discovered that they had changed the thickness of the rubber strings that held the meteorograph's pen. It had a slightly weaker elasticity, making it an unreliable pen lifter.[53]

This fieldwork was characteristically meteorological, not only in its scientific content, but also in its organization. Bjerknes filled a number of roles, but none of them brought him out into the field: he was the planner, organizer, and leader of the fieldwork. Bjerknes also played key roles in the practical set-up of the work, including modifications to the instruments in use. However, the actual balloon releases seem not to have been his job. Fieldwork to him was a matter of maintaining an infrastructure, not of going out in nature to observe for himself.

The International Commission for the Investigation of the Upper Air worked as a body of international, simultaneous meteorological fieldwork. The initiatives for international simultaneous balloon releases came both from Bergen and from other places – Jacob was eager to give back to the network as well as enjoy its services. "We would like to back up the undertakings that are being started from initiatives abroad, just as foreign countries have helped us."[54]

Starting in 1932, meteorographs were sent up in Norway annually, mostly from Ås. In 1937, radiosondes were attached to the instruments, reporting the measured values of air pressure, temperature, and humidity to the ground station as the balloon ascended. Even if the instrument was lost, the measurements were saved, which enabled releases from Bergen. The Finnish inventor Vilho Väisälä was present, and the radiosonde signals were detectable "far up into the stratosphere."[55]

Based on the evidence, the "research travelers" that D'Arcy Thompson met at the Geophysical Institute in the late 1920s were probably all oceanographers. The staff at Section B: Meteorology in the years around 1928 consisted of Professor Harald Ulrik Sverdrup and Assistant Professor Håkon Mosby. With his six years as scientific leader of the *Maud* expeditions to the Arctic (1918–1925), Sverdrup was at that time one of the most highly respected field scientists in the geophysical world. The younger Mosby was also already an experienced expeditioner due to Antarctic expeditions carried out in cooperation with the Norwegian whale hunter and businessman Lars Christensen. However, both Sverdrup and Mosby were soon to turn to oceanography, and meteorologists Jacob Bjerknes and Sverre Petterssen took their place. Thomson's impression in 1928 of the prevalence of oceanographers in meteorological research would probably not be valid much longer.

6
The oceans under surveillance

When the Second World War ended in 1945, the geophysicists in Bergen found the cards stacked against them: the building, equipment, and research vessel were in dire need of renovation and repairs, and the most prominent researchers had either moved away or were on the verge of retiring. While geophysical institutions elsewhere, particularly in the United States, had proven their worth in war and were rewarded with generous long-term funding, the German occupation of Norway had led to a standstill. But it was not all bad news: the Bergen legacy was strong, with oceanographic methods and tools, such as the Nansen bottle, being used all over the world. Bjørn Helland-Hansen was appointed president of the International Union for Geodesy and Geophysics (IUGG) in 1946, the same year that the Norwegian parliament decided to establish the country's second university with the Geophysical Institute as one of the cornerstones.

Since there was little or no collaboration between oceanographers, meteorologists, and researchers at "Section C," this chapter will focus solely on the physical oceanographers and how they faced the new reality of no longer being part of a leading scientific center. We will show that postwar physical oceanography was characterized by new technologies, international collaboration, a dramatic increase in observations, and increased disciplinary specialization. In Bergen, too, the most successful postwar oceanographic research project focused on developing technology – a niche with funding from NATO's Subcommittee for Oceanographic Research, led by the head of the Geophysical Institute, Håkon Mosby. What research interests did Mosby and the other oceanographers in Bergen pursue? On which parts of the oceans did they focus? Who did the oceanographers collaborate with during the Cold War, and in what ways did the quest to calculate the world continue into the postwar era?

After outlining the institutional situation after the war and the opening of the University of Bergen, the chapter will focus on the career of Håkon Mosby, who succeeded Bjørn Helland-Hansen as head of the institute from 1947 to 1970.[1] Mosby personified the age of institution-building pioneers, and had a central position domestically as well as in international oceanography. He was elected dean of the Faculty of Science (1954–59) and then rector of the University (1966–71).

Internationally, he was a leading member of a number of committees in the IUGG, UNESCO, and the International Council of Scientific Unions (ICSU); president of the International Association of Physical Oceanography (1954–60); and head of NATO's Subcommittee for Oceanographic Research (1960–65). To the next generation, Mosby came to represent the transformation from the generation of pioneers to a "modern oceanography." How did oceanography change, both in Bergen and in the rest of the world, during the reign of Håkon Mosby?

A new dawn

During World War II, the Geophysical Institute was characterized by occupation, stagnation, and overcrowding.[2] Since the German occupiers labeled weather observations as classified information, the meteorologists were reduced to analyzing old observations.[3] For the oceanographers, lack of fuel and other restrictions confined the research vessel *Armauer Hansen* to operations near Olden in the inner parts of Nordfjord. In the summers of 1940 and 1941, Mosby carried out a study of surface currents in the Tromsø strait based on observations of 50,000 floating objects moving through it, but otherwise he spent as much time as possible in Olden.[4] Next, Chr. Michelsen Institute had expanded to do research on the BCG vaccine against tuberculosis, and with limited space due to the occupation, a room in the basement was filled with cages for rabbits and guinea pigs. The staff was told to stay clear of the animals to avoid infections, but in a time of food shortages there were accusations of people stealing rabbit food for subsistence.[5] After Norway was liberated on May 8, 1945, it took many years before the rationing of foodstuff and other consumables was lifted.

On April 9, 1946, the Norwegian parliament voted to establish the University of Bergen, the second university in the country. Although the parliament had already formed a committee to address the question in 1938, it was the geophysicists who made the final push on behalf of the Bergen Museum. During the winter of 1945–46, Bjørn Helland-Hansen, director Bjørn Trumpy, and board chairman Wilhelm Mohr invited the Norwegian government to a dinner at the home of Helland-Hansen's son, Eigil, in Oslo. After a feast with roast beef brought from Bergen, the three gave an informal introduction to the plans for the University. According to Mohr, it was at this dinner that the government became convinced that the time was ripe for moving the decades-old university plans forward.[6] The three pointed out that the Bergen Museum already had ten divisions headed by professors with strong track records – three of them in geophysics – and offered an education to around 400 students.

The Bergen Museum had worked to become a university since the 1890s, but at every crossroads the plans had stalled due to costs. The city hospital at Haukeland had for almost four decades argued in favor of offering medical education, and was in dire need of expansion. In 1936, a business school (NHH Norwegian School of Economics) had been added to the academic community in Bergen, and the three pointed out that Denmark's second university in Århus, opened in 1928, had been a success. The last will and testament of engineer, officer, and businessman Lauritz Meltzer added to the urgency: in 1939, Meltzer had bequeathed his fortune of 5 million NOK to the University of Bergen, on the condition that it would open within ten years. Otherwise, the fortune would go to the town of Fredrikstad where Meltzer had grown up.

The Geophysical Institute was, quite literally, a cornerstone of the new University. When King Haakon VII put down the foundation stone on October 25, 1946, the first building project consisted of adding two new wings to the crowded Geophysical Institute. These would later host a botanical laboratory, as well as institutes for physics, chemistry, mathematics and biochemistry. On August 30, 1948, the University opened with faculties of natural sciences, medicine, and the humanities.

For the geophysicists, the university was both a blessing and a curse: rather than expanding the Geophysical Institute, the priority was to establish new institutes. Until 1960 the number of employees at the Geophysical Institute remained constant at 17 positions, suggesting that the record of excellence was taken for granted.[7] In comparison, the staff at Trumpy's neighboring Physics Department, which opened in 1948, had in the same time grown from zero to 26.[8] Although access to science funding improved with the establishment of Norway's first research councils in 1946 and 1949, there were also more researchers competing for funds.[9] Besides, Norwegian oceanographers did not benefit from close collaboration with the navy, which was the case in other postwar countries, especially the United States. Instead, Norwegian state funding for relevant military research went to the Norwegian Defense Research Establishment (FFI, formally established 1946), financed generously by defense budgets. The Norwegian researchers and engineers who during the war had worked in British laboratories and workshops were offered positions, and FFI quickly became the largest research institute in the country. By 1949, the staff had passed 100 employees and it kept growing, reaching 500 by 1967.[10]

Norwegian geophysicists were still held in high regard internationally. At the meeting of the 1939 IUGG meeting in Washington, members decided that the next 1942 General Assembly should be held in Oslo. Although the war postponed those plans, in 1948 Norway hosted the first postwar IUGG assembly. At a meeting of the IUGG executive committee in Oxford in 1945 in preparation for the event, Bjørn Helland-Hansen was elected president. In meteorology, Theodor Hesselberg, director of the Norwegian Meteorological Institute, was president of the International Meteorological Organization from 1935 until 1946, and headed the work to reform the organization under the United Nations umbrella as the World Meteorological Organization (WMO). However, increasingly the geophysicist's reputation was linked to Bergen and to Norway's legacy as a cradle for the geophysical dream of calculating the world, rather than as sites where cutting-edge research still took place.

In 1947 Bjørn Helland-Hansen retired after 30 years as head of the Geophysical Institute, and the following year he retired as the IUGG president. The era of pioneers was over.

A new leader for a new age

Helland-Hansen's heir as organizer and research administrator, locally as well as in international organizations, was Håkon Mosby. Fellow oceanographer Odd Henrik Sælen aptly described Mosby as "the nexus between the generation of pioneers and modern oceanography."[11] One example of this was his tenure as head of the Geophysical Institute, which he retained until his retirement in 1971, during which he started a tradition for shorter tenures when he stepped aside in 1958–62 due to international commitments. Like Helland-Hansen, Mosby was part of the absolute elite in establishing and administering international research agendas for oceanography.

In contrast to most postwar oceanographers, Håkon Mosby was recruited to geophysics through field practice rather than education. He began his career as an assistant to Fridtjof Nansen in 1923, when he, a 20-year-old student at the University of Kristiania, inherited the position from his older brother, Olav. In 1927 Mosby was appointed associate professor in meteorology at the Geophysical Institute in Bergen, where he started off as the scientific leader of the first of four *Norvegia* expeditions to the Southern Ocean.[12] During the expedition, Mosby mapped the distribution of water masses off the inhospitable Antarctic shelf. Explaining how Antarctic water masses form and what role they play in the global ocean circulation system became a lasting research interest for Mosby and the Geophysical Institute.[13] During the two months it took to sail to the Southern Ocean, Mosby also gained a reputation as an accomplished violinist, another lifelong passion.

After studying the interactions between wind and the ocean surface based on Sverdrup's observations from the *Maud* expedition, Mosby earned his doctorate on the properties, origins and movement of surface and deep waters of the Southern Ocean.[14] In addition to his own *Norvegia* observations, his dissertation rested on observations from a series of previous expeditions (the *Deutschland*, the *William Scoresby*, the *Meteor* and the *Vikingen*). While surface currents varied distinctly with the seasons, he found that the lower strata remained remarkably stable. To Mosby, this suggested that the southwestern part of the

Weddell Sea played a vital role in the physics of all oceans, namely that the extreme cooling produced bottom water that slowly spread to the abyss of all the world's oceans.[15] In 1936, he published a much-used equation on the solar energy reaching the ocean surface as a function of cloudiness and solar elevation.[16]

Although formally hired as a meteorologist, Mosby's research focused on the oceans. When oceanographer and mathematician Jonas Fjeldstad accepted a position at the University of Oslo in 1939, Mosby switched to oceanography. Where Fjeldstad had been a theoretician, Mosby put a premium on timely and accurate observations from the field. In his doctoral work he pointed out that by using observations taken several years apart, variations could give the appearance of permanent structures. One example of this was what appeared to be a division between two branches of the current off the coast of Antarctica at 65° S, 5° E. Mosby believed the current was merely a fluke caused by combining observations from different years, a view that was contested by British oceanographer George E.R. Deacon.[17]

To Mosby, variations and other questions of ocean dynamics could only be resolved through rigorous fieldwork. Working to improve oceanographic observations would be a common thread throughout his career, which included a total of three years at sea. Prior to World War II, when Mosby wrote a review of the history of oceanography for Norwegian sailors, it was not theoretical contributions, but expeditions, the construction of new instruments, and the development of new field methods that he highlighted as the main Norwegian contributions to physical oceanography.[18] Instead of calculations, emphasis was on observations – on putting numbers on the oceans. The pioneering collaboration between Helland-Hansen, Fridtjof Nansen, and Vagn Walfrid Ekman was mentioned, but his only comment about oceanographic theories was that they "in no way are developed to perfection":

"Theories for ocean currents, both those that apply to wind-driven currents and those caused by differences in density, are developed based on very simplified assumptions that are not met in nature. If one removes the simplifications and takes into account the depths of the oceans, that the ocean floor is uneven, that the earth is round, and that seawater is not homogeneous (even salinity and temperature), the calculations become insurmountable."[19]

Likewise, in a lengthy review of the history of Norwegian oceanography from the Enlightenment to the Second World War written by Mosby in the 1970s, emphasis was on expeditions, equipment, and observation techniques. Again, theoretical contributions were mentioned only in passing.[20] To Mosby, the most important events at the beginning of the 20th century were Martin Knudsen's hydrographic tables linking seawater density to salinity, temperature, and pressure (1901) and Vagn Walfrid Ekman's tables of the compressibility of ocean water (1908). "These two works (...) make up our complete knowledge of the 'nature' or state of ocean water. These are empirical results, and no one has managed to derive them theoretically and explain why ocean water behaves in this way."[21] In a short historical overview published ten years later, he repeated the same point, but gave more generous nods to Bjerknes's circulation theorem, the solenoids, the theory of dead water, and the Ekman spiral.[22] This reflected a shift in oceanography that Mosby himself had played an important part in bringing about, namely that the problem of how to produce time series from the deep was now considered solved, and the time was ripe for more calculations (modeling).

The need for new kinds of observations was inspired by Austrian oceanographer Albert Defant, who in 1941 published the first attempt at compiling an absolute topography of the physical sea surface level and the isobaric surfaces in the depths of the Atlantic Ocean. [23] The map of water masses with similar pressure (as a function of gravity, density, and depth) made it possible to put values on currents and volume transports in the deep sea. This opened the door to a conceptual shift from Helland-Hansen and Johan Sandström's simplified solenoids to calculating a complete picture of actual water movements in different parts of the Atlantic Ocean. An important research interest for Mosby became working out the budgets of transportation through different seas.[24] He also focused research on identifying and quantifying both stable currents and their variations. Rather than the synoptic snapshots that Helland-Hansen had encouraged, this approach to calculating the oceans required direct measurements over time in order to identify variations. The budgets became an important motivation for oceanography moving from sampling to surveillance. But first Mosby needed a new vessel.

Of ships and men

After several years in the brackish waters in Nordfjord during the war, the research vessel *Armauer Hansen* was in desperate need of repairs. In a country rebuilding after five years of foreign occupation, with limited access to labor and materials, it took several months to make the vessel seaworthy. In April 1946, just months after the first round of repairs, both masts had to be mended. That year only 19 of 182 hydrographical stations were taken on the open sea, the rest in sheltered fjords. The following years, the *Armauer Hansen* would do weekly observations in the fjord just outside Bergen, and visit other fjords every second month. From 1952 onward, the vessel was mostly left in port during the winter months between more repairs.[25] The wooden vessel that once had been the hallmark of a unique brand of oceanography – small, cheap and offering close contact with both nature and fellow crew members – was increasingly seen as backward and outdated.

A top priority for Mosby when he replaced Helland-Hansen as head of the Geophysical Institute in 1947 was to regain access to the oceans. Starting in the early 1950s he petitioned the newly established Norwegian General Research Council (NAVF) for funds for a new and more modern research vessel, arguing that it would be less expensive than the much-needed upgrades and constant repairs, that the ship did not satisfy standards for health and safety, and that it was impossible to educate "real oceanographers" without experience in the field.[26]

Starting in 1954, the Geophysical Institute received 1.5 million NOK per year from NAVF to build a new replacement research vessel, and in 1957, the finished ship was given the name *Helland Hansen* as a tribute to the first head of the institute.[27] The vessel was 113.8 feet long, weighed 187 tons, and had a 400-horsepower diesel engine. It was equipped with hydraulic winches, radars, Loran and Decca navigation systems, radios, sounders, and an autopilot. Onboard were three two-man cabins, as well as nine single ones, cooling systems for provisions, a laboratory, and workrooms. Fully stocked, it could operate on the open sea for up to four weeks with a range of 5000 nautical miles before resupplying. In comparison, the *Armauer Hansen* was 80 feet long, weighed 57 gross tons, had a 40hp engine supplemented by two sails, and only the captain had his own cabin.[28]

When Håkon Mosby proudly presented the new vessel to the readers of the popular science magazine *Naturen* in 1958, it was a leap into the modern age – on economy class: "Practical oceanography can hardly get any cheaper."[29] The *Helland Hansen* had a permanent crew of three, and hired three additional sailors for each cruise. In addition to equipment, winches, and anchors for hydrography, the vessel was prepared for attaching specialized equipment for sedimentologists to use gravity corers, for zoologists to trawl, and for meteorologists to send up weather balloons, so that the ship could be rented out. The income was used for upkeep, new equipment, and smaller expenses, and was a welcome buffer for the institute's budget. As owners of a ship with staff, it was seen as cheaper to use the vessel than to leave it in port. Steinar Myking, who was hired in 1969, recalls that many cruises were decided on the spur of the moment. "We were often told: The weather is nice, we'll go out tomorrow! The cruises often lasted for about a week. Today we often have to plan the expeditions a year or more in advance, but the vessels are larger and more stable. The *Helland Hansen* was small, and in rough seas, seasickness always lurked. This was never an excuse for remaining ashore."[30]

One of the first cruises the *Helland Hansen* took part in was the Atlantic Polar Front Programme (1958), alongside vessels from eight other countries: Canada, Denmark, France, Germany, Ireland, the Soviet Union, Spain, and the United Kingdom. The program was an extension of the International Geophysical Year 1957–58, which had involved 60,000 scientists and technicians from 67 countries and various geophysical fields. It was aimed at expanding the area of study farther to the north through numerous sections across the Atlantic Ocean. This was the first time the Soviet Union took part in an international expedition alongside oceanographers from the Western Hemisphere. The Soviets brought three vessels that all dwarfed the ship from Bergen:

the *Vitiaz* (5700 tons), the *Mikhail Lomonosov* (6000 tons), and the *Ob* (an icebreaker of 12,600 tons).[31] This illustrates how the superpowers had access to resources that went far beyond those available in Bergen, and how Mosby had focused on his vessel being cost-effective and down-to-earth.

Increasingly, it was participating – not having the largest vessel – that mattered in the international geophysical community. The observations from the Atlantic Polar Front Programme were collected in the World Data Centers and made available for the cost of copying and postage. The first data centers were established in the United States and the Soviet Union, while a third was subdivided between Western Europe, Australia and Japan. The acquisition of standardized observations through international research programs, and the way they were made available to the geophysical community internationally, greatly improved access to large quantities of quality data. This was big leap in making oceanography, not just the oceans, truly international. It did, however, require active participation in international collaborations. This was a landscape Mosby navigated with skill.

The Helland Hansen took part in numerous similar collaborations over the years that followed, such as the Overflow Program to the Iceland-Faroe Ridge in 1960 organized by ICES in which nine research ships took part, and in 1961 along with five other vessels to investigate the currents in the Gibraltar Strait, organized by NATO.[32] The vessel made frequent returns to these areas, in addition to sections in the Norwegian Sea, the North Sea, the Barents Sea, the Greenland Sea, and the Mediterranean, as well as to Norwegian fjords and coastal waters. The cruises had several aims, the most common being to produce synoptic overviews in collaboration with other vessels; to gain insight into specific phenomena, mechanisms, and variations; to test instruments; to measure currents; and to collect data to produce budgets of water transport. Several cruises set out to compare dynamic calculations with measurements.

From 1948 to 1986, the weather forecasting unit at the top of the Geophysical Institute also administered the two Norwegian weather ships *Polarfront I* and *Polarfront II*. The weather ship program was established by the Provisional International Civil Aviation Organization (PICAO) in 1946, and initially included 13 ships stationed in the North Atlantic to aid weather forecasting.[33] The two *Polarfront* vessels, two rebuilt British corvettes, operated Station M in the Norwegian Sea at 66 °N, 2° E. They were operated by the Weather Forecasting Unit for Western Norway, but were mainly funded by Sweden (43 percent) and Great Britain (35 percent). The weather ships were primarily meant to aid transatlantic flights and weather forecasting in Europe, but before becoming operational in October 1948, a committee consisting of Håkon Mosby, Jens Eggvin from the Institute of Marine Research in Bergen, and marine biologist Johan T. Ruud from the University of Oslo organized an oceanographic measurement program. In collaboration with Sverre Petterssen, who represented Norway in the negotiations, Mosby arranged for the Norwegian weather ships to be equipped with hydrographic winches in Britain so that the meteorologists would cover the expenses.[34]

The oceanographic observation program included taking weekly stations to determine temperature, salinity, and oxygen down to 3000 meters, and daily stations down to 150 meters' depth. The ships were also used to collect biological samples, such as plankton and pollen. Starting in the late 1950s they took water samples that were sent to the Norwegian Defense Research Establishment to measure radiation, and from 1966 onward the monthly samples were sent to the International Atomic Energy Agency (IAEA).[35] The oceanographic observation program lasted until Station M, the last remaining in the Atlantic Ocean, was decommissioned at the end of 2009.[36] Station M produced the world's longest time series of oceanographic data from the deep sea. The observations have been used in several publications on topics such as bottom water, heat exchanges between the ocean and the atmosphere, and climate studies.[37] Observations at the same location continue, but now from an instrumented mooring – an anchored surface buoy with oceanographic and meteorological instruments and satellite communication – and occasional ship visits in a collaboration between the Geophysical Institute and the Institute of Marine Research.

Oceanography internationally

During and after the Second World War the international geophysical research community grew significantly. In the United States, the "big two," Scripps Institution of Oceanography in California and Woods Hole Oceanographic Institution in Massachusetts (WHOI), closely liaised with the US military by doing research for the war effort, including research into sonar and sound interference, investigation of underwater explosives, and wave and surf forecasting for amphibious assaults. The collaboration continued into the Cold War. An overview of oceanographic research in the United States in 1958 shows that the six largest laboratories employed 598 scientific staff, an average of almost a hundred oceanographers per institution. In addition, nineteen laboratories at smaller universities employed a total of 228 scientific staff, and the US Navy employed 232 oceanographers.[38] The institutions also had technical staff, and there were oceanographers at the thirty US fisheries institutions. Similar institutions were established, or expanded, elsewhere, including the United Kingdom's National Institute of Oceanography that opened in 1949; its building opened at Wormley in 1952/53, employing 45 scientists and technicians.[39] In comparison, Mosby's section for oceanography in Bergen had three oceanographers, one instrument maker, two laboratory technicians and an office clerk.[40]

The growth of geophysics was reflected in the number of attendees at the IUGG General Assemblies. The 1936 assembly in Edinburgh, where Helland-Hansen was elected president of the International Association for Physical Oceanography, had 344 participants, up from 200 at the Lisbon meeting in 1933. The meeting in Oslo in 1948 was attended by 368 persons. In the 1950s, around 900 attended each assembly, and at Helsinki in 1960, the number increased to 1375 geophysicists. The expansion continued for another four decades, reaching almost 4500 in the 1990s.[41] Important drivers of these increases were the military and strategic importance that geophysics had proven to have during the war, and avenues for peacetime scientific collaboration offered by geophysics.

To coordinate their efforts, the oceanographers established several new international organizations. In addition to the International Association of Physical Oceanography (IAPO, since 1967 IAPSO), established in 1919 as one of seven IUGG associations, the most notable were the International Advisory Committee on Marine Sciences (IACOMS), established by UNESCO in 1955, and the Special Committee on Oceanographic Research (SCOR), established by the International Council of Scientific Unions (ICSU) in 1957. The three had overlapping missions and global ambitions. IAPO focused on scientific collaboration to be carried out with the aid of mathematics, physics, and chemistry; IACOMS sought to coordinate scientific information for application in improving the living conditions of mankind; and SCOR focused on fundamental research, establishing working groups to focus on narrower topics, and sending expeditions to distant and little-known areas.[42] In 1960, UNESCO created the Intergovernmental Oceanographic Commission (IOC) to coordinate activities that SCOR was not equipped to carry out, highlighting that oceans cover 70 percent of the earth's surface and are a crucial source for food for the world's increasing population.[43] Organizations of more regional scope also existed, the oldest being the International Council for the Exploration of the Sea (ICES), established in 1902. The International Hydrographic Bureau (IHB, since 1970: IHO), established in 1921, worked to support safe navigation and uniformity in nautical charts, and to develop methods for descriptive hydrography.

Historian Jacob Hamblin has argued that international geophysics in the first decade after the war worked to ease tensions and increase understanding between countries, but that the International Geophysical Year (1957–58) and the Soviet launch of the first man-made satellites, Sputnik I and II, changed the rhetoric: IGY was as much a geopolitical year as a geophysical one.[44] In December 1957 NATO responded by establishing a science committee with one representative from each member country. The Science Committee supported three kinds of activities: scholarships aimed at research exchanges, summer schools, and support for larger research programs. Historian John Krige has argued that the establishment of a Science Committee reflected a change from an arms race to a competition between civilizations, where the committee's goal was not primarily military, but gaining prestige in proving the West had a better way of organizing society.[45] According to Krige, this led to a strong emphasis on basic research, and both individual members and those who were funded had a large degree of academic freedom. Historian Ronald E. Doel has argued that military patronage of the earth sciences shaped the questions that researchers asked and valued, and limited their interactions with colleagues studying biology. Military interests were motivated by new weapon systems that required geophysical knowledge from the field, especially polar regions and other strategic areas: "Military thirst for geophysics intelligence was intense."[46]

In 1960, NATO's Science Committee established a Subcommittee for Oceanographic Research. The argument was that the initiative would lead to cooperation among member states, and that basic scientific research was also valuable to the military. NATO was also concerned that the West was falling behind the Soviet Union in technology and science, and that a shift in the balance of scientific power could lead to a shift in military power. The first leader of the subcommittee was Håkon Mosby.

Officially, Mosby welcomed collaboration in oceanography as a natural consequence of the oceans being worldwide and between countries: "The study of the sea is by nature of a true international character."[47] Additionally, Mosby highlighted that the oceans' capacity as receptacles for nuclear waste, pollution and carbon dioxide made them vitally important to the future of humanity. Although Mosby did not highlight this himself, the international organizations also provided the means for representatives from smaller states and institutions to overcome their lack of resources, which very much applied to Mosby.

"It may, however, also with good reason be asked if the establishment of so many different organizations is really needed and should be recommended," Mosby pointed out in 1959.[48] Yet, he added, the groups were well coordinated. SCOR and IACOMS usually arranged their annual meetings in the same place and about the same date, while IAPO met at the IUGG general assembles. About a third of the executive board members were also members of one of the other two main organizations. In 1959, this amounted to six people: Günther Böhnecke

(West Germany: IAPO, SCOR), Anton Frederik Bruun (Denmark: IACOMS, SCOR), George Deacon (UK: IACOMS, SCOR), Marc Eyriès (France: IAPO, IACOMS), Håkon Mosby (Norway: IAPO, SCOR), and Lev Zenkevitch (Soviet Union: IACOMS, SCOR). "Each group will, of course, have to pay interest mainly to plans that correspond to the desires and obligations of their sponsoring authority, but by the close contact any suggestion may easily be transferred to the proper group."[49]

As president of IAPO from 1954 to 1960, a founding member of SCOR, and from 1960 to 1965 head of NATO's Subcommittee for Physical Oceanography, Mosby was intimately involved in shaping and organizing postwar oceanography.[50] Being equally fluent in German, French, English and Norwegian was not a drawback, nor was representing a peripheral NATO country with a social democratic government. Mosby had two priorities: to organize international collaboration, and to channel more funding to reputable oceanographic research communities outside the United States and the Soviet Union. Between 1940 and 1960, his institution in Bergen had stagnated with no new positions.[51] Unlike Roger Revelle, director of Scripps Institution of Oceanography, and Harald Ulrik Sverdrup, director of the Norwegian Polar Institute, who actively stimulated oceanography in developing countries, Mosby argued that research funding should go to countries with strong oceanographic research traditions, such as his own: "The simplest, the most effective and also the cheapest way for UNESCO to increase our knowledge of the sea must be to further oceanographic research within countries where such research does already exist."[52]

Mosby had, as head of the NATO subcommittee, a large say in deciding what research projects to fund, and in Bergen several promising students were given NATO scholarships. Between 1960 and 1973, the number of employees at the Geophysical Institute in Bergen increased from 17 to 53. The section for oceanography increased from seven to 19. Some of the new positions came with the introduction of a new study plan in 1960; most were paid by external funding. To the oceanographers, NATO was by far the most important source of external funding: Mosby's emphasis on international collaboration had paid off.

Tools for a new age

From the late 1950s onward, oceanographers became increasingly interested in the instruments they used. At the 1957 IUGG General Assembly in Toronto, German oceanographer Günther Böhnecke summarized existing methods and principles for measuring currents, and the IAPO decided to produce a global survey of oceanographic instruments. At the next 1960 IUGG Assembly in Helsinki, his colleague, Hartwig Weidemann, presented a booklet detailing 67 instruments, which was organized as index cards, and was intended to be expanded as information on more devices (especially from the Soviet Union) came in. Each card described one instrument, including its name, purpose, methods for use, site-specific requirements, measurement principle, accuracy, dimensions, power supply, depth limitations, how often it needed service, relevant literature, contact person, and whether it was for sale. The information was organized in different categories, represented by punched holes on the sides for easy reference.[53] While the content demonstrated a widespread culture for institutions and individual oceanographers developing and using their own tools, the booklet itself exposed a new desire for standardization.

Håkon Mosby was among those who contributed input to the booklet. One of the instruments was "the Mosby thermo-sound," which produced a continuous plot of temperature throughout a water column and could be used to great depth. The concept was first presented at the 1939 IUGG General Assembly in Washington, and then developed in collaboration with Odd Dahl during the Second World War. The main mechanism was a framed brass wire that changed its length with changing temperatures. The wire was attached to a pen that mechanically recorded a scratching on a circular glass plate that rotated as the instrument descended through the water column. The rate of revolution could be adjusted to accommodate different depths. In the late 1950s, it was praised for being both simple and robust.[54] A major drawback, however, was that the glass plates needed to be taken ashore and analyzed under a microscope, as a 1° C change in temperature corresponded to only 0.15 millimeters.

Among the motivations for developing new tools was making

oceanographic observations faster, more practical, and more accurate. Only months after returning from taking part in the Atlantic Polar Front Program, a new electric salinity meter was installed on the *Helland Hansen*. The new instrument increased the precision of the observations by a factor of three, and reduced the time needed for processing by a factor of four. A decade later, the research vessel *Helland Hansen* was equipped to use CTDs (conductivity, temperature, and depth) in real time, invented by Neil Brown from the Woods Hole Oceanographic Institution in the 1960s.[55]

As the first head of NATO's Subcommittee on Oceanographic Research, Mosby funded several projects aimed at producing new instruments. In 1962, one group began the construction of a "Mare-graphe," an instrument for measuring variations in sea level, and another began the construction of an instrument for taking bottom samples 20–30 meters below the seafloor. The largest project, which began in Bergen in 1960, was the construction of autonomous current meters that could be anchored in fixed locations, and make continuous measurements for months at a time. The "oceanographic buoys project" was the only project fully financed by the NATO subcommittee, and instrument maker Odd Dahl at Chr. Michelsen Institute was put in charge.[56] Illustrating both how the geophysical center of gravity had moved to the United States and how smaller institutions had to carve out niches for themselves, Dahl's first task was to tour eleven different US institutions that all had relevant experience in producing "gadgets." The final product was based on a design from WHOI in Massachusetts. Woods Hole also took part in the first field survey aimed at finding where to anchor the buoys in the Faroe-Shetland Strait with its research vessel the *Chain*, a repurposed navy salvage vessel.

As Mosby put it in a speech to the NATO Parliamentary Congress in 1963, the potential for the instrument developed in Bergen was nothing less than to herald a new age for oceanography:

"The climate of Northern Europe is known to be abnormally mild for the latitudes, and the primary cause is sought in the branch of the Atlantic Current or the Gulf Stream entering into the Norwegian Sea through the Faroe-Shetland Channel. More than 50 years ago indications were found of fluctuations of this important water transport, and many efforts have later been made to elucidate the causes of these fluctuations. But clearly, the first condition for doing so, with any hope of success, must be a fairly reliable knowledge of the fluctuations themselves, in other words a method by which the fluctuations can be measured (...) As will be understood, the oceanographic buoys project is something different from most other projects taken up by the Subcommittee. If it comes out successfully it will mean a technical achievement which could hardly have been thought of in any single European country, a tool for oceanography not yet to be really understood."[57]

The primary objective for the current meter was that it had to operate autonomously and make precise and reliable measurements to great depths. Secondly, it had to be cheap, easy to deploy, and easy to retrieve at the desired time.[58] When the finished instrument was presented in 1964, it was about 40 cm tall with a 13-centimeter diameter, and was enclosed in a metal cylinder protecting it from the elements, with a rotor and other measuring equipment sticking out. The memory unit was a tape recorder, and the instrument was powered by six 1.5 volt batteries of the same kind used in flashlights.[59] Using a unique combination of mechanics and electronics, the power consumption was kept very low.

Many of the oceanographic cruises in the early 1960s focused on field-testing the current meter and numerous other prototypes. The cruise to the Faroe-Shetland channel in 1962 in collaboration with the British *Discovery II*, for instance, tested the NATO buoy, a submerged floating buoy, a geomagnetic electro-kinetograph, an acoustic current meter, and a five-rotor current meter. While the instruments operated on different principles, and had different strengths and weaknesses, it was also a race to create an instrument that would become the new standard in its defined niche. When finished, the buoys were used to supplement ordinary oceanographic cruises. During two spring months in 1967, for instance, three vessels from Norway and Britain made observations along a 187-nautical-mile-wide area in the Atlantic current at 66 degrees north 17 times. The observations were supplemented with data from four permanent buoys.

The buoy research was at the core of a larger development effort at the Chr. Michelsen Institute. The side projects included a release mechanism that, on an acoustic signal, would release anchored buoys, as well as a surface buoy that could receive data from instruments at depth and transfer them via radio or satellite. The latter part of the project involved technicians from Belgium, Great Britain, Germany, and the USA.

The buoy project can be understood as part of a larger project in postwar geophysics, namely the emergence of a collective ambition to use technology to put the earth under constant surveillance. The transformation was driven by geophysicists developing and adapting new tools for observation, communication, and computation, subtly but irreversibly changing oceanographic practices. As Mosby noted in the annual report from 1965–66 after anchoring six buoys over the continental shelf outside Greenland: "The extensive use of oceanographic buoys as part of a cruise represents a new phase in oceanography. The work with the buoy observations has now become a routine to the extent that it is possible to decode and analyze the observations through electronic computations."[60]

The NATO buoys proved themselves internationally in July 1967, when they were deployed in a competition arranged by WHOI, as part of SCOR Working Group 21's calibration of oceanographic instruments outside Bermuda. In addition to the Bergen current meter, a British, a German, and a US instrument were used. Three buoys of each kind were placed at 500 meters' depth for eight days, and the Bergen current meter was the only one that finished with three full sets of data.[61] For a science relying on observations from expensive expeditions, reliability was a major selling point.

At the 1966 SCAR symposium on Antarctica in Santiago, Chile, Mosby suggested making direct observations of the bottom water current on the shelf and under the sea ice in the Weddell Sea in Antarctica. During the International Weddell Sea Oceanographic Expedition (IWSOE) in 1968, oceanographer Thor Kvinge from the Geophysical Institute and Jan Strømme from the Chr. Michelsen Institute brought four instruments onboard the US icebreaker *Glacier*.[62] The instruments were placed at the edge of the continental shelf, between 400 and 600 meters below the surface. The goal was to test Mosby's theory regarding bottom water formation.[63]

When returning to retrieve the buoys a year later, the area was covered by sea ice. In 1970, the acoustic release mechanism did not work. Only in 1973 were two of the buoys retrieved by using a trawl. Incredibly, after five years deep in Antarctic waters, the instruments were unharmed and had observation series of 269 and 474 days (9 and 15 months), which was how long the batteries had lasted.[64] The water circulation and transformation underneath the floating ice shelves in Antarctica became a lasting research topic at the institute.

Based on the data from the NATO buoys and a number of new observations made in the 1970s, several new theories on the sub-ice shelf circulation were proposed. Herman Gade at the Geophysical Institute also formulated a theory on when water freezes or ice melts underneath the ice shelf as a function of salinity, the latent and specific heat of the ice, its in situ freezing point, and the ice shelf core temperature. The equation, known as the Gade line, is still in use.[65]

Another early adaptor of the Bergen Current Meter was mathematician Martin Mork, who began work at the Geophysical Institute in 1965. Mork was interested in the process of wind-driven waves, for which measurements from a single point using Nansen bottles were not sufficient.[66] Mork theorized that when persistent winds over an ocean suddenly die down, inertial oscillations result. Whether this phenomenon existed was a question that could only be answered by an instrument that could make time series. Since such oscillations in the Atlantic would be indistinguishable from tidal waves, the experiment was conducted in the Mediterranean: Mork had two colleagues rent a van and drive the measuring equipment, including four kilometers of anchoring wire, to Monaco.[67] After sufficient time series were collected, he had to develop new methods for analyzing the observations before he could confirm the existence of wind-driven inertial oscillations.[68] This was but one example of how the instrument could answer new research questions.

By the early 1980s, the Bergen Current Meter (BCM) was one of the most popular current meters in the world, known for being reliable and simple to use.[69] The only change from the initial design was that the nickel coating on the copper pressure case was removed, because under certain conditions it created a magnetic field that interacted with the enclosed compass needle.[70] Another improvement was new software that made it possible to import the observations directly to a computer. Earlier, the process had been cumbersome. First, the tape had to be connected to a printer, which resulted in long series of numbers on paper. Then, the numbers had to be entered manually onto punch cards, which were fed into the university's computer, a process that often had to be repeated several times to weed out errors.[71] Only then could the actual analysis begin.[72]

In 1966, Ivar Aanderaa, a physics graduate from Bergen who was in charge of developing the buoy, formed his own company, Aanderaa Instruments, with the current meter as its cornerstone.[73] The payment for the patent consisted of donating six current meters to the Geophysical Institute. Aanderaa was not the only one to commercialize oceanographic instruments developed in close collaboration with oceanographic institutions. That same year Shale Niskin, of the University of Miami's Institute of Marine Science in Florida, patented the "Niskin bottle" and founded the company General Oceanics. Over time these bottles came to replace the more cumbersome Nansen bottles.[74] Similarly, in 1974 Neil Brown from WHOI set up his own company to mass produce CTDs, Brown Instrument Systems.[75]

Oceanographic tools, increasingly being produced by private companies and available off the shelf, had unforeseen consequences. On the one hand, although the instruments remained expensive, they provided much-longed-for standardization, the quality of the observations improved, and less time was spent on development and field testing. On the other hand, technicians in the workshops were either demoted from instrument-makers to caretakers doing maintenance and minor adjustments, or made obsolete altogether. Likewise, the introduction of onboard electronic equipment for measuring salinity in the late 1950s – and then CTDs in 1976 – meant an end to the need for manual titration in the laboratories. Steinar Myking recalls:

"The shift from mechanical to electronic tools happened very quickly, but there was little attention to the implications for us technicians. For those of us who found new tasks, or were curious to learn new skills such as using computers, this was fine. Others were left in a vacuum. As long as you showed up at work, no one reacted if you had no tasks. Some had a very hard time."[76]

Hunting submarines?

In a study of the NATO Sub-Committee of Oceanographic Research published in the journal *Centaurus* in 2012, historian Simone Turchetti argued that the prime motivation for establishing the subcommittee was to gather military intelligence from specific areas in order to detect enemy submarines.[77] In 1952, there were military reports that the Soviets were developing a nuclear deterrent based on nuclear missiles onboard submarines, and if the Cold War turned hot, NATO needed to detect them before they reached missile range. This gave the straits that connected the Soviet Union and the Atlantic Ocean vital strategic importance, and created a pressing need for more information on their physical properties.

Mosby and other leading members of the subcommittee did indeed work closely with the military. From 1955 to 1961, the Geophysical Institute in Bergen hosted a "Water Office" (Farvannskontoret) with employees from the Norwegian Defence Research Establishment (FFI). The office worked with projects concerning the defense of Norwegian harbors, underwater mines, and submarine warfare. This included deciding the placement of underwater sonar along the Norwegian coast and field testing of listening equipment, detecting the first submarines in Korsfjorden in 1955. Mosby also took part in exchanging information with the US Navy on the physical conditions in the oceans through "Information Exchange Program No. 1 (IEP N-1)."[78] Likewise, the US representative to the subcommittee, WHOI Director Columbus Iselin, was influential in sponsorship schemes for the Office of Naval Research, the largest naval research agency in the United States. Both Iselin and his British counterpart on the Subcommittee, UK National Institute of Oceanography Director George Deacon, had been active in anti-submarine warfare work during the Second World War. Deacon had investigated the effects of currents on sound transmission and how they could be used to hide from enemy sonars.[79]

In the time leading up to the subcommittee being established, the British representative, who also represented the navy, had called for collaboration on developing technology to improve their "detection and kill capacity" in the seven straits that would serve as passageways if

Soviet submarines wanted to enter the Atlantic Ocean.[80] The military especially desired insights into environmental conditions that could affect NATO surveillance devices, namely temperature, current, and salinity. These were the same seven straits that the subcommittee later singled out as areas for joint oceanographic projects. Although Turchetti does not present documentation for his claim, it is unlikely that a top secret military motivation would have been written down in an unclassified document. Besides, despite being defensive rather than offensive, a NATO project aimed at detecting submarines carrying nuclear weapons would probably have been controversial. In Norway, NATO membership was a divisive issue causing a schism in the ruling Labor Party, and major protests in the late 1950s led to a ban on storing nuclear weapons on Norwegian soil.[81]

Rather than being motivated by warfare alone, a more reasonable interpretation is that the military and the oceanographers had overlapping interests. During a meeting in New York in September 1959, Mosby suggested that the NATO subcommittee should pursue research in two specific areas: direct current measurements in the Faroe-Shetland and the Gibraltar Straits, and an instrument development project aimed at developing an automatic buoy for current and temperature measurements.[82] His argument was not that the straits were of interest to military intelligence pursuing submarine surveillance and warfare, but that they play a vital part in the ocean's circulation system. The Faroe-Shetland Strait was the main point of entry of the warm Gulf Stream to the Norwegian Sea, while the salty water exiting through the Gibraltar Strait could potentially produce some of the bottom water in the Atlantic.[83] Bottom water, to Mosby, was "one of the most important features in the physics of the oceans."[84] In the years that followed, the NATO subcommittee started similar projects in the Alboran Sea in the western Mediterranean (1961), the Tyrrhenian Sea (1962), the Irminger Sea (1962), the Skagerrak (1962) and the Turkish straits of Bosporus and Dardanelles (1962), which were identified as focal points where oceanographic researchers from many countries could collaborate. The goal was to map, or find methods to map, the transport of different water masses through the straits.[85]

As historian Gunnar Ellingsen has pointed out, the research questions Mosby and the other oceanographers sought to answer far pre-dated the Cold War context: What happens below the surface, what physics are involved, and how can the state of the oceans be calculated and predicted?[86] In order to study the physical processes and currents, the oceanographers looked for the same parameters they had since Mohn's days, namely temperature, pressure, and salinity. Projects that were more directly defense related were pursued at the NATO undersea warfare laboratory in La Spezia, Italy, which was financed by the United States Department of State and opened in 1959.[87] In 1963, disappointed with the military relevance of its science committee, NATO established the Defence Research Directors Committee (DRDC) to conduct research more directly relevant to the military.[88] Although geophysicists, like Mosby, undoubtedly exploited the Cold War context to secure military funding, the buoys that Turchetti argues were used to detect submarines were sold commercially with no restrictions. For the oceanographers, NATO provided the structural conditions necessary to fund the research they wanted to pursue regardless. Or, as historian John Krige has put it: "NATO was now more than simply a military alliance. It was there to weld a community of nations together around a core of set values and to confidently proclaim the scientific and technological strength of the Free World against a program of techno-scientific seduction by an ebullient Soviet Union."[89]

Guests and destinations

Under Mosby's leadership, the oceanographers were by far the most internationally oriented of the three divisions at the Geophysical Insti-tute. Between 1945 and 1965, more than 90 percent of all foreign guests to the Geophysical Institute were visiting the oceanographers. This peaked while Mosby was head of the NATO subcommittee, averaging more than 22 guest researchers per year (see Figure 5 next page). Some stayed for a year or more; others spent only a day or two as an extension of international meetings or to give lectures. During the interviews for this book, several people described friendships with colleagues from other institutions and their families, with home dinners and return

Figure 5

Number of foreign guests to the Geophysical Institute in Bergen based on the
institute's annual reports.

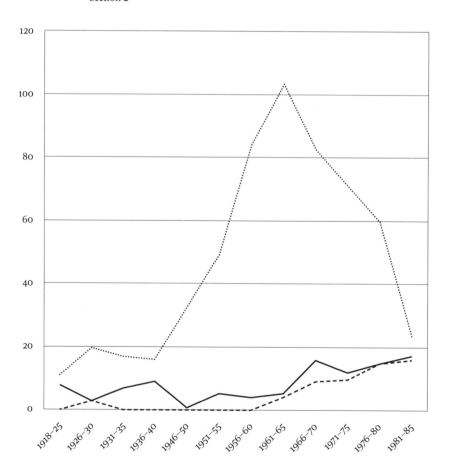

visits. After Mosby resigned as head of the NATO subcommittee in 1966, the number of guests dropped. The following decade, three out of four guests were oceanographers; by the 1980s the proportion had fallen to 40 percent.

In the first years after the war, most international visitors came from Sweden, but by 1950, there were more British and American guests. While the number of visitors from Britain declined over time, the number of US guests increased until 1980. There were also frequent visitors from Germany, Denmark, France, and Canada. In the first half of the 1970s, the Geophysical Institute had 21 guests from the Soviet Union, mostly thanks to a cultural exchange program between Norway and its Soviet neighbor.

There were also numerous guests from other countries, but the connection to Brazil stands out. In 1967, the Bergen shipyard Mjellem & Karlsen launched the research vessel *Prof. W. Besnard*, Brazil's first and only purpose-built oceanographic research vessel until 2008. Before and after its maiden voyage back to Brazil via the Canary Islands – the joint Brazilian-Norwegian *Vikindio* expedition – oceanographers from Bergen worked as visiting professors in São Paulo.[90] The vessel, named after the founder of the Oceanographic Institute in São Paulo, Professor Wladimir Besnard, was 49 meters long, weighed 700 tons, had a crew of 23, and had space for 16 scientists. While the oceanographers in Bergen returned ashore to do their analysis, the *Besnard* specialized in doing physical, chemical, geological, and biological analysis in situ. In 1968, the vessel took part in producing the first geological charts of the Brazilian coast.

While the collaboration with Brazil was relatively short-term, the engagement in southern waters was not. Over the years, geophysicists from Bergen have made numerous return visits to Antarctica. The Weddell Sea and surrounding waters play an important role in bottom water formation and the global ocean circulation system, and geophysicists from Bergen have produced what is likely the world's longest oceanographic time series from Antarctic waters.[91] In 2010, after melting through 1000 meters of the ice shelf, they also observed the coldest ocean water ever measured, -2.6° C.[92] The area has also been of special interest for climate studies.

In addition to Norwegian Roald Amundsen's conquering of the South Pole in December 1911, a main reason for Norway's interest in Antarctica has been that it annexed almost a fifth of the frozen continent in 1939, an area seven times the size of the Norwegian mainland. The area was given the name Queen Maud Land. When Norway, Sweden, and Britain organized a joint expedition to prove that European science had not been broken by the war (1949–52), they set sail for Antarctica.[93] Being present in Queen Maud Land became a political goal in itself, and meant access to research funding through the Norwegian Polar Institute. In 1957, the International Council for Scientific Unions (ICSU) established a committee to coordinate scientific activities on the continent, the Special Committee on Antarctic Research (SCAR), and two years later Norway was among the twelve original signatories to the Antarctic Treaty, which froze any territorial claims and ensured that the continent would be used exclusively for scientific exploration and other peaceful purposes. Expeditions to Antarctica are exceedingly expensive due to its remoteness, inhospitable climate, and huge distances, and the majority of the Norwegian research expeditions have been conducted from foreign research vessels, which required sustained participation in international networks.

While the rationale for understanding Antarctic bottom water formation is that it affects the whole of the Atlantic Ocean, other postwar oceanographic research had a more regional scope. One of Mosby's colleagues, Herman Gade, focused much of his career on investigating fjords. Formed by retreating glaciers, most fjords have shallower bedrock at their mouths, known as fjord sills. This, according to Gade, led to the water in the fjords having three distinct layers. The bottom contains "basin water," acting like large containers in which seawater remains stagnant for extended periods of time before being flushed in massive exchanges with the adjacent sea. Above the fjord sill, fjords have a layer of "intermediate water" in direct communication with the ocean. Finally, the fjords receive runoff from rivers, especially in summer, which creates a distinct layer with reduced salinity in the top 50 meters of the water column. The three water masses exchange on different time scales, and have different impacts on the dynamics of the costal currents.[94]

While dividing the fjords into layers provided new insights, it also gave rise to a host of new questions. What prompted the exchanges of different layers, and could they be predicted? What happened between the layers? In addition to identifying seasonal variabilities, the research focused on identifying the mechanisms involved in the different exchanges, such as prevailing wind, tides, or seasonal temperature changes, as well as the question of mixing. Next, a contested topic was what happened when flushed fjord water met the coastal current: Did the different water masses simply mix, or were they – as Gade argued – forced by the more powerful coastal current into vortices along the coast? The fjord research had clear practical implications, such as how much pollution a fjord could survive. In the early 1960s, Gade was put in charge of physical oceanographic research in a large interdisciplinary study of water exchanges in the Oslo fjord, south of the Norwegian capital, headed by Hans Munthe-Kaas from the Norwegian Institute for Water Research (NIVA). The project was prompted by concerns about water pollution.[95]

Local collaboration – or lack thereof

In his historical writings, Håkon Mosby put biological and physical research on the sea on an equal footing. The direct inspiration seems to have been Harald U. Sverdrup, Martin W. Johnson, and Richard H. Fleming's *The oceans: Their physics, chemistry, and general biology*, first published in 1942. The book was initially meant to fill a gap in the curriculum at the Scripps Institution of Oceanography, but after the war, *The Oceans* was translated into many languages, and came to be known simply as "the Bible" for oceanography. Throughout the 1960s, the book was updated and reprinted every one to three years.[96] In addition to bringing together insights from European and American oceanography, this first comprehensive textbook sought to reunite the physical properties of the oceans with its biology. The stage was thus set for close collaboration with other institutes in Bergen studying the sea, in particular the Institute of Marine Research which focused on fisheries research, and the Biological Station, which focused on marine biology more generally.

In December 1949, Mosby and several of his colleagues from the Geophysical Institute participated in founding both the Bergen Geophysical Association (Bergens Geofysikeres Forening) and the Norwegian Association of Marine Scientists (Norske Havforskeres Forening). The former focused on collaboration in Bergen, and arranged shared colloquia for the town's meteorologists, oceanographers, and weather forecasters, as well as researchers from the Chr. Michelsen Institute and the Institute for Marine Research. The latter had members from the University of Oslo, the Norwegian Polar Institute, the scientific museums in Trondheim and Tromsø, and the Ministry of Fisheries. In addition to giving several lectures to both groups, Mosby was the head of the Bergen Geophysical Association (1956–57), and the leader of the Association of Marine Scientists (1959–62).[97] But while geophysicists participated in most of the monthly meetings, and reported the number of members in both associations in their annual reports, there was no formal research collaboration until after Mosby retired in the early 1970s.[98]

Instead of the oceanographers reaching out to the marine biologists, the Institute of Marine Research established its own section for physical oceanography. Its first leader, Jens Eggvin, remained in the position until he retired in 1969. Eggvin, hired in 1931, was tasked with building expertise on physical oceanography in relation to fisheries and fisheries research. During the 1930s, he established a number of permanent oceanographic stations along the 2700-km Norwegian coastline and began a systematic collection of ocean temperatures and water samples taken by coastal steamers and liners. The goal was to identify the movement of water masses with different temperatures and to use these to aid the fisheries through forecasts for when, where, and at what depths cod and spring-spawning herring were to be found. Eggvin's doctoral thesis, *The movement of a cold water front* (1940), followed an outbreak of cold water from the Skagerrak and its propagation along the Norwegian southern and western coasts and discussed its implications for the herring fisheries.[99] But according to historian Vera Schwach, poor personal chemistry was an obstacle to collaboration with the geophysicists.[100] Eggvin also clashed with colleagues representing more traditional fisheries biology at his own institution.[101]

In addition to the lack of partnerships with other local institutions, there was little collaboration among the staff at Mosby's Geophysical Institute. Apart from sharing lunch, and occasionally popping in to see the weather forecasters to get the latest forecasts, oceanographers, meteorologists, and weather forecasters worked as different units under the same roof. When the institute was asked to analyze what local consequences the development of hydropower and rerouting of rivers had on fjords, such as in the Fuglesett fjord north of Bergen in 1967, no meteorologists were involved. When the meteorologists studied the impact of hydropower development, the oceanographers were not involved.[102] Nor did the oceanographers collaborate with each other. According to researcher Svein Østerhus, who came to the Geophysical Institute in 1980 on external funding, there was simply not a culture for teaming up: "Collaboration was seen as defeat, as proof you were unable to work independently. There were some exceptions, but the culture was to do everything on your own, from observations to analysis to publication. The oceanographers preferred their own instruments and their own models."[103] In effect, there was more emphasis on gathering empirical data from the field than on developing theory or seeing data and theory in connection.

One notable exception was a project headed by mathematician Martin Mork. Following up his interest in waves, Mork involved himself in the Joint North Sea Wave Project (JONSWAP), an international measurement program that began in 1967 aimed at measuring the growth of waves and how they propagate into shallow water. As we will see in chapter 9, this led to Mork taking the initiative to organize a large national collaboration to study the Norwegian coastal current from different disciplinary angles, starting in 1975.

The lack of local collaboration can be seen as an unintended consequence of Mosby's philosophy of science. In *Quo vadis universitas* (1971), a reflection on the relationship between university and society that Mosby published the same year he retired as rector of the University of Bergen, he argued that the role of the university was to "seek truth for the truth's own sake, and to achieve this it needs autonomy."[104] Autonomy was to extend from the institution to its researchers. The premise for good science, as Mosby put it, was that each scientist was given freedom to pursue his interests, and had undisturbed time to work.[105] In this perspective, the role of the leader, the scientist-administrator, was to facilitate and gather funds, and not to apply pressure that would stifle the curiosity that scientific progress relied on. With increasing administrative obligations both locally and internationally, a politically conservative viewpoint, a formal demeanor, and the strict head of administration Karen Sofie Olsen serving as a gatekeeper in the anteroom, Mosby has been described by colleagues as a skilled and inclusive leader, but also autocratic and uninvolved in research conducted by his colleagues.[106]

Publication practices

The lack of collaboration was reflected in publication practices: in the 1950s and 1960s, 85 percent of all oceanographic publications produced at the Geophysical Institute had a single author. But when the tide began to shift, it changed relatively rapidly. In the 1970s, the proportion of single-authored publications fell below 60 percent. By the 1980s, it had dropped to 30 percent. Although we have not managed to compile complete publication records from 1988 to 2000, it seems that the formation of formal research groups in the early 1990s cemented the trend. Since the turn of the millennium, only 22 percent of the publications have had a single author. The average number of authors of each publication has also increased dramatically: in the 1970s, each publication had an average of 1.6 authors. The following decade, the average number of authors was 3.4. From the turn of the century to 2015, each publication had an average of 4.8 authors. The same pattern is reflected in geophysical publication practices elsewhere, mirroring how research is conducted in groups and requiring different specialties working together.

The publications genres have also changed over time. Most noticeably, the popularizations that in the interwar period had made up a quarter of all oceanographic publications dropped to only 8 percent in 1945–59. In the 1960s, the oceanographers did not produce a single publication aimed at a general public, and in the 1970s and 1980s, the proportion increased only slightly, to 3 and 1 percent respectively. It was hardly a coincidence that this happened during a period with high attention to military funding, with student numbers more than doubling, and with emphasis on producing data series from the field. Instead, two new genres began to dominate: in-house booklets and reports. The booklets were mainly printed lecture notes for students, which in lieu of updated textbooks in the 1960s made up 19 percent of the publications. The reports were often addressed to external funders, and made up about a fifth of the publications in the 1960s. The proportion increased to slightly more than half in the 1970s (52 percent). This reflects perceptions about audiences: oceanography was seen as technical and of interest mainly to those with special interests. Since the turn of the millennium, however, reports of various kinds have made up 16 percent of the publications, while the booklets have disappeared completely.

The postwar drop in popularizations coincided with a drop in publications aimed at academic journals. During the interwar period, scientific journal papers made up about half of all oceanographic publications. Between the war and 1960, this had halved to one in four, and the number continued to drop to only one in eight publications (14 percent) in the 1960s. Instead, in the first fifteen years after the war, more than half the publications were in the form of conference proceedings and book chapters (30 percent), and yearbook publications (24 percent). There were two main reasons: first, as a cornerstone institute in a new university, publishing in the institution's yearbook was a priority. Second, oceanography was understood to have entered a faster pace than before the war, which meant that participating and publishing in conference proceedings was seen as more important than potentially having to wait for a year or more to get the same results in an academic journal. In the 1960s and 1970s, the proportion of anthologies and conference proceedings fell further, before making a brief comeback in the 1980s. Since the turn of the millennium, only 5 percent of publications are in the form of conference proceedings and anthologies, while yearbook publications make up less than 1 percent of the total. In return, the proportion of papers in academic journals has since the 1970s only increased. Since the millennium, scientific papers in international journals again make up more than half of the publications (54 percent). This, too, reflects a wider pattern in the field: papers in scientific journals, produced by groups collaborating, have become the new gold standard. Notably, the age of papers has coincided with an increase in popularizations, which since the turn of the millennium have made up 17 percent of the publications. This has to a large extent coincided with attention to man-made climate change. As core producers of knowledge about this challenge to humanity's future, geophysicists have seen communicating the nature of the threat, their research results, and how they have arrived at them, and the need for political action, as vital.

Table 1: Publication genres, oceanography (percentages).

Year	Conference proceedings/ anthologies	Populari- zations	Journals (Norwegian)	Journals (international)	Reports	Observation series	Year-books	Booklets
1917–1929	0	37	22	30	0	0	7	0
1930–1940	14	19	31	22	0	0	8	3
1945–1959	30	8	11	16	5	0	24	5
1960–1969	23	0	4	10	20	11	14	19
1970–1979	12	3	3	17	52	6	3	3
1980–1988	22	1	2	33	41	1	1	0
2000–2015*	5	17	3	54	16	1	0	0

Table: Publication genres (in percentages) from the oceanographers at the Geophysical Institute in Bergen. Based on publication records from the Geophysical Institute's annual reports (1917–1988), *Navn og Tall* (2000–2003), and Cristin (Current research information system in Norway, 2005–2015).
* The numbers for 2000–2015 also include publications by meteorologists and climate researchers.

The steamship *Vøringen* hosted the Norwegian North-Atlantic Expedition (1876–78). The expedition coined the name "Norwegian Sea," and was the first systematic study of the oceans outside Norway. Photo: UiB, Special Collections.

VØRINGEN.

The scientists onboard the *Vøringen* expedition. From the left: marine biologist Georg Ossian Sars, captain Carl Fredrik Wille, professor of meteorology Henrik Mohn, landscape painter Franz Wilhelm Schiertz, Captain-Lieutenant C. Petersen, physician and zoologist Daniel C. Danielssen, shipmaster Joachim Grieg, zoologist and businessman Herman Friele, and chemist Hercules Tornøe. Photo: Bergen Museum.

From the aft deck of the *Vøringen* in 1876.
Henrik Mohn (top, middle) is holding a
hand on the instrument he designed to
reproduce the ocean surface in miniature.
Photo: UiB, Special Collections.

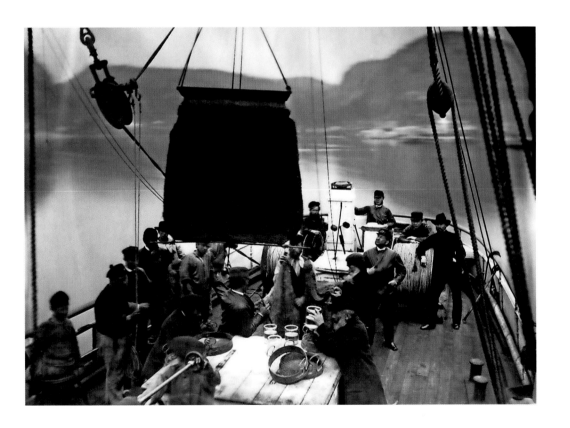

↓ The next major expedition to set sail from Bergen was Fridtjof Nansen's *Fram* expedition, which left port on July 2, 1893. The goal was to let the specially designed vessel freeze into the ice, and drift across the North Pole. What actually happened was even more dramatic. Photo: Johan von der Fehr. UiB, Special Collections.

→ The aurora borealis, the northern lights, as drawn by Fridtjof Nansen during the *Fram* expedition. The image was first printed in his 600-page history of polar exploration, "Nord i tåkeheimen" (1911: 376), also published in English under the title "In Northern Mists. Arctic Exploration in Early Times." Photo: UiB, Special Collections.

Michael Sars, the research vessel belonging to the Norwegian Fisheries Board, from which Bjørn Helland-Hansen and Fridtjof Nansen in 1900–1904 made their observations for the classic "The Norwegian Sea" (1909). Photo: UiB, Special Collections.

From 1892 to 1917, Bergen Museum's
Biological Station at Marineholmen was
the main institution for marine research
in Bergen. It was located some 200 meters
from today's Geophysical Institute. Photo:
K. Nyblin. UiB, Special Collections.

← The research vessel *Armauer Hansen* on its maiden voyage in 1913. The small wooden vessel was financed through a local fundraising campaign, and would for decades remain a hallmark for Bergen oceanography: economical, unpretentious, maneuverable, and with a large operational range. Photo: UiB, Special Collections.

↓ Fridtjof Nansen and Bjørn Helland-Hansen onboard the *Armauer Hansen*. For three decades, these two were the most prominent Norwegian oceanographers, and played a key role in establishing the scientific discipline of physical oceanography. Photo: UiB, Special Collections.

Onboard the *Armauer Hansen*, science and leisure could be combined. Here Håkon Mosby plays the violin, while his brother Olav Mosby plays the accordion. Photo: Aagot Borge. UiB, Special Collections.

Bjørn Helland-Hansen and a colleague attaching a Nansen bottle onboard the *Armauer Hansen*. As Olav Mosby wrote in his diary, again and again: "Stations and stations ..." Photo: Aagot Borge. UiB, Special Collections.

Swedish oceanographer Vagn Walfrid
Ekman doing maintenance on his current
meter in the laboratory onboard the
Armauer Hansen. The picture is probably
from the mid-1920s. Photo: UiB, Special
Collections.

Bjørn Helland-Hansen spent significant parts of his academic career onboard the *Armauer Hansen*; here among friends and colleagues after the Geophysical Commission's meeting at his cabin in Olden in 1931. Photo: Aagot Borge. UiB, Special Collections.

The return of the *Fram* expedition in
1896 has been described as Norway's
first national event, nine years before the
country became independent. In Bergen,
the town built a triumphal arch by the fish
market to celebrate. In the background to
the left, the Sandviken Mountain where
the Nansen monument was planned in
the 1930s. Photo: Gustav Emil Mohn. UiB,
Special Collections.

Ernst Müller-Blensdorf's model of the Nansen monument and the Sandviken Mountain overlooking Bergen. Photo: UiB, Special Collections.

From the left: polar researcher Adolf Hoel, polar explorer Magnus K. Giøver and the German sculptor Ernst Müller-Blensdorf with the scale model of the Nansen monument. Photo: UiB, Special Collections.

The painting "New Telegram" by Fredrik Kolstø, painted at the fishing community of Røvær in 1891, shows one of the temporary telegraph stations erected to safeguard the fisheries. The station opened in 1868, two years before Norway's first systematic storm warning service. Photo: Moss kunstforening.

The Geophysical Commission meeting at the Haldde observatory in 1920:
V. Bjerknes, H. Köhler, B. Helland-Hansen, O. Devik, C. Störmer, O.A. Krogness, O. Edlund, Th. Hesselberg and O. Stoll.
Photo: Rurik Köhler's album / Alta Museum.

The Bergen school of meteorology
consisted of a group of young men and an
assistant, who analyzed weather maps in
Vilhelm Bjerknes's attic. Here Tor Bergeron
and Jacob Bjerknes are working to solve the
puzzle of tomorrow's weather. Photo: UiB,
Special Collections.

The participants at the 1921 conference for the investigation of the upper atmosphere. The event was part of a systematic dissemination of the invsights from the Bergen school of meteorology, and a source of both pride and ridicule by local journalists. Little did they know that the methods developed by their local weather forecasters would spread worldwide. Photo: UiB, Special Collections.

Knowing the structure of cyclones made it possible to provide more accurate weather forecasts. This model of a polar front cyclone, and how a passing cyclone is experienced on the ground, was drawn by meteorologist Wiggo Hårvig and first used in Dannevig 1969: 321.

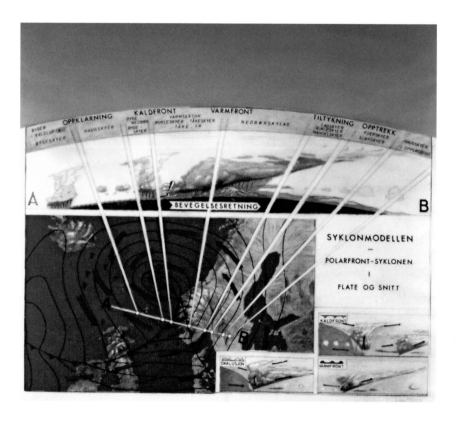

Meteorologist Sverre Petterssen overlooking the weather forecasters in Bergen in the 1930s. In 1944 Petterssen would famously postpone the Allied invasion of Normandy, Operation Overlord, most often referred to simply as D-Day. Photo: UiB, Special Collections.

Jack Bjerknes at Ås, supervising the first
of a series of weather balloon ascents in
1932. Three years later, Bjerknes organized
international synoptic (simultaneous)
weather balloon ascents in order to get
a three-dimensional image of passing
cyclones, a project in which 18 European
observatories participated. Photo: UiB,
Special Collections.

The measurements from the weather
balloons were scratched on sooted glass
plates, and had to be analyzed under
the microscope. Unfortunately, not all
balloons were recovered, and not all that
were recovered survived intact. Photo: UiB,
Special Collections.

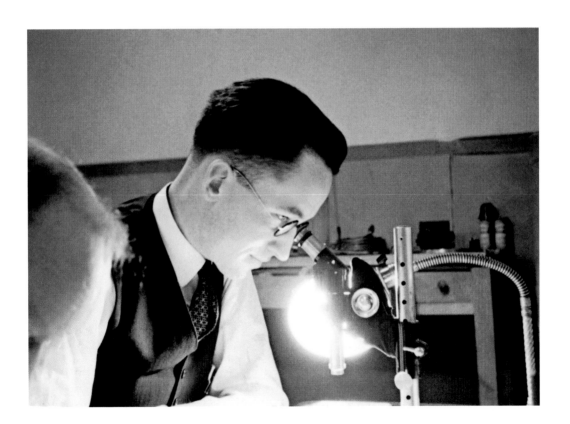

The final stages of the erection of the Geophysical Institute's building, "Geofysen," photographed in March 1928. Photo: Atelier KK. UiB, Special Collections.

From inside one of Geofysen's new laboratories. The instruments could be hoisted to
the roof by a set of pulleys. Metal inside the
walls and specially designed hatches over
the windows meant that the room could
be turned into a Faraday cage. Photo: UiB,
Geophysical Institute.

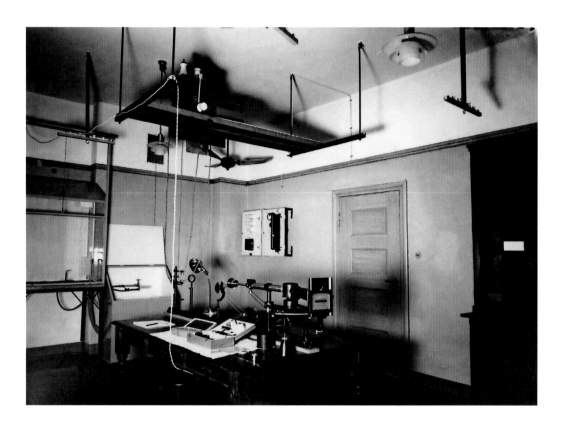

The Geophysical Institute in the 1930s as seen from the waterfront. The Biological Station the institute grew out of was situated on the waterfront some hundred meters to the left. Photo: UiB, Geophysical Institute.

2280.— BERGEN: GEOFYSISKE INSTITUT, FLORIDA.   K.K. BERGEN.

Enjoying the field without science. Tor Bergeron (lower left), Aagot Borge (middle) and friends enjoying the western Norwegian landscape, probably in the early 1920s. Photo: Aagot Borge. UiB, Special Collections.

October 25, 1946: King Haakon VII places the foundation stone for Bergen University. The first building project was to expand the Geophysical Institute with new wings to host a botanical laboratory and an institute for chemistry. Photos: Franz Blaha. UiB, Special Collections.

↓ The Geophysical Institute photographed in 2011. The weather forecasters still occupy the central "tower," while the Bjerknes Centre for Climate Research now occupies the newly refurbished wing to the left. Photo: Alf Edgar Andresen, UiB.

→ Sculptor Stinius Fredriksen and Vilhelm Bjerknes, probably taken in the latter's office at Blindern in the late 1940s. The bust is on display in the lobby of the Geophysical Institute in Bergen. Photo: UiB, Special Collections.

← The IBM-650 computer EMMA being installed at the Geophysical Institute in 1958. This was Norway's first commercially available computer, and was used for science, calculating tax returns and more. Embedded: a) The (human) computers in Bergen prior to EMMA used punch card machines. b) EMMA arriving at Bergen Flesland Airport. c) The computer arriving at the Geophysical Institute. In the background, the University's Van der Graff reactor. d) EMMA in use. Photos: Atelier KK. UiB, Special Collections.

↓ Guro Gjellestad, the first female professor at the Geophysical Institute, introduced the science of paleomagnetism: the study of the magnetic field imprinted in fossil records. The instrument on the picture, however, is used to analyze the spectral characteristics of stars and was taken in the basement of the Mount Wilson and Palomar Observatory's Library Building in Pasadena, California, in 1951. Photo: UiB, Picture Collection & the National Library.

Carl Ludvig Godske left weather forecasting behind and instead focused on computers, weather in landscapes and outreach. Here he is at the "Godske farm for wild ideas" at Kleppe north of Bergen, organizing one of many summer camps for young people. Photo: UiB, Special Collections.

In 1973, meteorologists organized an observation point at Mount Ulriken (643 m) overlooking Bergen. An identical set of instruments had been installed on the top of the Geophysical Institute in 1952. By comparing the two observation points, the goal was to gain new insights into the lower part of the atmosphere and the effects of pollution. From the left: Arild Guldbrandsen, Kjell Nytun and Herfinn Schjeldrup Paulsen. Photo: UiB, Special Collections.

ADVARSEL
INSTRUMENTENE ER DELVIS STRØMFØRENDE
BERØRING KAN MEDFØRE LIVSFARE

Berit Kjersti Bjørndal was one of the longest-serving executive officers at the Geophysical Institute. She was hired in the early 1970s, and retired in 2016. Here behind the most important tool for the administration before the computer – the typewriter. Photo: Arne Foldvik.

The Geophysical Institute has long and strong links to polar exploration. After spending seven years in the ice on Roald Amundsen's *Maud* expedition (1918–1925), Harald Ulrik Sverdrup took over Vilhelm Bjerknes's professorship in Bergen. Photo: UiB, Special Collections.

Håkon Mosby was the head of research at the first of four Norvegia expeditions to the South Polar Sea. The expedition made landfall at Bouvet Island in 1927 and claimed it for Norway. Photo: Norwegian Polar Institute.

Harald Ulrik Sverdrup (opposite) was
head of research for Hubert Wilkins's and
Lincoln Ellsworth's Arctic Expedition
using the submarine *Nautilus*. After the
expedition, the *Nautilus* was scuttled at
350 meters' depth in the Byfjorden outside
Bergen. Photos: Atelier KK. UiB, Special
Collections.

After the Second World War, the *Armauer Hansen*, the once proud landmark of a distinct branch of geophysics, was seen as outdated and old-fashioned. Photo: UiB, Special Collections.

In 1958, the oceanographers received funding for a new and modern research vessel, the *Helland-Hansen*. Two years later, the Geophysical Institute had its first staff expansion since before the Second World War. Photo: UiB, Special Collections.

In 1976, a strong wave came across the stern, and the research vessel *Helland-Hansen* capsized and sank at Stadlandet. Two persons died. Photo: Arne Foldvik.

From 1980 to 2016, the *Håkon Mosby* was
the main research vessel for the oceanographers at the Geophysical Institute. Photo:
Frank Cleveland.

↓ The Aanderaa current meter, an autonomous instrument developed under the auspices of NATO in Bergen in the 1960s. By the 1980s, 10 000 units had been sold worldwide. Photo: Frank Cleveland.

→ Norway is the only country that has dependencies in both the Arctic and the Antarctic. Here, the Norwegian Antarctic Research Expedition 1976 (NARE-76) is landing their equipment on the ice shelf. After the equipment was ashore, it became clear that the iceberg was young and unstable, and the equipment had to be hauled back onboard the research vessel *Polarsirkel* for a new landing elsewhere. Photos: Arne Foldvik.

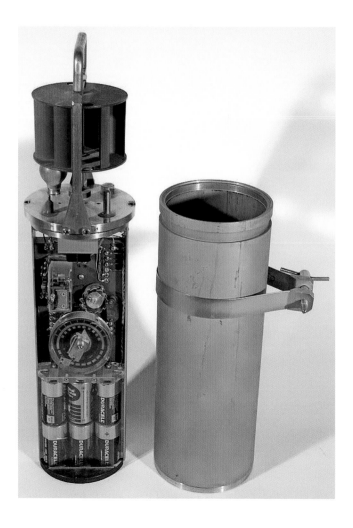

The mess hall onboard the research vessel
Polarsirkel during the NARE 76 expedition
to Antarctica. Photo: Arne Foldvik.

Ola M. Johannessen is briefing the participant on MIZEX 83 on the ice and meteorological conditions that should be expected in the field. The picture was taken in Tromsø in June 1983, at the beginning of MIZEX 83, and was borrowed from Johannessen's private collection.

↓ Johnny Johannessen (left) from the Nansen Center is using a winch to set out a CTD from the vessel *Polarbjørn* during the project NORSEX-79, while Steinar Myking from the Geophysical Institute assists. In the background is a Soviet icebreaker that kept a close eye on the expedition. The research took place in a strategically sensitive area north of Svalbard, at the height of the Cold War. Photo: Ola M. Johannessen.

→ A weather balloon being sent up from the deck of the German research icebreaker *Polarstern* during MISEX-84. Seven ships, eight airplanes and four helicopters took part in the Marginal Ice Zone Experiment in the Fram Strait between Svalbard and Greenland, again under the watchful eyes of the KGB. Photo: Ola M. Johannessen.

Geophysical fieldwork in the Arctic requires scientific instruments, tools, supplies – and guns to keep the polar bears at bay. The picture was taken at the "last outpost" on an ice floe during MISEX-83. From left to right: Arne Hansen, Miles McPhee, Johnny Johannessen (with the saw), and Jamie Morison. Photo: Ola M. Johannessen.

An anchored Seawatch Wavescan buoy is measuring wave direction, motion and temperature in both the atmosphere and the water column. Weighing almost a ton (1000 kg), it is still no match for a Norwegian storm. This instrument was recovered near Slettringen Lighthouse on Titran, north of Kristiansund, about one hundred meters from the shoreline. The underwater acoustic Doppler current profiler (ADCP) is still attached. Photo: Frank Cleveland.

↓ Eva Falck and Ilker Fehr are looking at real-time data from oceanographic instruments below the research vessel *Lance*. Photo: Frank Cleveland.

→ Since its establishment in 1986, the Nansen Remote Sensing Center has depended on external funding. After initially catering mainly to the oil industry, shipping and fish farmers, the center soon focused on environmental research, including "Environmental" in its name in 1990. The text on the front page of its annual report in 1996 reads "We are looking for more funding for research."

Vi ser
etter
mer
midler
til
forsk-
ning

Årsrapport 1996
Nansen Senter for
Miljø og Fjernmåling

Part of the "environmental turn" at the Nansen Center involved exploring shallow-water injection of CO2 for deep-water storage as a possible partial solution to the problem of global warming. After an initial paper in *Nature* in 1992, the research continued for about a decade. After repeated applications for field-testing for the possible environmental impacts were declined, the research was abandoned.

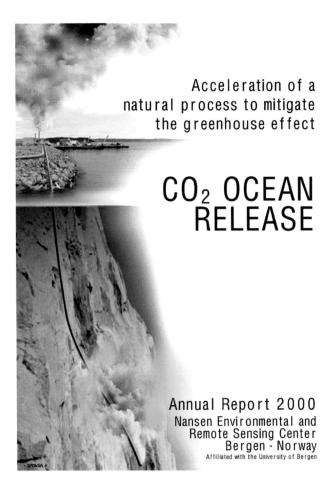

Acceleration of a
natural process to mitigate
the greenhouse effect

CO$_2$ OCEAN RELEASE

Annual Report 2000
Nansen Environmental and
Remote Sensing Center
Bergen - Norway
Affiliated with the University of Bergen

Researching wind flow, the optimal placement of offshore windfarms and individual
windmills is a relatively new research area
at the Geophysical Institute. This picture
is from the towing of the first windmill
from Stord, south of Bergen, to the Hywind
windfarm off the coast of Scotland. Photo:
Roar Lindefjeld / Statoil.

Staff and researchers keep a close watch on the ice from the bridge onboard the research vessel *Lance*. Their task is to warn the researchers on the ice about any approaching polar bears. The picture was taken during fieldwork for N-ICE2015 (Norwegian Young Sea Ice Cruise) in March 2015. Photo: Algot Kristoffer Peterson.

The Norwegian Coast Guard vessel *Svalbard* escorts the research vessel *Lance* into the ice north of the Svalbard archipelago on its way to conduct field observations during N-ICE2015. Normally, the *Svalbard* is up front breaking the meter-thick ice, but here the researchers onboard the *Lance* are stuck and need help breaking free from the grip of the ice. Photo: Algot Kristoffer Peterson.

A helicopter ferries scientists to the
research vessel *Lance*. The vessel was teth-
ered to sea ice to act as a field laboratory
for polar research. After six weeks adrift,
a 1 km-wide ice floe suddenly shattered
like glass, separating the researchers from
their instruments. The *Lance* served as the
scientists' only lifeline as they continued to
drift for another six weeks in inhospitable
polar conditions. The goal of N-ICE2015
was to investigate the interactions between
ice, sea, atmosphere and biology in an area
of the Arctic previously covered by ice all
year round, but where the ice now melts
during the summer. Photo: Algot Kristoffer
Peterson.

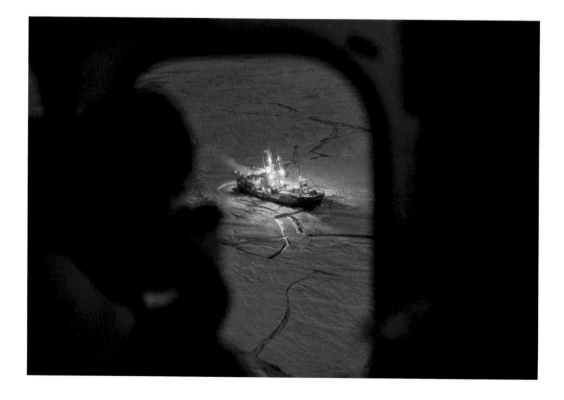

Recovering instruments in the Fram Strait between Greenland and Svalbard during a student expedition out of UNIS (the University Centre in Svalbard) in 2012. The instruments were put on the ice at the Russian base of Barneo, close to the North Pole, and had to be rescued before the ice melted. Photo: Algot Kristoffer Peterson.

A researcher is packing down the equipment from a field station on the ice north
of Svalbard in February 2015. After moving
everything back onto the research vessel
Lance, it was transported to a new ice floe
further into the Arctic, where the measuring continued throughout the winter.
Photo: Algot Kristoffer Peterson.

Equipment for field research in the ice is always placed at a distance from the vessel to avoid contamination. This remote camp was established during N-ICE2015. Photo: Algot Kristoffer Peterson.

Vilhelm Bjerknes's circulation theorem formed the theoretical basis for both physical oceanography and numerical weather forecasting. Bjerknes still has a strong presence among the Bergen geophysicists, often expressed in anecdotes, such as celebrating researchers with "the courage to commit stupidities." This painting by Rolf Groven (1983) is on display at the Geophysical Institute in Bergen.

Finally, the language used in the publications has changed, reflecting how the scientific center of gravity has changed over time. In the interwar period, the most commonly used language was English, comprising about half of all publications. The rest were split equally between German and Norwegian, with occasional publications in Swedish and French. After the Second World War, the use of German came to an abrupt halt, while Norwegian was briefly used for more than half the publications. Since the middle of the 1950s, however, around three out of four publications have been written in English.

Table 2: Publication languages (percentages).

Year	English	Norwegian	German	Other
1917–1929	46	24	26	4
1930–1940	50	24	24	1
1945–1959	55	38	2	6
1960–1969	76	23	0	1
1970–1979	70	30	0	0
1980–1988	74	26	0	0
2000–2015	77	23	0	0

Table: Publication languages (in percentages) from the Geophysical Institute in Bergen. Based on publication records from the Geophysical Institute's annual reports (1917–1988), *Navn og Tall* (2000–2003), and Cristin (Current research information system in Norway, 2005–2015).

The publication trends from the Geophysical Institute, especially the decline in texts aimed at general audiences, were not unique to Bergen. Historian of science Naomi Oreskes has shown that in the United States, postwar earth scientists, in particular oceanographers, "simply didn't engage the general public as they previously had."[107] Her argument is that popularization in the interwar period had been motivated by the need for funding, either directly through speaking fees, royalties from books, or commissions from newspapers, or indirectly through attracting private patrons. During and after the war, funding for geophysics in the United States, in particular for oceanography, came from military sources which "for decades were used to proceeding without public scrutiny, and the idea of the public as their ultimate patron was rarely if ever raised."[108]

Supporting her argument, Oreskes points out that popularization made a comeback only after the Cold War had ended and military funding started to dry up. Oceanographers quickly realized that "trust us, we're experts" or "trust us, our intentions are good" did not pass muster. This led to more publications aimed at a general public motivated by building trust, and sparked discussions of scientific "values," not just immediate scientific "value": "Forty years of military patronage were not just epistemically consequential, they were socially and culturally consequential as well."[109]

The lack of outreach in Bergen coincided in time with military patronage being at its most intense. While Mosby was head of NATO's Subcommittee for Oceanographic Research (1960–65) and the Norwegian member of the NATO Science Committee (1965–69), the oceanographers at the Geophysical Institute did not produce a single publication aimed at the general public, and outreach remained low throughout the Cold War. However, collaboration between geophysicists and the military was, in Bergen, not a Cold War phenomenon. As early as 1918, Vilhelm Bjerknes had approached the navy to help establish a meteorological observation network along the coast. And although the extent remained very limited in the 1970s, the oceanographers in Bergen did occasionally publish articles aimed at general audiences.[110]

Historian of science Theodore Porter's argument that science in the postwar era became increasingly specialized also has merit in the Bergen setting.[111] Porter describes science as being located on an axis between technical and common sense, where the technical was based on concepts and vocabulary that only have meaning for specialists. Common sense science had, in comparison, a goal of popularizing the findings to the general public. In the postwar period, outreach was seen as political, and driven by specialization and professionalizing, science became bureaucratic and technocratic: "While the scale and applicability of science advanced enormously after 1900, scientists have more and more preferred the detached objectivity of service to bureaucratic experts over the cultivation of an engaged public."[112] As described earlier, the research questions, which shifted from general features to details, time series, and mechanisms, were of necessity more technical in nature. Several of those we have interviewed have also pointed out that while many of the researchers were skilled and respected, and had strong international networks, they opposed the bourgeois formalities that Mosby embodied, and did not see value in seeking attention that could possibly undermine their scientific credibility.

Lastly, in Bergen the decline in publications aimed at general audiences coincided with the "Sverdrup Plan" for higher education, and increased the attention given to teaching. The Sverdrup plan, a major reform for Norwegian higher education inspired by the US academic system, was introduced in Bergen in 1959.[113] It seems likely that teaching was understood as fulfilling the university's mandate for public outreach. In Mosby's reflection on the university and society, published the same year as he retired as rector of the University of Bergen in 1971, he did not remark on what role science should have towards the general public. To Mosby, scientists' audiences were fellow scientists, indoctrinating students to become part of the scientific community, and providing unbiased knowledge to politicians. Mosby's reasoning explicitly referred to the development of the atomic bomb during the Second World War: scientific curiosity produced insights that could be applied for various and powerful ends, but scientists themselves could not be held responsible for how politicians or others chose to apply their insights.[114]

The Helland Hansen sinks

The oceanographic cruises were never without danger. On October 17, 1955, steward Alfred Nilsen from Telavåg died during a fjord cruise on the *Armauer Hansen*.[115] On Friday September 10, 1975, after 19 years of operations, the *Helland Hansen* sank at Stadlandet, 16 nautical miles west of Svinøya. The area is known for its dangerous waters: in addition to being open to the Atlantic, currents meet and create large and unpredictable waves.

"We were doing lunar day observations, hourly observations for about 25 hours. The weather got rough, but we wanted to finish our measurements before seeking shelter. A strong wave came across the stern, and the vessel capsized. Before the ship had a chance to stabilize, we were hit by another wave," Steinar Myking recalls. He was in a cabin with his wife, a laboratory technician, when the ship capsized.

"We immediately realized something was very wrong. The boat lay sideways, the engine stopped, and it became quiet. We threw on some clothes and ran out to the hallway. There we saw water come gushing in. When we got to the deck, we could already see much oil in the water. The captain helped us the last leg to the lifeboats. One was already under the ship, the other was stuck. We also had rafts, but one was submerged, and the other would not inflate. Suddenly the other raft popped to the surface. We jumped ship and swam. I remember someone yelling: The ship is going down! The raft was still connected by a cord, and was dragged down. It was cold, maybe ten degrees, and we all had frozen fingers. We had a knife, but it was wrapped in plastic. The steward broke a tooth trying to unwrap it. We failed. Thankfully, the cord snapped. Only then did we notice not everyone was accounted for. The other laboratory technician, a mechanic and the captain were missing. Just then we heard a knocking on the side of the raft – it was the mechanic, an old war sailor: Let me in, boys!

"We sat in the raft submerged to our waists; ten degrees in the air, ten degrees in the ocean. We were freezing. Then we heard the sound of a helicopter. Before abandoning ship, the first mate had managed to send a mayday signal over the radio and give the position. Unknown to us, someone had notified the Rescue Coordination Center. The helicopter was dispatched from a nearby oil platform. I remember it hovering above us, letting down blankets and a rope. The steward tied the rope around his waist and was pulled up by hand. He gave a status report, how many we were; that two were missing, injuries. Shortly thereafter we heard the sound of a rescue helicopter. After getting onboard, we searched for the missing. We found the laboratory technician floating in the water, dead. He left several young children behind. The captain, a man in his 60s with grown children, we never saw again."[116]

The investigation after the capsizing, which used a scale model of the vessel in a model tank, showed that the ballast was positioned too high, and that the stability requirements set forth at the Torremolinos International Convention for the Safety of Fishing Vessels, adopted in 1977, were inadequate.[117] In Norway, the shipwreck prompted new and stricter regulations for ballast. However, three years later, on a cruise with the rented fishing boat *Siljan* investigating the Moskenes current in Lofoten, the oceanographers experienced a similar situation: a wave over the stern capsized the vessel. Fortunately, the vessel had lower ballast and a deep keel, so the ship straightened up on its own. "One of the laboratory technicians was scalded by a pan of boiling water and was badly injured, and we lost the trawl, but we did not sink and could set course toward land and the closest hospital."[118]

The University of Bergen built a new research vessel in 1980, and named it the M/S *Håkon Mosby*. Again the size increased, to 155 feet (47.2 meters) and 493 gross tons. The *Håkon Mosby* was equipped with a 1500-horsepower engine and had 15 single and two four-man cabins. Unlike the two previous vessels, the ship was owned by the university, and not the Geophysical Institute. To the oceanographers, having to share the vessel with other researchers meant the end to the rental fund and an end to the spontaneous cruises. Now, ship-time often has to be reserved more than a year in advance.

An oceanographic transformation

The reign of Håkon Mosby entailed a transformation of oceanography both in Bergen and elsewhere, and in more ways than one. On the local institutional level, he remained the undisputed head of the Geophysical Institute, but by stepping down in 1958–1962, he paved the way for shorter tenures, which became the norm after Mosby retired. He introduced larger, multi-purpose research vessels constructed to be used for more than just physical oceanography.

The research questions were fundamentally the same as those of his predecessors, but new instruments led to a shift in focus from sampling to surveillance. Mosby also personified the new age characterized by military patronage, and the development of new technologies aimed at putting the oceans under surveillance. On the other hand, Mosby did not have co-authors, and like most of his colleagues, many of his publications were in the form of reports, rather than in peer-reviewed journals. Simultaneously, the oceanographers stopped publishing texts aimed at general audiences.

Despite his publication record and representing a small institution in a growing field, Mosby was a central member of several of the leading international organizations that were established after the war. Under his leadership, the oceanographers in Bergen had a host of international visitors, and Mosby traveled extensively. His career thus reflects a vital change in postwar geophysics: rather than individual geniuses working and publishing alone, the leaders in postwar geophysics were the administrators coordinating research efforts. Whether in fieldwork, in analysis, or developing new instruments, geophysics was changing from individual curiosity to concerted efforts where international coordination was a matter of course. Although it seems paradoxical that Mosby, who put such emphasis on international collaboration, appears to have done little to encourage collaboration at home, this was due to his belief in autonomy as a premise for good science. Mosby actively introduced his younger colleagues to international guests and activities elsewhere, and frequently hosted dinners at home, but it seems he took it for granted that his researchers would be as self-driven and ambitious as himself, which was not necessarily the case. As the drop in the number of foreign guests illustrates, when Mosby retired the next generation focused its attention not primarily on the international stage, but on national collaboration.

In the next chapter we will show how his meteorological counterparts chose a very different strategy when faced with being a small group in a growing field.

7
Matters of scale in meteorology and earth physics

The Eighth Assembly of the International Union of Geodesy and Geophysics (IUGG) in Norway in 1948 was meant to be a celebration of Bergen's continued importance as a geophysical world capital, but instead it marked a departure. When Norway was awarded the meeting at the 1939 IUGG assembly in Washington, the plan was to start the event in Oslo, continue on a ship along the Norwegian coast, and end in a grand finale in Bergen. Instead, the event took place in Oslo, with Bergen included only as an optional three-day excursion for meteorologists and oceanographers.[1] The ailing Helland-Hansen was unable to attend. During the assembly, the new professor of meteorology in Bergen, Carl Ludvig Godske, gave a lecture in which he argued that weather forecasting had received more attention than it deserved, and raised doubts about its empirical basis at a time when weather forecasters were celebrated as instrumental in winning the war and when optimism about further improvements prevailed. Instead of continuing the research program that had made the Bergen school famous, Godske would leave weather forecasting behind. His new focus on climatology at a local level brought new clients, tasks, and goals, but it was perceived by colleagues to be a backward approach to meteorological research.

In this chapter we will examine the departure from the Bergen school of meteorology, and investigate the new research programs that were put in its place, which ranged from field studies and youth camps to computers and automated equipment, as well as new worries about humanity's impact on the environment. It all began with a book twenty years in the making.

The meteorologists were not the only group for which a new professor meant a shift in research interests. At the end of the chapter, we will investigate how Section C, when its leadership changed, moved from exploring particle physics to investigating how rock records of the geomagnetic field could be used to answer questions about the origin and development of Earth through geological time.

A farewell to forecasting

Carl Ludvig Godske was both the last member of the Bergen school of meteorology and the first of what was to follow. When Godske was a graduate student at the University of Oslo in 1928, his lecturer in physics and mathematics, Professor Carl Størmer, had singled him out as the most promising science student in his cohort and recommended him as an assistant to Fridtjof Nansen.[2] Godske declined, explaining that he was more interested in mathematics and meteorology than oceanography, and he was instead hired as a Carnegie assistant to Vilhelm Bjerknes.[3] Among other tasks, Godske assisted on the equations in *Physikalische Hydrodynamik mit Anwendungen auf die dynamische Meteorologie* (1933), a 797-page tome written by Vilhelm Bjerknes, Jacob Bjerknes, Halvor Solberg and Tor Bergeron that summarized the Bergen school's latest findings.[4] In 1934, Godske defended a doctorate in hydrodynamics at the University of Oslo, and the following summer he worked on a mathematical theory of cyclone formation in collaboration with Jacob Bjerknes.[5] He also took courses on pure and applied mathematics with Richard von Mises in Istanbul, followed by studies in mathematics and fluid mechanics in Paris and atmospheric radiation in London, all topics Godske later pursued in Bergen.

In 1937, Godske became involved in what was to become the final monograph from the Bergen school of meteorology: the textbook *Dynamic Meteorology and Weather Forecasting*. When he became a research fellow at the Chr. Michelsen Institute and moved from Oslo to Bergen permanently in fall of 1938, Godske was appointed lead editor of the book written in collaboration with two Bergen school originals, Jacob Bjerknes and Tor Bergeron. On April 1, 1940, the 33-year-old Godske was appointed as a substitute for Bjerknes, who had been given a year-long leave to head the Meteorological Association at the IUGG's congress in Washington and visit various US research institutions in 1939. However, the outbreak of war had stranded him in the United States. On Godske's ninth day as professor, Germany invaded Norway, beginning its five-year occupation.

With Bjerknes in Los Angeles and Bergeron in Stockholm, the book collaboration was put on hold until the war ended. In December 1945,

Bjerknes announced that neither he, Sverdrup nor Holmboe would "block the road for other well qualified Norwegians who have carried the burdens of war" by returning to Norway.[6] Therefore, Godske retained the professorship. Bjerknes did, however, spend almost a full year of sabbatical leave in Bergen starting in the fall of 1946 to finish his part of the manuscript. At this point, the book had already been nine years in the making. The visit went well and ended with Godske and the head of the weather forecasting unit for western Norway, Finn Spinnangr, recommending Jacob Bjerknes for the distinction "Knight 1st Class of the Royal Norwegian Order of St. Olav" for his contributions to science, highlighting his description of the structure of cyclones: "This discovery has completely revolutionized weather forecasting, which today all over the world is done by means of the 'Bergen methods.' In further developing these methods, J. Bjerknes has played a greater role than any other."[7]

Summarizing the insights of the Bergen school turned out to be more time consuming and frustrating than anticipated. In February 1947, Bjerknes invited one of his former students from UCLA, Robert C. Bundgaard, to write up the section on practical weather forecasting. Starting in 1944, Bundgaard had been head of the US division for upper-air forecasting in Europe, and after the war, as a lieutenant colonel, he was a leading forecaster in the United States Air Force. However, when Bundgaard started delivering drafts, the collaboration collapsed. Godske was already annoyed at Bjerknes for taking months to reply to even short letters, and at Bergeron for being an eternal perfectionist who spent months on simple proofreading. But that was peanuts compared to Bundgaard. In a series of increasingly angry letters to the other authors, Godske complained about incomprehensible language, misplaced manuscript pages, tasks not being carried out, and agreements not being kept. As Godske put it in a letter to Bjerknes in April 1949:

"You have no idea how often I have felt the need to curse you for bringing me Bob [Bundgaard] as a colleague – for it is I who have to deal with him, despite having nothing to do with that part of the book. He is a stately fellow in all ways, good mannered and pleasant, this is not the problem. You have no idea what a brick he is when it comes to writing. Not a single page, practically not a single sentence, could be used without major revisions – larger paragraphs I have had to rewrite repeatedly to get a basic understanding of what he actually meant to say."[8]

Some months later, Godske wrote to Bundgaard that he was "tired of the book – dead tired – and I hope to make a flight into other fields as soon as possible."[9] Even after the manuscript was finally complete in 1951, the conflict continued: despite Godske having authored the majority of the book from scratch, Jacob Bjerknes sent it to the publisher with a title page explaining that the book was authored by "V. Bjerknes and collaborators," and merely "rewritten and extended by L. Godske with contributions from T. Bergeron, J. Bjerknes and R.C. Bundgaard."[10] Godske probably felt as though he was being treated as the Bergen school's eternal assistant. In a letter to Finnish meteorologist Erik Palmén, who had collaborated with Bjerknes on the weather balloon ascents in the 1930s, Godske explained that the book had broken his "interest in this field."[11] As a result of further delays on the part of the publishers, it was not published until 1957, six years after the authors' last edits, and more than two decades since writing had begun. According to the reviewers, it was published too late.

"That such a delay in printing should have occurred at a time when theoretical meteorology was developing rapidly can only be regretted. Turning the pages emphasizes how rapidly some branches of meteorology have developed since 1951. (...) This is not to say that the book is wholly outmoded, but its main value will be as a reference work regarding the state of the science at the end of World War II. (...) Clearly this is a book which should be available to all meteorologists – it represents the culmination of an epoch in scientific meteorology – but appearing at the present time it is likely to be little read."[12]

The collaboration's disintegration was not the only reason Godske abandoned synoptic and dynamic meteorology. After Vilhelm Bjerknes had moved from Bergen in 1926, he had established a strong research group for theoretical and dynamical meteorology at the University of Oslo under the leadership of Halvor Solberg and Einar Høiland, Bjerknes's last Carnegie assistant. After the war, Høiland's Institute for Weather and Climate Research was the main national research community to continue Vilhelm Bjerknes's dream of using hydrodynamics to calculate the weather.[13] In a country with limited resources for research, Godske considered it meaningless for Bergen to establish a research group with the same specialty.[14] He was not inherently negative toward using hydrodynamics to calculate tomorrow's weather, but argued that the study of atmospheric movements was much more than merely forecasting weather for the benefit of shipping and aviation. As he put it at the 1948 IUGG meeting in Oslo, "Now, meteorology is not only a servant to aviation and navigation, but also to agriculture, forestry, horticulture, town-planning, technique, hygiene etc."[15]

Postwar expansion of weather forecasting

Godske's argument that forecasting was given too much attention was shaped in a postwar context where rebuilding and extending the weather forecasting service became the number one priority. Just a month after liberation, forecasts from Oslo and Bergen were back on the airwaves four times a day, with Tromsø following suit shortly thereafter. To the public, familiar voices presenting the forecasts were a symbol that society was returning to normal.[16] But instead of merely rebuilding, the forecasting service was expanding rapidly: before the war the Norwegian Meteorological Institute had 116 employees, but by 1948 the number had increased to 293.[17]

The expansion was organized by Sverre Petterssen, who had returned to Norway a war hero. From 1931 to 1939, Petterssen had succeeded Jacob Bjerknes as head of the weather forecasting service in Bergen, and he had then moved to work with Carl-Gustaf Rossby at MIT. After producing two widely used textbooks, he was recruited to head the Norwegian meteorologists who had enrolled in the Allied war effort in Britain. Petterssen relocated to London and became head of the department for upper-air forecasting at Dunstable, where he made forecasts for bombing raids over the European continent and for a range of Allied landing operations. Petterssen played a crucial part in the forecast for D-Day as head of one of three forecasting centers giving advice on weather developments for the invasion of Normandy. The

invasion was planned for June 5, 1944, and ships were already creep-
ing south along the British coast, when Petterssen, at 3 AM on June
4, managed to get the invasion postponed, avoiding an amphibious
landing during a storm. On Petterssen's advice, the invasion took place
on June 6, in a window between two cyclones. Meanwhile, the German
Navy had sought shelter in port, their air force was grounded, Field
Marshal Erwin Rommel had traveled to Berlin to celebrate his wife's
50th birthday, and several leading officers were given leave due to the
bad weather.[18] The D-Day experience, not to mention his vivid telling
of the story after his return to Norway, made him a national hero.[19]
After the war, Petterssen was appointed National Chief Forecaster
(Riksværvarslingssjef) and second in command at the Norwegian
Meteorological Institute, and he represented the country in several
international meetings focused on rebuilding and expanding meteorological networks.[20] All Norwegians who had worked for the Allied
meteorological service during the war, some thirty in all, were offered
positions in the forecasting service. Rather than simply being a sign
of gratitude, this exemplifies how the Bergen school techniques had
changed from a thriving research field to a set of methods that practi-
tioners could use regardless of location.[21]

In addition to expanding the weather forecasting services in Oslo,
Bergen, and Tromsø, Petterssen established a meteorology service with
forecasters for all airports that served both military and civil aviation
needs. Petterssen also represented Norway in the negotiations organized by the Provisional International Civil Aviation Organization
(PICAO, from 1947: ICAO), which established an international weather
ship service in the Atlantic Ocean.[22] Finally, Petterssen attached Norway to the international radiosonde network that was established
shortly after the war, again organized by PICAO, with aviation forecasting as its main objective.[23]

The postwar expansion of the meteorological service was not
unique to Norway, and was characterized by a strong emphasis on
weather forecasting using the Bergen school methods. In the United
States, between 7,000 and 10,000 meteorologists and 20,000 observers
had been educated during the war, and the number of US meteorologists with higher education is estimated to have increased by a factor of

twenty.[24] Four of the five schools offering training used Bergen school methods: MIT under the leadership of Petterssen until 1941, followed by Henry Houghton and Hurd Curtis Willett; the University of California, Los Angeles, under the leadership of Norwegians Jørgen Holmboe and Jacob Bjerknes; the University of Chicago, under the leadership of Carl-Gustaf Rossby; and New York University, under the leadership of Athelstan Spilhaus, one of Rossby's MIT students.[25] The main argument for using the Bergen school methods was, as Rossby put it in 1943, that they offered a scientific approach to weather forecasting, based on physics and mathematics that could be universally applied:

"Earlier methods of training meteorologists, particularly in the United States Weather Bureau, were based entirely on the accumulation of experience. A man trained over a number of years in, say, San Francisco, would in that fashion become a good forecaster for our West Coast but would have to start all over again if he were transferred to another part of the country.

"We do not have the time to give our students adequate basic training and also a large amount of experience within the short period of time at our disposal. Hence, we *must* concentrate on the application of fundamental principles of analysis and forecasting which can be used in any part of the world."[26]

Also Russian and German forecasters had by the Second World War adapted the Bergen school's air mass analysis. In the early 1930s, Tor Bergeron had made two longer stays in the Soviet Union, and his lecture notes became the basis for a textbook that shaped Soviet weather forecasting.[27] The textbook was also translated into German, and starting in 1935, Hermann Göring demanded that all meteorologists in the Luftwaffe should adopt the Bergen school air mass-analysis.[28] The Bergen school methods were also the basis for a number of textbooks published in the United States, most notably Horace Byers's *Synoptic and Aeronautical Meteorology* (1937) and *General Meteorology* (1944), and Petterssen's *Weather Analysis and Forecasting* (1940) and *Introduction to Meteorology* (1941). However, rather than developing the methods further, the emphasis was on dissemination.

After the war, many wartime forecasters returned to work in the expanding meteorological institutes back home, and many meteorologists worked to make forecasting a cornerstone of peacetime internationalism. In addition to weather forecasts having been of vital importance during the war, weather forecasting was seen as a way to show that humanity, like weather, could unite beyond national borders. This was made possible by standardization, largely thanks to all forecasters having adopted the same Bergen school methods, as well as the development of a global infrastructure for rapid exchange of weather observations. In Norway, the teleprinter network that was set up by the German occupiers during the war was integrated into the exchange of weather data. In 1946, Ludwig Weickmann, who had headed the German meteorologists in Norway during the Second World War, was made the first president of the German Meteorological Service (Deutsche Wetterdienst, DWD) for the area controlled by the Allies that became the Federal Republic of Germany (West Germany) in 1949.

In 1951, the International Meteorological Organization was formally reborn under the auspices of the United Nations and named the World Meteorological Organization. In the decades to follow, WMO played a part in several large-scale projects aimed at improving international collaboration, such as the infrastructure program World Weather Watch established in 1963 as a continuation of the International Geophysical Year's (1957–58) meteorological program; education and training programs; and global research efforts aimed at putting the atmosphere under constant surveillance. The largest research program was the Global Atmospheric Research Program (GARP), which lasted from 1967 to 1982, and was aimed at advancing the range of weather predictions and understanding the physical basis for climate variations. While weather forecasting and observation technology took precedence, the observations from GARP form an important basis for our knowledge of anthropogenic climate change, and the GARP data are still used to calibrate and test forecasting and climate models. However, since membership in WMO was restricted to national meteorological institutes, meteorologists in Bergen had little impact on these developments. The only exceptions were taking part in a handful of workshops; hosting a weeklong symposium on radiation and satellite techniques in 1968, arranged in collaboration with both WMO and the IUGG; and field-testing meteorological buoys in the 1960s and 70s.[29]

Figure 6

Carl Ludvig Godske defined meteorology as the study of movements of the atmosphere at different scales, and often used this illustration to argue that weather forecasting (synoptic meteorology, "Syn.") was but a small part of meteorology. Throughout his career, Godske warned that forecasting received a disproportionate share of resources. (Godske, C.L. *Statistics of Meteorological Variables: Final Report.* Research conducted for the United States Air Force under contract no. AF 61 (052)-416. Geofysisk Institutt, Bergen. 1965: 8; Godske, C.L. "The Future of meteorological data analysis." *WMO Technical Note* No. 100. Geneva, 1969: 53.)

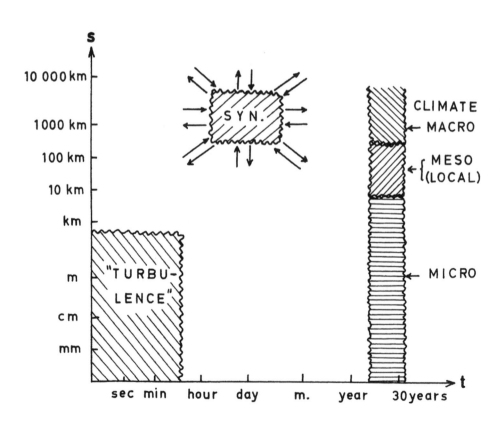

The expansion of the international meteorological observation network was reflected in the number of daily weather observations that reached the country from abroad. From 1948 to 1949 the number of daily weather observations that arrived in Norway via telegraph wire, radio signals, and the teleprinter network grew from 30,000 to 80,000. By 1950, the number of daily observations had grown to 175,000 per day.[30]

A main point for Godske at the 1948 IUGG meeting was that different clients had different needs and required different observations. Instead of having as many observations as possible, it was more important that the observations were representative of what they were supposed to be used for. To Godske, it all boiled down to the question of scale. Weather forecasters (synoptic meteorologists) were interested in changes taking place from hour to hour or day to day over hundreds or thousands of kilometers. By contrast, studies of turbulence focused on centimeters and seconds, while local meteorologists were interested in changes taking place over years or decades in areas covering only a few kilometers. Throughout his career, Godske argued that when the scales in time and space change, methods and observations must also change.[31] During the war he had shown that meteorological observations could differ greatly over small distances.[32] Rather than simply *more* observations, Godske asked for better knowledge of the local conditions for each station, how weather behaves in different landscapes, and more knowledge about the behavior of the lowest two meters of the atmosphere – where plants grow and animals live. These were arguments he had developed during the Second World War, and they led to a research program that differed distinctly from the methods developed by the Bergen school of meteorology.

Turning to the local

Weather forecasting was banned during Germany's occupation of Norway from April 9 1940 to May 8 1945. Tomorrow's weather was considered of military significance. In a compromise with the occupiers, Norwegian meteorologists were allowed to keep their posts in order to be ready for when the war ended.[33] The forecasters in Bergen were instructed to analyze past meteorological shipping observations, and the head of the weather forecasting unit, Finn Spinnangr, started a project analyzing weather maps over western Norway found in the archives. Spinnangr thought that by focusing on one and only one weather phenomenon (wind, fog, thunderstorms, temperature changes, or precipitation), it would be possible to statistically show the likelihood of different weather phenomena appearing under certain conditions, and thereby improving forecast accuracy.

Spinnangr's approach was both a continuation of and a departure from Bergen school meteorology. In the 1920s, emphasis had been on the structures of weather systems – fronts, cyclones, and air masses – and on identifying models to aid in the analysis of synoptic weather maps. Instead of focusing on only one phenomenon, all observation data were put on the same map, and the Bergen school treated forecasts as "puzzles" where all the pieces had to fit together. In the 1930s, Petterssen had focused on developing equations and procedural methods for calculating how features on the weather map moved over time. But, as Petterssen's successor Spinnangr pointed out, the premise for Pettersen's equations to work was that the weather continued to develop according to idealized models. Along fjords and mountains in western Norway, this was seldom the case. To Spinnangr, empirically identifying what happened when weather systems hit specific locations on the coastline was a logical step to improving the forecasts, more so than attacking them through theoretical idealizations.[34] His idea became of lasting research interest to forecasters. In the 1960s, Spinnangr's successor, Harald Johansen, developed a classification of "weather types" based on prevailing wind direction and pressure.[35] In the 1970s, Johansen's weather types were used in several empirical studies.[36]

Godske's interest in local meteorology had also developed during the occupation. In 1942 and 1943, he took the initiative in Norway's first micrometeorological survey. In collaboration with Spinnangr, he recruited some 30 volunteers, equipped them with thermometers, and sent them on routes in the Bergen valley to map the distribution of minimum temperatures. Unlike weather forecasting, this effort was considered harmless by the occupiers.[37] Based on 3300 measurements gathered over three clear, calm spring nights, Godske found that the temperatures could differ by up to ten degrees Celsius from the permanent meteorological station, which was positioned at one of the warmest points in the valley:

"The atmospheric variables for the region as a whole are then only inadequately described by the observations taken at the station, which can thus be characterized as '*synoptically representative*' and '*microclimatologically irrepresentative.*' (...) the temperatures measured at a well exposed station 2 m above the ground, cannot uncritically be used, say, for the prediction of frost phenomena occurring at 2 m or lower in sheltered hollows in the vicinity."[38]

Godske's study of minimum temperatures in the Bergen valley was decidedly useful, especially for agriculture. A main finding was that cold air formed in hillside marshes and then ran like streams or rivers down the hillsides. By building small hills or stone fences, these invisible rivers could be diverted from crops, extending the harvest season and preventing frost damage. Godske established a close collaboration with Olav Skard, professor of horticulture at Norway's Agricultural College at Ås, and before liberation in 1945 he tried to establish an Institute for Agricultural Meteorology in Bergen. When the board of the Bergen Museum rejected the proposal, he was advised to approach the Rockefeller Foundation to secure funding, but the plan was abandoned when it became clear that Godske would keep Bjerknes's professorship.[39] However, he continued to lecture, give interviews, and frequently publish for gardening and farmers' associations along the coast, and he published popular books on gardening that were widely read.

With Godske at the helm in a country where various foodstuffs were rationed until the early 1950s, meteorological research focused on local meteorology to the benefit of agriculture, forestry, and horticulture. In the 1950s and 1960s, he organized a series of field investigations in southern Norway, financed mainly by the Norwegian Research Council for Agriculture, aimed at investigating the distribution of parameters linked to plant growth: temperature, precipitation, humidity, evaporation, evapotranspiration, and surface textures' effects on soil temperatures.[40] In 1962, after five years of planning, a 2500 m^2 permanent field station was established at Stend in Fana, outside Bergen, as part of an international network of research gardens. Each garden had 50 species of trees and bushes, used to investigate the relationship between plant growth and climatic conditions. Additional field stations aimed at improved agricultural output were established at Nes in Hedmark (1958–60), along the Hardanger and Sørfjord fjords (1959–61), in Sogn (1962–67), and in Aust-Agder (1969–72). However, rather than addressing fellow meteorologists, the results were published as reports tailored to local clients.

The field research focused on small areas, and was supplemented by research in micro-meteorology aimed at growing conditions in the top centimeters of the soil and evaporation rates from various surfaces. The latter approach was especially important to the test field on the mountain plateau Hardangervidda (1968–72), which was set up as part of the International Biological Program's Section for Production of Terrestrial Community, a research program that combined big science and ecosystem ecology from 1964 to 1974. The field station at Hardangervidda used automatic measuring equipment attached to the data logger Aanderaa had developed for the oceanographers' NATO Buoy project. As environmental concerns came to the fore in the late 1970s, the meteorologists produced a series of reports on the local climatological impacts of hydropower dam construction.[41] Undoubtedly inspired by his oceanographic colleagues' new research vessel, Godske successfully petitioned the Norwegian General Science Fund for a "rolling observatory": a car equipped for measuring solar radiation and minimum temperatures, with space for accommodations. Starting in 1959, the "research car" provided easy access to the field.

The reports from the meteorological field research were highly empirical. Apart from an initial goal of using statistics to cut the observation times from 5–10 years to about three, emphasis was on the results of the data, not on how they fit into a larger picture, theory, or methodology, or what lessons were applicable elsewhere.[42] Instead of convincing scientific colleagues elsewhere, which had been characteristic of the Bergen school in the 1920s, the main audiences were the clients who could directly benefit from the results. This was reflected in the publication strategies: like the oceanographers, the share of reports grew from the 1960s to the 1980s, while the proportion of scientific publications in academic journals fell. Meanwhile, the proportion of publications in Bergen Museum's yearbook (from 1948 onward: Bergen University's yearbook) aimed at colleagues fell from 60 percent in the 1940s, when few other outlets had been available, to around 3 percent in the 1970s and 1980s. Unlike the oceanographers, who published reports for international colleagues, the meteorologists increasingly wrote in Norwegian (from 40 percent in the 1950s, to 60 percent in the 1970s), reflecting the local aspirations of their research. However, as international colleagues became more important audiences in the 1980s, English has increasingly been the language of choice in scientific publications.

In a situation where the number of researchers at the Geophysical Institute remained static, while the surrounding university expanded, hands-on experience from fieldwork became vital to Godske's approach to recruitment. In a letter from 1946, in which he discussed whether his research assistant, Herfinn Scheldrup Paulsen, should accept an offer for a permanent position at the forecasting office or pursue an uncertain future in research, Godske wrote: "[Paulsen] is in great doubt as to what to do. So am I. He has a point regarding an uncertain future. But I still dislike the idea of 'letting him go,' since this in many ways will mean I would return to square one – with new and inexperienced assistants."[43] Godske convinced Paulsen to stay. He remained for almost 40 years doing research on physical meteorology, in particular solar radiation, using instruments to continuously measure the solar rays that hit the ground, air transparency, and surface cooling as a function of temperature, solar radiation, humidity, and wind.

The local meteorological station at Kleppe on the island Osterøya illustrates how Godske combined meteorological research with outreach, collaboration, recruitment, and novel ideas about the role of academia in society. Starting in the summer of 1945, only months after liberation, Godske began arranging summer camps for youths. Godske was an avid participant in the Scouting movement, and the camps introduced the participants to the wonders of nature. In 1950, he convinced the Bergen Rotary Club to buy a farm at Kleppe, on Osterøya, an island about 30 kilometers northeast of Bergen. "The Godske farm for wild ideas" became the base for both summer courses and a meteorological field station. Guided by the slogans "research and outdoor life" and "nature and culture," young people were introduced to fieldwork in geology, botany, zoology, meteorology, and archeology; there were trips, plays, lectures, discussions, stage plays, and singalongs.[44] Instead of adjusting individuals to fit society, Godske argued that the goal of the university was to provide new members of society with the necessary tools to develop rich personal lives and make the world a better place: "The young must not be considered raw material for forging directors."[45] Godske believed everyone was born with scientific curiosity, but for many this was lost in the school system. He promoted filling "leisure time" with adventure and discovery, and argued that this was especially important in a future where machines and automatization would do most of the work.[46] In addition to offering educational activities for the summer holidays, the station provided observations for a series of scientific papers, investigating how different surface textures affected soil temperatures.[47] As he had recruited Paulsen, Godske offered a permanent position to his field assistant, Kåre Utaaker, once he had a new opening.

The meteorologists' goal of being "useful" was manifested in different ways. When Bergen decided to expand with a new suburb, Fyllingsdalen, in 1955, Paulsen investigated the hourly sunlight at different locations to aid town planners.[48] Around the same time, he was asked by the industrial town of Sauda, host to Europe's largest smelting plant, to map whether smoke from the factories had reduced the light intensity to such an extent that it had contributed to the decline in the surrounding forest.[49] Another practical use for meteorological studies presented itself in 1962, when the Committee on Traffic Safety Research asked Paulsen to draw up a report on road safety when driving in darkness or with reduced visibility due to weather conditions, research that continued for seven years and included a statistical analysis of traffic accidents in Bergen and Oslo, in nighttime and daytime conditions.[50] Local studies, both those focusing on growth conditions in the 1940s-1960s and reports on the consequences of hydropower regulations in the 1970s and 1980s, offered findings useful to clients.

The meteorological research interests were inspired by the local setting. To the Bergen school, the cyclones repeatedly hitting the Norwegian west coast and the tight network of surface observations had held the key to the cyclone model. For Godske, his daily morning walks between his Geofysen office and his home in Professorveien, two kilometers up the Bergen valley, led him to investigate how minimum temperatures differed on a local scale. In 1973, a permanent observation station was established at Ulriken, the highest of the seven mountains surrounding Bergen. Since 1961, the 643-meter-tall mountain three kilometers away from the Geophysical Institute had been accessible by cable car, and Paulsen installed equipment for hourly automatic measurement of solar radiation. The instruments were connected via telex to an identical set of instruments at the Institute's tower, which had been assembled in 1952. The idea: by comparing the two observation points, it would be possible to gain new insight into the lower part of the atmosphere and the effects of pollution, which, according to the Geophysical Institute's Annual Report (1954–55) was "a core issue in today's discussions on possible causes of climate variations."[51]

In 1963, the meteorologists installed an instrument to automatically record the observations: the MADAM AIR (Multichannel Analog and Digital Automatic Measuring And Integrating Recorder). The instrument was developed in collaboration with a local company, Rieck & Co., and every thirty minutes it automatically read the instruments and stored the observations on punch cards. The following year, Paulsen presented MADAM AIR at the International Radiation Symposium in Leningrad, organized by the WMO. This resulted in an equipment order from the Meteorological Institute in Bratislava, Czechoslovakia, and also in solar radiation observations being sent to the WMO's Data Center in St. Petersburg, which was part of its postwar efforts to organize a constant surveillance of the atmosphere.[52] The radiation data were collected alongside measurements of the chemical composition of precipitation, CO_2 levels, radioactivity and ozone.[53] Starting in 1966 the Geophysical Institute also published a "Radiation Yearbook" with hourly values of radiation, diffractive sky radiation, direct sunlight, illumination, atmospheric long wave radiation, sunshine duration, and cooling index from the instruments on the Institute's roof.[54] In

1972, the recording instrument was upgraded and given the name "MADAM AIR II," and supplemented automatic dust collectors for analyzing air pollutants.

MADAM AIR was part of the same shift towards putting the atmosphere and oceans under constant surveillance as the oceanographers' Aanderaa Buoys. Manual observations were replaced by autonomous observation instruments, and then new instruments were needed to manage and analyze the increasing amount of observational data. In the process, the instruments changed from tools for specialists to sources for scientific authority.[55]

Let the machine do the calculations?

The climatological research program relied on huge numbers of observations, and Godske was in the forefront of introducing new tools for analyzing the data. In 1952, Godske rented an IBM-602A multiplier punch card calculator, which was used to statistically analyze meteorological observations. Three years later, Godske, Kåre Fløisand, and J.B. Hannisdal established a "punch card division," which also offered computing time to other institutes at the university. This became the base for the company EMMA (acronym for Elektronisk Matematikk MAskin [Electrical Mathematics Machine]), a collaboration of the Geophysical Institute, the municipal administration, and ten local businesses. EMMA rented an IBM 650, which was installed in the basement at the Geophysical Institute in May 1958. The computer could read 200 punch cards per minute, and write to a hundred cards per minute. It could also make 1000 additions or subtractions, or 60 multiplications, per minute, and the main storage unit could store 2000 ten digit numbers. It was used in research, calculating tax returns, and commercial applications, and was Norway's first commercial computer. After the IBM 650 was replaced by newer models, it was used for several years in undergraduate teaching and by graduate students working on their dissertations in physics, mathematics, geophysics, and chemistry.[56]

In Norway, the history of computing is intimately linked to meteorology. The first analog computer, a differential analyzer which solved equations using wheel-and-disc mechanisms, was built on the first floor of the Institute of Astrophysics at the University of Oslo. This was the same building in which Vilhelm Bjerknes, Solberg and Høiland worked, which hosted Høiland's research group in theoretical meteorology, and where Godske had his office before he moved to Bergen. The machine was designed and built by a former Bjerknes assistant, Svein Rosseland. Rosseland had started out as a weather forecaster in Bergen in 1919, but when it turned out he was mediocre at drawing weather maps, he was sent to physicist Niels Bohr in Copenhagen in 1920. Before being made professor of astrophysics at the University of Oslo in 1928, Rosseland was a Rockefeller Fellow at the Mount Wilson observatory in Pasadena, California. Back in Norway, he was a key figure in establishing both the Institute for Theoretical Astrophysics (1934) and the Harestua Solar Observatory outside Oslo (1954). The Institute, the Observatory, and the differential analyzer were financed largely by the Rockefeller Foundation. For four years, the Oslo analyzer was the most powerful computer in the world, and was used for hydrodynamic calculations, until Rosseland dismantled key parts in 1941, buried them in the Institute's back yard, and fled to the United States. After the war, Rosseland was instrumental in promoting the purchase of the first Norwegian digital computer, NUSSE, which was placed in the Institute of Physics.

Although the computers increased the speed of calculations, the technology did not dictate what calculations they should be used for. The world's first successful numerical weather prediction had been made in 1950, using the ENIAC computer at the US Army's Aberdeen Proving Ground in Maryland, as part of the numerical weather prediction program at the Institute for Advanced Study, Princeton, New Jersey. A team including the Oslo-based meteorologists Arnt Eliassen and Ragnar Fjørtoft calculated how the geopotential height (altitude) of the 500-millibars pressure surface would develop in the next 24 hours, and the calculations took nearly 24 hours to complete.[57] The prospect of using computers for numerical weather prediction was an important part of weather forecasters' optimism in the 1950s and 1960s. At a lecture at a Scandinavian-American conference in 1958, where some 120 researchers celebrated the 40th anniversary of the Bergen school of meteorology, Godske argued that the computers were the future for resolving "the problem of weather forecasting," but that the final solution was to be found in combining numerical and statistical methods. At the University of Oslo meteorologists had experimented with numerical methods using NUSSE since 1953, and in 1962 the Norwegian Meteorological Institute installed the computer FACIT EDB. To avoid duplication, Godske decided that the computer in Bergen should focus on the use of statistical methods.[58] Rather than numerical forecasting, EMMA was used as a direct continuation of how the punch card machines had been employed: to crunch numbers using a statistical approach.

From 1960 to 1968, the research into statistical meteorology received funding from the US Air Force, but this had little impact on the research agenda. While the oceanographers operated a NATO research center with its own budget, Godske received a grant from US Air Force Cambridge Research Laboratories through the European Office of Aerospace Research. The funding, $20,000 every 18 months, made it possible to hire two graduate students as temporary research assistants.[59] The main objective of the contracts was to identify "inter-relations which possess a certain amount of permanency so that they are valid also in the future."[60] Godske used the concepts of "signal" and "noise": The signals were regular periodic variations, such as seasonal patterns, or temperature changes over the course of an average day during different parts of the year. Weather systems, like fronts and cyclones, were considered "noise": short-term changes that only served to hide what he assumed were underlying regularities. The research focused in particular on analyzing the Norwegian Meteorological Institute's time series on air temperatures. The funding also made it possible for Godske to make several visits and lecture tours to the United States, and to attend a number of conferences organized by the WMO's working group for climatology. However, the statistical approach did not produce any "breakthroughs." In the final report in 1969, Godske commented with candid honesty:

"The work carried out by myself and my collaborators has been more 'romantic' than 'classical,' different methods of attack being adopted and attempted and no systematic procedure aimed at. Consequently, most of the results are still of a provisional type. It is, however, our intention to continue the work along the same lines as far as possible with the modest resources available at my institute."[61]

By the time the contract with the US Air Force had begun, statistics had been a focus of teaching for a decade. In 1951, Godske and his colleague, Oddvar Bjørgum, started a course in applied mathematics, with a strong emphasis on statistics.[62] Two years later, Bjørgum accepted a position at the Department of Mathematics, and left meteorology.[63] After the curriculum was reorganized in 1959 as part of a national effort to facilitate mass education, Godske continued to lecture in statistics and practical mathematics, including coding and programming, and was very popular among the students. Between 1958 and 1962, he also functioned as head of the Geophysical Institute. Starting in 1963, the Department of Mathematics offered the first courses in computer science, numerical analysis, and programming, followed by a graduate program. Computer use grew rapidly, and by 1966 Kåre Fløisand had expanded the punch card division from three to fourteen people, ten of whom were employed year after year on temporary contracts. The division was relocated, and as an independent entity it became the basis of the University of Bergen's IT department.

A matter of prestige

Bergen school meteorologists disdained the statistical approach. In a lecture in 1974, four years after Godske died, his colleague of 21 years, Kåre Utaaker, argued that the climatologists had failed to utilize the collection of observations to solve specific problems, while forecasters' single-minded belief in progress had led to an irrational contempt for the past:

"It is said that a student who in the 1930s asked J. Bjerknes what climatology was got the answer: 'It is to add 30 numbers and divide the sum by 30.' Whether the story is true, I do not know, but it illustrates the condescending attitude many meteorologists working with synoptic and dynamic approaches usually have had towards climatology, a branch of meteorological science that has been neglected and belittled."[64]

In the first decades after the Second World War, climatology was also held in low regard elsewhere. In his first public lecture as President for the Commission on Climatology at the WMO in 1953, US geographer and climatologist Charles Warren Thornthwaite stressed: "I hope that we may soon rise up from our inferior position in the hierarchy of meteorology."[65] To the extent that climatology received attention in the *WMO Bulletin* in the 1950s, it was to repeat that the specialization should have higher ambitions than providing statistical descriptions of the past, and to allow subsequent presidents of the Commission to argue that the specialty deserved more respect from other meteorologists.[66]

One spurious reason for the contempt was that the statistical approach had clear similarities with "the analogue method." The analogue method assumed Earth's atmosphere to be a closed system, and that eventually the weather pattern would start repeating. Its method for weather forecasting hinged on making archives of weather maps, searching the archives for similar weather situations in the past, and basing the forecasts on what happened the last time the situation was the same. In a worst-case scenario, when the forecast was wrong, one could add the case to the archives and thus be better prepared for the next time the situation arose. During the war, analogue forecasting had been the curriculum at the California Institute of Technology (Caltech), the fifth and final school educating meteorologists in the United States. The program was headed by Irving P. Krick, who had visited Bergen for two weeks in 1934, but ended up dismissing the Bergen school's air masses, fronts, and research on the physical structure of cyclones as overly complicated. He also saw them as quite useless when making forecasts for the southern United States where the weather, rather than being dominated by cyclones formed in a constant battle between cold and warm air masses, is characterized by seasonal variations and the El Niño cycle. Although the statistical approach was based on probabilities and not individual cases, it assumed that a key to improving the forecasts lay in answering what "normally" happened. Harald Johansen's concept of "weather types" further underlined the connection: Krick saw the origins of his analogue method as "weather typing," an approach first used by the US Army Signal Service in 1872.[67]

During the war, Krick had been Pettersen's counterpart in the forecast for D-Day, and Krick had been the one to argue that the invasion should take place on June 5, 1944. That day turned out to have had strong on-shore winds that would have made amphibious landings impossible, and continuous low cloud that would have hindered naval and aerial bombardment.[68] It was this forecast that Petterssen, at the last minute, had managed to overturn, postponing D-Day and avoiding disaster. However, in 1954, Krick would claim that he was the American hero who had made D-Day happen and presented Petterssen as a gloomy and uncooperative pessimist, failing to mention anything about the invasion being postponed.[69] By then Krick had been kicked out of Caltech for running a private weather forecasting bureau out of the university campus. Presenting himself as a war hero was part of promoting his private weather consulting business that specialized in long-term forecasting, including for movie productions and presidential inaugurations, and weather modification. To the Bergen school meteorologists, this was quasi-scientific.

The lack of new positions at the Geophysical Institute until 1960, while the university grew around them, was an apt metaphor for the international role of the meteorologists in Bergen: rather than an absolute decline, the meteorological stagnation was amplified by growth elsewhere. While the postwar expansion of meteorology elsewhere revolved around the Bergen school weather maps, Godske's turn to the local meant that meteorological research soon became disconnected from the weather forecaster's work on the Geophysical Institute's top floors. The forecasters shared lunchrooms and were invited to the annual Christmas parties, but neither local nor statistical meteorology had much impact on the daily practice of drawing weather maps at high speed to meet short deadlines. Research aimed at using computers to aid the daily weather forecasting took place mainly in Oslo, where Ragnar Fjørtoft and Arnt Eliassen worked. Eliassen was a close friend of the US meteorologist Edward Lorenz, who showed that minor changes in the initial conditions could have large consequences for the numerical forecasts.[70] Aptly nicknamed "the butterfly effect," chaos theory puts an absolute limit to predictability. Exactly how far into the future the weather could be calculated was a matter of controversy. Danish meteorologist Aksel Wiin-Nielsen, who was a professor at the Geophysical Institute from 1972 before becoming the first director of the European Centre for Medium-Range Weather Forecasts (ECMWF) in 1974, aimed at producing 10-day forecasts. Fjørtoft and Eliassen put the limit closer to five days, and on their advice Norway did not join ECMWF until 1988.[71]

Although statistics was not the only approach to meteorology in postwar Bergen, those using a hydrodynamic approach did not stay long in Godske's meteorology department. From 1939 to 1953, Oddvar Bjørgum had used hydrodynamics to study turbulence, but he then departed for a chair in the new Department of Mathematics. In 1960, as a consequence of the Sverdrup plan leading to increased demand for teaching, Arne Foldvik was hired in a lecturing position. For eight years, this strengthened the expertise in hydrodynamics. Since his days as a student, Foldvik had been attached to Einar Høiland's "Oslo school" of meteorology based at the Institute of Astrophysics at the University of Oslo, which had continued Vilhelm Bjerknes's most theoretical work. Foldvik's research focused on mountain waves, waves that can appear downwind of high ground, in a similar way to how waves can form behind stones in a river stream. In the atmosphere, the waves create lenticular clouds and are a hazard to aviation. The international radiosonde network set up in the aftermath of the war was spread too thinly to offer relevant observations, and in 1963–64, Foldvik visited J. Bjerknes and Holmboe's Department of Meteorology at UCLA to study the phenomenon using computer calculations. However, the computing power turned out to be too limited. After returning to Bergen, he built an experimental tank, 6x1 meter, and attached a 16-mm camera to an artificial mountain that could be dragged along the bottom. The tank produced wonderful waves, and was used in several student theses, but it had one fundamental flaw: one meter above the "mountain," the waves reflected back from the tank's surface. This does not happen in the atmosphere. In 1968, Foldvik switched from meteorology to oceanography.

After collaborating with fellow oceanographer Thor Kvinge on a research program in the Weddell Sea in 1973, which resulted in the recovery of two Aanderaa current meters that had been placed there five years earlier, Foldvik focused mainly on polar research. He took part in organizing several international expeditions to the Barents Sea in the north, and the Weddell Sea in Antarctica, and he did research on tides, continental shelf waves, mixing, and bottom water formation. In the 1980s, this led to research on the ocean climate on a global scale:

"We observed changes in the deep sea I had not believed could exist. We had assumed that the depths of the Norwegian Sea had constant temperature and salinity, and used samples from fixed locations to calibrate our equipment. But when the instruments improved, we noticed that also the climate in the ocean is changing. The oceans are also very important for the climate in the earth's atmosphere."[72]

However, both the global perspective and the international collaborations took place after Foldvik had left meteorology, and for the most part after Godske and institute head Mosby had retired.

Introducing geological time

The meteorologists under Godske were not the only ones to change focus in the postwar period. Section C for Geomagnetism and Cosmic Physics, under new leadership, embarked on new research questions with new timescales: Why does the surface of the earth look the way it does, and have the continents changed over time?

After the University of Bergen had established a Department of Physics in 1952, Section C at the Geophysical Institute became little more than the Magnetic Bureau. Bjørn Trumpy remained the leader of Section C until 1959, but for the last seven years both his professorship and his research concentrated on physics in his new department.[73] The Magnetic Bureau carried out smaller projects on the interaction between Earth's magnetic field and cosmic radiation, for instance after a solar flare in February 1953, but focused on working with geomagnetic mapmaking and publishing the annual yearbooks from the magnetic station at Dombås. Besides, until 1953 Trumpy had been the first rector at the new university. Like Håkon Mosby, Trumpy frequently opened his home to guests, colleagues, and students. While "Professor Mosby" had a formal style, and "Calle" (Godske) took pride in being informal, "Trumpy" is described by his peers as outgoing and inspiring: "Among other things, he regularly invited us to festivities at his home at Muséplass. Even though we were then a small institute, it could get youthful and crowded on the dance floor, far into the small hours of the night."[74]

The research interests changed markedly after Guro Gjellestad entered the scene. Gjellestad had graduated as a physicist in 1950 at the age of 36, and had then received a Fulbright scholarship that took her to various astronomical observatories in the United States. At the Smithsonian Institution in Washington, D.C., she was introduced to the new science of paleomagnetism, the study of the magnetic field imprinted as fossil records in many volcanic and sedimentary rocks. By measuring the direction of the magnetic imprints in rocks of different ages, it was in theory possible to investigate polar wandering and how continents have changed over geological time. In 1955, Gjellestad was hired as a lecturer at Section C in Bergen, and four years later she was made head of the section after outcompeting four male rivals.[75]

With a global geophysical perspective in the 1950s, paleomagnetism was linked to renewed interest and fierce debates about the continental drift theory. In August 1920, German polar explorer, geophysicist, and meteorologist Alfred Wegener had presented his theory of continental drift at an evening lecture in Bergen, held during his visit to learn Bergen school methods of weather forecasting. Although Wegener had worked on his theory since 1912, this was the first time it had been presented outside German-speaking Europe. Wegner's theory suggested that all continents at one point had been connected in one supercontinent, Pangea, centered on the present South Pole, but that they had since drifted apart. His arguments were that the continents fit together almost like a jigsaw puzzle, that the fossil plant records seemed to continue uninterrupted from one side of the Atlantic Ocean to the other, and the geological structures and rock types formed similar patterns on both sides.[76] At the time, the dominant theory was that the continents had always been fixed, that the fossil records were explained by similarities in the climate, and that the continents at different points in geological time had been connected by land bridges caused by rising and sinking sea levels. Although the initial response in Bergen seems to have been positive – thanks to the British mathematician Sydney Chapman, who attended the lecture, it had led to an invitation to the University of Manchester – continental drift was generally dismissed among geologists. Historian Naomi Oreskes has argued that in Britain the debate focused on whether the evidence fit the theory; in the United States the critique was that the theory was unscientific: Wegener had set out to prove his theory rather than to test it. Comparing the drift theory with the later theory of plate tectonics, Oreskes argues that the crux of the matter was not that Wegener lacked formal training in geology, or a mechanism for how continents could move, but the nature of the evidence used to demonstrate it. Wegener's argument rested on similarities in patterns in rocks in the field, homologies, described in words and pictures, and with hammers, lenses and notebooks from the field as the main tools, whereas the later plate tectonics rested on the geophysical properties of the earth, measurements of numerical data, and the use of laboratory tools: seismographs, magnetometers, and computers.[77]

That continental drift received renewed interest in the 1950s was prompted by new data sonar mapping of the ocean floors showing the presence of mid-ocean ridges, suggesting that in areas where continental plates were moving apart, new oceanic crust was formed through volcanic activities. This "seafloor spreading" was a new kind of evidence, indicating that the continents are floating plates, pushed by convection from the hot mantle below. Furthermore, techniques from the new field of paleomagnetism made it possible to analyze the magnetic field imprinted as rocks and sediments. It was known that magnetic polarity shifted over time, and samples from the seafloor sediments showed patterns of magnetization that suggested it had indeed been formed over long time periods. Paleomagnetic studies also showed that the direction of the magnetic curves differed from continent to continent, indicating that they had moved independently.[78] Third, the network of sensors set up in the 1960s to monitor compliance with the Nuclear Test Ban Treaty of 1963, which also recorded earthquake activities, showed that almost all earthquakes and volcanic activities occur at the edges of what increasingly were referred to as "continental plates." This led to the birth of the model of plate tectonics, a scientific theory that would soon become Gjellestad's main focus for Section C at the Geophysical Institute in Bergen.

Gjellestad soon reached out to King's College in Newcastle (since 1963, Newcastle University), where the British physicist Keith Runcorn was building up a research group for paleomagnetism. A three-month stay there in 1961 became the beginning of a long and fruitful collaboration. By the next summer, Gjellestad was traveling around Norway with David Collinson and Jim Parry from Newcastle to find suitable sites for red sandstone, a sedimentary rock containing the iron oxide hematite, to bring to the laboratories to investigate their fossil magnetizations. During the decade that followed, Gjellestad, colleagues, assistants, and students were frequent users of the ferry between Bergen and Newcastle. However, as her student Karsten Storetvedt has pointed out, Gjellestad started out with no knowledge about geology, no laboratory, and no equipment.[79]

One strategy Gjellestad used to establish paleomagnetism studies in Bergen was to build prestige by inviting prominent international guest researchers. In 1961, Runcorn came to Bergen for the first of several visits. His lecture "Rock Magnetism, Polar Wandering and Continental Drift" filled the auditorium to the brim. The prestige gained from this and later international visits, such as when the first man in space, Yuri Gagarin, visited as a goodwill ambassador from the Soviet Union in 1964, was important capital when Gjellestad started applying for funding to establish and equip paleomagnetic laboratories in Bergen.[80] In collaboration with Runcorn's group, Gjellestad set up a magnetic laboratory in a house near the Geophysical Institute, and one at Espegrend Biological Station, about 20 kilometers south of the city center. As soon as the laboratories were equipped with astatic magnetometers, microscopes, precision declinometers, and demagnetizing instruments, they were put to use, establishing polar wandering curves and dating rock formations.[81] By comparing the magnetic imprints in rocks of similar ages, and creating comparative timelines, it was possible to show how Europe and North America had moved in relation to one another through geologic time, then a highly controversial topic.[82] By the time Gjellestad died in January 1972, at age 57, her research group and laboratories were attracting a number of international guests every year.[83] The section also continued the tradition established by Trumpy of placing a strong emphasis on publishing in respected international journals like *Nature*, *Astrophysical Journal*, *Tectonophysics*, and *Earth and Planetary Science Letters*. While the meteorologists and oceanographers increasingly produced reports, the proportion of publications in international journals in the section for geomagnetism increased from 40 percent in the 1960s to 60 percent in the 1980s.[84]

In the 1960s, Section C in Bergen was both the first in Scandinavia to educate students on continental drift and plate tectonics, and the first to question the new paradigm. The suspicion that the theory was wrong was led by one of Gjellestad's first students, Karsten Storetvedt, who started his career as her research assistant in 1962. As a student, Storetvedt had been introduced to Runcorn's group in the Newcastle Physics Department at the height of the controversies surrounding the revival of Wegener's continental drift theory. By 1969, his investigation into secular oxidation processes of rock samples from the Orkney Islands, Corsica, and Portugal led him to conclude that "The results have given new confirmations that many paleo-geophysical conclusions rest on an unrealistic basis."[85] Rather than a stable imprint, Storetvedt asserted that the paleomagnetic record might have changed over time. He also pointed out several ad hoc-hypotheses that were used to make findings fit plate tectonics, and later argued that if Wegener's theory of continental drift had not been the only alternative available to interpret the data, "today's geoscience would probably have been fundamentally different."[86]

Then, in the spring of 1989, while on sabbatical leave at Newcastle University, Storetvedt had a "eureka moment": plate tectonics was wrong because it had misinterpreted the formation of Earth itself.[87] Dismissing the traditional idea of planet Earth beginning as a hot, convecting globe, Storetvedt asserted that the proto-earth consisted of a fast-spinning cold mix of mineral components and gases. When the earth's crust formed at an early stage, it was a thick, dry blanket, similar in composition to the lunar highland crust. Inside the earth, driven by gravity, dense materials fell toward the core, while lighter materials rose toward the surface. The lighter materials included water. Instead of the idea that oceans have been present from the very beginning, or arrived with asteroids, Storetvedt argues that seawater has been expelled slowly from Earth's core and formed as the sea floors sank toward the planet's core and mantle. In this perspective, what are generally recognized as meteor crater lakes were not formed by impacts from asteroids, but were the products of gases being released from the earth's deep interior. Similarly, this process of "degassing" produced hydrocarbons: oil and natural gas. According to Storetvedt, "degassing" is an ongoing process, and is a culprit behind earthquakes. At some point in the future, when the initial shell finally decompresses, he contends that the earth will be completely covered by relatively deep oceans. According to Storetvedt, his "new global model has the latent capacity of becoming a general evolutionary theory for terrestrial planets."[88]

One implication of Storetvedt's new planetary genesis is that plate tectonics is fundamentally wrong. Plate tectonics relies on an inner warm core on top of which plates are moved due to heat exchanges. To Storetvedt, Earth's core is heating up, but remains much colder than ordinarily assumed. Next, the continents have deep roots and relatively fixed locations. Volcanoes are explained by hot magma patches heated by local radioactive processes near the surface. Magnetic differences between continents are explained not by lateral movements, but by rotation. Storetvedt, therefore, named his new theory "Wrench Tectonics." Equipped with his new history of Earth, Storetvedt went on a crusade against what he saw as dogma, indoctrination, and irrational science, with an explicit goal to "decanonize the patron saints of the plate tectonics establishment."[89] He has written two textbooks and a scientific autobiography in which he describes the scientific community as dogmatic and non-rational, where the pressure to conform is greater than the dedication to truth.[90] According to Storetvedt, his rebellion "against the tyranny of scientific orthodoxy" has made him increasingly unpopular among his peers.[91] None of the books are included in any university curriculum, which to Storetvedt is a confirmation of his bleak view of science as unable to see past its own paradigm.[92]

Another research interest at Section C was magnetic resonance of materials in different states and at different temperatures. Starting in the 1970s, Einar Gjøen focused on pulsations and interactions between particles and waves in the plasmasphere, the inner part of the magnetosphere that surrounds Earth. Reidar Løvlie, who graduated in 1970 within the new field of paleomagnetism and returned to Section C in 1972 after spending time at Columbia University's Lamont Doherty Geological Observatory north of New York, focused on magnetism at ultra-low temperatures, both in laboratory experiments and on several field trips to Antarctica and the Arctic. In 1980, he took the lead in upgrading the laboratory with a SQUID magnetometer, an instrument that allows for extremely precise magnetic measurements but relies on cooling the sample with liquid helium or liquid nitrogen.[93] Løvlie also collaborated with the oil company Statoil in analyzing core samples from offshore drilling, and with local geologists and chemists in Oslo. Lastly, Section C worked to digitize and fully automate the magnetic station at Dombås, and took part in annual gatherings where geomagnetic researchers from all five Nordic countries packed up their instruments and gathered to compare and calibrate their equipment.

However, apart from a brief collaboration between Helge Dalseide and meteorologist Arild Guldbrandsen on solar X-rays, and the work to automate the Magnetic Station, Section C remained the odd one out at the Geophysical Institute. Although paleomagnetism and cosmic radiation concern geophysical phenomena in perpetual motion, it was increasingly difficult to see the relationship between the magnetic field, the oceans, and the atmosphere. And despite Godske having raised the issue of time scales, his "long periods" were still only 30 years. In comparison, the paleomagnetic timescales could span up to hundreds of millions of years.

In 1989, when the Faculty of Mathematics and Science reduced the number of departments from 18 to 14, Section C left the Geophysical Institute and merged with the University's Earthquake Station to form the Institute of Solid Earth Physics. The idea seems to have been that there would be synergistic possibilities between seismology and tectonics, but the marriage was not successful: their research perspectives had little in common, they shared no goals, and although the employees at Section C moved their offices to a different building, their laboratories remained at Geofysen.[94] In the 1990s, when planning the Bjerknes Center for Climate Research, paleo perspectives became relevant to questions of climate change, in particular as a measure for natural variability and the relationship between temperatures and CO_2. In 2003, in a new merger, the institutes for Solid Earth Physics and Geology were united to form a new Department of Earth Science (Institutt for Geovitenskap).

8
New ways of organizing
geophysical research

During the last three decades of the 20th century, new ways of organizing geophysical research were introduced in Bergen alongside the government-funded research conducted by the three departments of the Geophysical Institute. Geophysicists started to collaborate more frequently with scientists from other disciplines and institutions in short-term interdisciplinary projects. Furthermore, new institutions were set up outside the formal structures of the university, gathering scientists from several disciplines to specialize in the application of a particular technology or on particular topics. The first of these was the Nansen Remote Sensing Center, set up in 1986 to specialize in the application of satellite-based data in studies of the sea and the sea ice in Arctic waters. The second was the Bjerknes Centre for Climate Research, formally established in 2000 to focus on climate change with the explicit goal of providing knowledge to political decision makers, industries and the public.[1] Lastly, geophysical research became increasingly dependent on external funding provided by industry and by domestic and international research councils.

These changes were not restricted to geophysical disciplines, nor to scientific institutions in Bergen. Several scholars have observed that starting in the 1970s universities in many parts of the world experienced increasing demands to produce *useful* knowledge, and that the universities responded by introducing reforms to encourage interdisciplinary collaboration to address practical, political problems, to collaborate closer with industry, to attract external sources of funding, and to promote economic growth.[2] How did the new ways of organizing geophysical research in Bergen come about? Since similar changes occurred in many other places about the same time, we cannot focus exclusively on local context to understand how and why new institutions and ways of organizing research materialized in this period. In this chapter we show that a variety of actors, in Bergen and elsewhere, contributed to facilitating changes in the ways of organizing geophysical research in Bergen between the early 1970s and the early 2000s.

Interdisciplinary collaboration and contract research

In 1975, oceanographers at the Geophysical Institute in Bergen (GFI) started collaborating with scientists from a range of other disciplines and Norwegian institutions in a five-year project called the Norwegian Coastal Current Project, the first project to engage the major share of Norwegian oceanographic research institutions in a coordinated collaboration. Among the other participants were the Institute of Marine Research (IMR), the Institute for Geophysics in Oslo, and the Institute for Biology and Geology at the University of Tromsø.[3] The project was coordinated by Martin Mork, mathematician, theoretical oceanographer, and head of the GFI.[4]

The emergence of interdisciplinary project research conformed with international trends, and both oceanographers and meteorologists at GFI started to engage in international and interdisciplinary projects in the early and mid-1970s. In 1976, Norwegian Coastal Current Project's participants joined forces with a large international consortium known as the Joint North Sea Information System (JONSIS), which had been formed by British and German oceanographers in 1971, to develop a network of moored ocean data stations to collect data for the modeling of the North Sea's circulation. The consortium conducted two experiments in the North Sea, JONSDAP 73 and JONSDAP 76 in 1976. The Norwegian oceanographers participated in the latter, and by this time the consortium had expanded to include the International Council for the Exploration of the Sea (ICES) and several other countries.[5]

In 1971, GFI meteorologist Aksel Wiin-Nielsen became a member of a working group in the large international project known as the Global Atmospheric Research Program (GARP), initiated by the World Meteorological Organization (WMO) and the International Council of Scientific Unions (ICSU) in 1967. GARP's main goal was to conduct field experiments to make further progress in numerical weather prediction.[6] Wiin-Nielssen contributed to planning how data gathered from the project could contribute to the production of long-term weather prognoses.[7] In 1974, Mork and the meteorologist Arne Grammeltvedt became members of the Norwegian National GARP committee. Mork later coordinated a GARP subproject on coastal current dynamics and ocean climate.[8]

In addition to participating with scientists from other disciplines, institutions, and countries, in the mid-1970s scientists at GFI experimented with industrial collaboration. In 1976, the new leader of GFI's Department of Oceanography, Arne Foldvik, signed a contract with the Norwegian oil company, Statoil. The institute was to conduct a series of measurements of wind and waves at two stations located at Sotra outside Bergen, in order to provide insights that would be useful for Statoil's North Sea operations.[9] In retrospect, Foldvik recalls that this sort of contact with industry was met with resistance by several scientists at the institute, who argued that industrial collaboration was not appropriate for a university. Yet Foldvik welcomed the collaboration because it enabled the institute to invest in new instruments for its ongoing investigations in Antarctic waters. The lack of good instruments had been an obstacle to progress in oceanographic research in Bergen for a while, Foldvik recalls. Although the Aanderaa Current meter buoy had been a success, the research at the institute had lagged behind that of other countries, particularly the USA, due to a lack of investments in measurement devices. The budgets covered salaries, but not upgrading instruments for conducting field observations. The Statoil project was an opportunity to address this problem and to catch up with the research conducted in more well-funded countries.[10]

While scientific collaboration became commonplace in the late twentieth century, contract research for industrial companies did not really take off until the latter part of the 1980s, when the newly established Nansen Remote Sensing Center started appealing to private companies for research funding.[11] Still, the almost simultaneous initiation of collaboration among scientists from different disciplines, among scientific institutions, and among scientific institutions and industrial companies in the "outside world" in the mid-1970s suggests that something was beginning to change at the Geophysical Institute in Bergen in this period. Historians and sociologists who have observed similar changes in several countries in the 1970s and 1980s do not agree on what triggered the emergence of new ways of organizing and conducting scientific research. While some scholars have emphasized changes in economic conditions, others have pointed to changes in people's values, beliefs and attitudes toward science and toward universities.

Economic realities versus economistic ideologies

In the book *The new production of knowledge. The dynamics of science and research in contemporary societies* (1994), sociologist Michael Gibbons and colleagues observed that a new mode of knowledge production was emerging in Western academia. They proposed the concept "mode 2" science to designate the new mode of knowledge production, and contrasted it to a traditional "mode 1" science. While mode 1 science was conducted in universities and addressed disciplinary problems set and solved in a context governed by the academic interests of the scientific communities, mode 2 science was conducted in a variety of new sites, by interdisciplinary teams, and addressed problems formulated in a "context of application." The latter mode of knowledge production was often funded by industry and other external agencies.[12]

Gibbons and his colleagues explained the emergence of mode 2 science mainly in terms of changes in economic conditions. The core of their thesis was that the new mode was brought about by a "massification of education" on the supply side, and "increasing demands for specialist knowledge" on the demand side.[13] While the massification of education was the product of various "more or less independent forces" at work in the post-WWII era, changes in market conditions in the 1970s and 1980s were portrayed as the real "eliciting factor" that brought about changes in the mode of knowledge production.[14] Increased demand for specialist knowledge, they argued, was driven by a more intense international competition in business and industry, brought forth in part by the emergence of many new players on the international scene who challenged traditional industrial countries like the UK, the USA, Germany, and Japan. In this new situation, the industrialized countries started to search for safe niche markets. During the 1980s, when the newcomers' capabilities increased and challenged these niche markets as well, the industrialized countries started investing in Research and Development (R&D) to remain competitive. This led to a situation in which in-house research by industrial companies was no longer sufficient. Knowledge now had to be drawn from a wide range of sources.[15]

In his article "On the historical forms of knowledge production and curation. Modernity entailed disciplinarity, postmodernity entails antidisciplinarity" (2012), historian of science Paul Forman argues that changes in people's values, beliefs, and ideologies played a far greater role than changes in the economy. Forman suggests that an entirely new cultural-historical epoch – "postmodernity" – substituted for the former age of "modernity" in the late 1960s or early 1970s. He identifies a set of values and beliefs that he considers to have been shared by most people during modernity, and which the modern institutions and research practices depended on. These values and beliefs, he argues, constituted "cultural presuppositions" for the existence and maintenance of these institutions and practices. One of these values was *disinterestedness*. According to Forman, the capacity to think and act in disregard of one's personal interest was the "most highly respected quality of mind and character in modernity." In the current age of postmodernity, he argues, most people consider disinterestedness to be unrealizable as well as undesirable. The "disinterested scientist" has lost his former cultural rank to the "intensely self-interested entrepreneur."[16] Forman explicitly made the case that changes in ideology were more decisive than changes in economic realities:

"I, however, do not think that the [fall] of disciplinary science can be explained economically. I am quite certain that the world-historical transformation from modernity to postmodernity that caused the Icarian fall of disciplinarity was too broad, too fast, and involved too radical a reversal of modern perspectives to be explained by economic circumstances. I welcome the recent work of [an historian of economic thought, who] assigns the determinative role not to economic realities but to an economistic ideology."[17]

What brought about this sudden change in ideology, and in people's values, beliefs and "state of mind"? Forman suggests that it had something to do with student riots, and a more general, anti-authoritarian "revolt of the client," in the late 1960s and early 1970s. The intellectual leaders of these movements contributed to spreading the new values – including a thorough devaluation of scientific institutions – to other parts of the world.[18]

In the book *The history of the University of Bergen* (1996), Astrid Forland invokes changes in both economic realities and in ideology, beliefs, and values to account for reforms and changes at the University of Bergen from the 1970s onward. Forland speaks of a transition from an "expansion phase" to an "integration phase" at the university in this period. Following a period of expansion in student numbers, employment of academic and administrative personnel, and state funding starting in the 1960s, a period of stagnation and integration set in. The term "integration" refers partly to the university's integration within the broader society, and partly to the integration of different scientific disciplines within external research institutions.[19]

Forland observed that the university was exposed to heavy-handed economic treatment, and pointed to a general economic decline, exacerbated by the OPEC countries' oil embargoes in 1973 and 1979, as an important explanation. Yet, she added, echoing Forman, a devaluation of the university in the eyes of broad segments of the population probably contributed to the heavy-handed treatment of the university.[20] Also echoing Forman, she ascribed this devaluation of the university to the actions of student protests and "radical elements" at the university, which were inspired by the international wave of student riots. The "society-critical university," she suggested, triggered the emergence of a "university-critical society." The university campus in Bergen received the nickname "the Lenin height," and it became a widespread notion that the university had turned into a remote, academic "ivory tower." In this context, the University of Bergen was faced with a "crisis of legitimacy." Forland frames the chapter on the integration phase as a story about the university's battle to regain legitimacy in society. "Outside" society, and the government "above," leaned on the university, demanding reforms and useful science. The University of Bergen responded to the increasing external demands by prioritizing applied science over basic science and vocation-oriented education over theoretical education, and by facilitating externally funded contract research.[21]

Some historians have identified a different source for the emergence of new values, beliefs, and attitudes toward science and universities. In *The University of Oslo 1975–2011. Towards a new societal contract?* (2011), historian Kim G. Helsvig suggests that the Organization of European Cooperation and Development (OECD) contributed to spreading a new attitude toward science and universities among Norwegian politicians, bureaucrats, and the wider public from the 1960s onward. In addition to the international wave of student riots, OECD personnel and like-minded people in Norway helped to introduce the legitimation crisis of Norwegian universities. OECD, Helsvig argues, contributed to a "science-political regime change."[22] Political-administrative strategists were busy formulating and circulating a new science policy to replace the traditional "basic research ideology." In 1966, OECD set up a Committee for Science Policy with the explicit goal of forging a transition from "policy for science" to "policy through science." That is, scientific research and research institutions were to become means towards political and practical ends rather than mere ends in themselves. As shown by Helsvig, OECD encouraged member governments to formulate clearly to which national goals scientific research should contribute to reaching.[23] Helsvig suggested that a "complex dialectic" materialized between these political-administrative ambitions to repurpose science to promote technological and economic development, and the student rioters' critique of positivism, capitalism, and the "technocratic-materialistic" society.[24]

An important source of Helsvig's observation of OECD's role is Edgeir Benum's article "A new science policy regime? Basic research, OECD and Norway 1965–1972" (2007). Benum's article has the advantage that it renders visible some of the work involved in producing and circulating a new discourse about how scientific research should be legitimated. First, he shows that OECD personnel, in the early- to mid-1960s, started to formulate and circulate reports discussing concerns about a "technological gap" between European and US industries, and suggesting that European research institutions should be reformed as a means to address this concern.[25] The OECD, Benum argues, became a center for dissemination of the idea that universities should be turned into vehicles for economic growth and policy formation.[26] While

OECD initially limited itself to promoting the initiation of applied research *in addition to* the basic research traditionally conducted at universities, in the latter part of the 1960s it started to advocate the introduction of "demands for relevance" for research conducted at the traditional basic research institutions as well. At the same time, OECD started to advocate a broader objective for the governments' science policy: in addition to facilitating economic growth, governments should formulate plans for their science policy so that it could be used to address "politically defined societal problems."[27]

Benum's "OECD thesis," maintaining that the OECD played an important role in introducing new ways of legitimating and organizing research, is more convincing than the "student riot thesis" for four reasons. First, it identifies the production and dissemination of a discourse more intimately connected to the observed changes in the organization of research. The assumed causal connection between the student protests and the subsequent external demands about useful science appears, in contrast, to be more speculative. The student riot thesis may explain why universities introduced reforms to "democratize" their decision-making organs, but it does not have the same explanatory power on the question of why universities started to promote interdisciplinary, project-based research, external funding, and the formation of new, external research institutions.

Second, Benum shows that people at a variety of sites and at different moments in time contributed to preparing the stage for the introduction of new ways of organizing scientific research in Bergen and elsewhere. He observes that the Norwegian government set up the Main Committee for Norwegian Research in 1965, and that representatives from this committee participated in OECD meetings. This committee contributed to disseminating the OECD discourse to a Norwegian audience, including the Norwegian government and various university institutions. Indeed, an important mandate for the committee was to provide the government with a solid basis for the formulation of science policies.[28]

Third, the OECD thesis might contribute to explaining why similar changes in the organization and funding of science occurred in several countries at about the same time. As pointed out by Benum, OECD was an intergovernmental organization engaged in formulating and disseminating the new discourse to several member governments. Lastly, Benum's thesis gains some support from another source. In 1975, the Norwegian government presented a White Paper entitled *On the organization and funding of* research, a document which in many respects echoed the OECD discourse. [29] We will investigate the content of this White Paper in more detail below.

A problem with Benum's account is that he fails to clarify why the new discourse was produced and circulated in the first place, and how it gained the force to influence science policy debates and actual science policies on a national level. The ambiguity on this point is evident in a discussion of two reports ordered and discussed by the OECD secretariat in the late 1960s. One report discussed whether European countries lagged behind the USA in technological innovation. While the report did not identify a significant "technology gap" between the two regions, Benum suggests that the OECD secretariat was able to use the report "for its own purposes." Another report, written about the same time, concluded that the technology gap was real, and recommended that European research institutions be reformed in order to close the gap. On this occasion, Benum questions the quality of the report: "Methodologically, the report's comparison of European and US universities appears dodgy."[30] Benum could have discussed whether there was a connection between changes in economic realities, as presented in the latter report, and the new discourse, or ideology, produced and circulated by OECD personnel and members of the Main Committee on Norwegian Research, but he did not exploit this opportunity. Instead, he suggests that the new discourse was produced and disseminated by people pursuing their own "dubious" goals. These goals apparently came into being in complete isolation – with no connection to changing economic conditions. This way, he seems to place himself firmly in the "ideology" camp together with Paul Forman.

Even though Benum shows that other people contributed to disseminating the new discourse in the late 1960s and early 1970s, he allows the new discourse to become an autonomous force capable of triggering effects during subsequent decades. "Over time," he argues, "the spreading of the new discourse had considerable consequences for science policy thinking and the framework conditions of science."[31] A similar deficiency is detectable in the contributions of other scholars. Forland presents her story about the University of Bergen during the integration phase as a story about "*the legitimation crisis* that characterized the major portion of the 1970s and 1980s, and the ways in which the development of the university took a new path in this period."[32] Forman exposes himself to the same critique when he suggests that the "self-consciously held ideology" of postmodernism, popularized in the 1970s, turned into "unreflectively held cultural presuppositions" with consequences for the course of events during the age of postmodernity.[33] These quotes indicate that the authors treat actions and events in the late 1960s and early 1970s as autonomous "causes" that produce far-reaching and long-lasting effects. Yet, as pointed out by the social theorist Bruno Latour, it requires additional work to make particular events have a bearing on other, subsequent events.[34] Once we *add* this additional work to the account, the simple cause-effect relationships become more complex.

A call for useful research

In the early 1970s, the Norwegian government was busy formulating a new science policy along the lines advocated by OECD. The White Paper *On the Organization and Funding of Science*, completed in November 1975, invited the Members of Parliament to a broad discussion of the organizational and financial aspects of research in Norway.[35] The explicit reason was that both basic and applied science were spreading to new areas and "now played a crucial role to the development of our society." The government observed that most Western countries had been addressing similar concerns for a while.[36]

The government pointed out that Norway had a long tradition of using science as a means to promote industrial production, and suggested that the four independent research councils set up at various moments after WWII had played a crucial role in facilitating good connections between the government, the universities, and various industries. While the government wanted to maintain this apparatus more or less unchanged, it argued that a new research council was required: a "Research Council for Societal Planning." Echoing the OECD discourse identified by Benum, the government explicitly made clear that the goal was to use this research council to promote research on issues considered deserving of scientific inquiry for "political reasons." The list contained issues like the organization of the production of goods and services, and concerns about the use and management of the environment and natural resources.[37] The government also made explicit that it was crucial to promote interdisciplinary and problem-oriented research to address these issues:

"Most of the research needs identified [here] transgress established disciplinary, institutional and administrative boundaries. The government wishes to integrate problem-oriented, interdisciplinary research on important societal problems in the research council structure, and thus in the overall research sector, more forcefully than today. The proposition to set up a new research council for societal planning must be seen in the light of this goal."[38]

The White Paper did not simply "echo" the OECD discourse. It also made explicit references to advice from the Main Committee for Norwegian Research, the forum that had sent representatives to OECD meetings from the mid-1960s onward.[39] Yet the question remains whether this indicates that the "spreading of the new discourse," in Benum's terms, was beginning to have "considerable consequences for science policy thinking and for the framework conditions of science" in Norway. Once we recognize that additional work was required to bring elements of this discourse into the White Paper, we realize that this White Paper itself was an event in which the discourse was spread, or transported, to other sites. The "spreading of the new discourse" then appears as a "consequence" of this work, rather than a cause that explains the emergence of new science policies and science policy thinking. This does not rule out the possibility that the Norwegian politicians were inspired by the work going on in OECD, but it raises the question of whether the Norwegian government brought in this discourse because it was inspired by OECD or because of its own, more pragmatic, purposes. Recall that Benum insinuated that the OECD secretariat produced and disseminated the new discourse for "its own purposes," and that they were not really inspired by the reports discussed among the OECD representatives. This is certainly possible, but Benum does not provide any evidence to support it. In the same way, it is possible that changes in Norwegian science policies were a product of inspiration from OECD. Yet it is also possible that the new discourse was brought in due to purposes, or goals, that had been defined elsewhere.[40]

Another reason to leave the question about such "external" influences more open is that scientists in Bergen about the same time argued the case for reorganizing scientific inquiries for reasons that do not echo the OECD discourse. In 1976, two participants in the Norwegian Coastal Current Project suggested that interdisciplinary collaboration was necessary to achieve "further progress" in oceanography.

How to achieve "further progress" in ocean studies

In the report "Some preliminary results from a synoptic experiment in the Norwegian Coastal Current" (1976), written shortly after the participants in the Norwegian Coastal Current Project had conducted their first field experiment, scientists Gunnar Furnes and Roald Sætre established that the explicit objective of the Norwegian Coastal Current Project was to obtain "a better understanding of the structure and dynamics of the coastal current and to establish the relationship between biological, physical, and marine geological phenomena in Norwegian coastal regions."[41] The two scientists, members of the project's steering group and active project participants, also elaborated on why oceanographers had started to search for new ways of organizing scientific research. An important argument was that collaboration with other disciplines was necessary to achieve further progress in oceanographic research:

"Oceanographic research within the fields of physical, chemical, and geological oceanography, and marine biology in the coastal current have long traditions in Norway. However, a common factor in the earlier investigations is that they have been limited to a relatively narrow framework, with respect to geographical region, period of time, or discipline. The investigations have for the most part been directed at special phenomena, and have therefore been difficult to combine. Further progress in oceanographic research demands that problems within one discipline should not be regarded separately, but must be seen in the light of those conditions dealt with by the other disciplines. This having been recognized, new forms of cooperation between oceanographic research and reporting institutions have been sought."[42]

Furnes and Sætre elaborated on how interdisciplinary collaboration could help address questions requiring skills and insights from more than one discipline. One such problem regarded the relationship between biological and physical conditions in the sea. Coastal regions generally had high biological production, which made them important sites for fishing fleets. There were indications of a close correlation between physical and biological phenomena, but the relationship between them was "not at all clear." Another interdisciplinary question regarded the relations between topographical features, currents, and the distribution of sediments on the sea bottom, from which marine geologists could benefit from input from physical oceanography.[43]

Meeting demands for "further progress" was not the only reason to initiate a large, interdisciplinary, and inter-institutional study of the Norwegian coastal current. The authors mentioned three more reasons. First, a large-scale collaboration could help optimize the use of available resources. One immediate advantage was that the scientists could conduct "synoptic experiments," using several ships simultaneously and making measurements "with a reasonably tight observation net" along the entire Norwegian coast.[44] While synoptic experiments were surely a means to achieve progress in oceanography, the point about optimizing the use of available resources also contains an element of the economic advantages of collaboration. Second, the project could also contribute to a better understanding of biological processes of interest due to their "importance to the world's fisheries." Lastly, it could provide insights about man's influence on the marine environment. It was clear that industrial waste, pollution from densely populated areas, heat pollution from power plants, and regulation of fresh water flows all influenced the marine environment, but the full consequences of their influence were unknown.[45]

The latter arguments about optimizing the use of available resources and producing knowledge for the fishing industry and to institutions responsible for the management of the marine environment conform well to the ideas and concerns articulated in the OECD discourse. However, Furnes and Sætre presented these potential practical insights of the investigations more as 'bonuses' than as important reasons for joining forces with scientists from other disciplines. The main argument presented by these "insiders" of the local scientific community was that new ways of organizing research were demanded in order to make further progress in oceanography. Apparently, to the oceanographers, making progress in oceanography was the main goal.

This observation does not conform well to the literature discussed above. Most of those accounts ascribe changes to "external pressures" from a society that did not consider knowledge production "for its own sake" a legitimate goal. From the 1970s onward, the story goes, universities were conceived of as "ivory towers," and scientific research was considered valuable if it served practical purposes.[46] Furnes and Sætre's paper indicates that the emergence of new ways of organizing research was not simply an outcome of pressures from "outside" society or from the government "above." Local scientists also articulated complaints about the traditional mode of knowledge production, and advocated change. Furthermore, the authors presented "further progress" in science as a goal, not just a means to obtain useful knowledge. We do not know how important this argument was to the people involved in realizing this project, but the mere presence of this discourse suggests that explanations invoking external demands and a postmodern devaluation of science should be qualified.

A new site for geophysical knowledge production

In 1986, the establishment of Nansen Remote Sensing Center introduced another change in the mode of organizing geophysical research in Bergen. The center specialized in remote sensing – the use of satellites and satellite data – in studies of the ocean and the sea ice in Arctic waters. Organized as an independent foundation, its prime mover was the GFI oceanographer Ola M. Johannessen, who became the first Professor in Remote Sensing in Norway with the center's opening. Johannessen had obtained a degree in oceanography at GFI in 1965, and subsequently spent nine years abroad. After a short stay at the University of Sao Paulo in 1966, he spent four years at McGill University in Canada (1966–1970) and another four years at the NATO Center in La Spezia, Italy (1970–1974). At the two latter sites, Johannessen gained experience with remote sensing. Returning to Bergen and the Geophysical Institute in 1974, he soon started assembling a group specializing in validation and use of satellite data in oceanographic research.

In 1984, Johannessen started carving out the plans for setting up an independent center for remote sensing in Bergen. According to Johannessen, the initial trigger was an invitation to evaluate the Scott Polar Research Institute in Cambridge, UK. If the British oceanographers had an institute named after the British Polar Expedition hero Robert F. Scott, he reasoned, then certainly Bergen should have a research institution named after Fridtjof Nansen. An important step was involving the director of Rieber Shipping, Christian Rieber, in the plans. After a successful visit at Rieber's office, Johannessen had closed a deal guaranteeing free accommodation for the Nansen Remote Sensing Center during its first two years of operation.[47] Next, Johannessen formulated a formal application to the Dean of the Faculty of Mathematics and Natural Sciences, in which he argued the case for setting up a center for remote sensing in Bergen. One point was that the use of satellite data would enable a "radical improvement in the description and understanding of the marine environment." The satellites, he argued, had already contributed to uncovering processes in the sea and the atmosphere, and many of those processes could best be studied with a combination of satellite data and conventional measurements.[48] This argument echoed Furnes and Sætre's arguments in 1976 that interdisciplinary studies would facilitate "further progress" in oceanography.

Unlike Furnes and Sætre, Johannessen strongly emphasized the many practical applications of the center's research. Observations from satellites could provide insights to a variety of industries, including the oil, shipping, and fishing and fish-farming industries. Of particular importance was the potential for developing offshore warning systems for ocean circulation and ice conditions. The Nansen Center would contribute to research and development of operational models capable of processing vast amounts of satellite data, which would improve both short- and long-term forecasts of ocean and ice circulation of particular usefulness to the oil industry. Satellite observations could also enable detection of harmful algae from Skagerrak that occasionally led to high mortality in fish farms along the Norwegian coast. Lastly, studies of the internal structures of the earth could potentially lead to useful insights for the oil industry.[49]

A third argument was that Bergen was a "center of gravity" in oceanography, and had significant competence in satellite-based remote sensing. He listed eleven scientists and a secretary who had worked together on various remote sensing projects since 1978. "A considerable competence has been built up," he argued, and the group should constitute a "natural core in the proposed Nansen Remote Sensing Center."[50]

A final argument was that the research center would facilitate interdisciplinary collaboration. Satellite-based remote sensing was an "essentially interdisciplinary" activity, Johannessen argued. It involved high-tech space research; in-situ verification of measurements with conventional instruments on- and offshore; knowledge about electromagnetic radiation; skills in instrumentation; development of algorithms; geophysical, biological, and geological interpretations; numerical modeling; real-time analysis for input in models applied to forecasting; and interactive databases to be applied in research and management. The new research center, Johannessen suggested, would be a site for coordinating interdisciplinary projects in which remote sensing constituted a crucial element. The center was to employ a relatively small staff that would collaborate with "regional, national, and international institutions and industrial companies."[51]

In retrospect, Johannessen recalls that the university leadership did not immediately embrace the idea of setting up an independent center for remote sensing, and demanded that another sponsor in addition to Rieber be included to ensure its financial base.[52] Johannessen turned to Tenneco Oil, a US oil company operating in Norway with which he had become acquainted through Bjørn Landmark, leader of the Technical Science Research Council's (NTNF) space division.[53] Eventually, Rieber Shipping and Tenneco Oil became co-founders of the Nansen Center together with the Science Foundation at the University of Bergen (UNIFOB), a foundation recently set up to promote and administer externally funded contract research.[54]

Johannessen played a decisive, proactive role in setting up the Nansen Remote Sensing Center, but he did not operate in a vacuum. Johannessen's application contains several clues about the work already conducted by people in a variety of sites contributing to preparing the stage. First, the arguments that satellite-based remote sensing would lead to "radical improvement" of the description and understanding of the marine environment, and that research based on satellite data would have many practical applications, would have appeared far more questionable 20 years earlier. Johannessen would certainly not have been able to point to important discoveries to substantiate this claim. As shown by historian Eric M. Conway, in the mid-1960s a representative of the US National Aeronautics and Space Administration (NASA) made efforts to recruit geophysicists to space science. By this time, however, it was an open question whether, and how, satellite data would be valuable, particularly to oceanographers. At a conference convened jointly by NASA and the Woods Hole Oceanographic Institution in 1964, the participants identified a variety of research areas where satellites could be useful, but many participants were skeptical about the satellites' potential for providing data of satisfactory quality. By 1969, Conway argues, the oceanographers' confidence in the satellites' potential usefulness had strengthened.[55] Yet a lot of work remained before the satellite gained the status of an indisputable asset to oceanography. Two satellite launchings in the early 1970s provided important "proofs of concept," according to Conway. Data retrieved from the Skylab satellite, launched in 1973, and the GEOS 3 satellite, launched in 1975, showed that satellites could indeed provide information of great value to oceanographers.[56]

Second, when Johannessen argued that remote sensing would have many practical applications, he referred to a Norwegian Official

Report (NOU) on satellite remote sensing, written in 1983.[57] This reference points to other sites in which people had been busy preparing the ground. In October 1982, the Norwegian government had set up a committee to investigate the potential of satellite remote sensing and to formulate policy advice. The director of the Norwegian Defense Research Establishment (FFI), Finn Lied, was appointed to lead the five-member committee. Lied, a veteran of satellite-based research and monitoring for military purposes, had played an active role in establishing the Norwegian Andøya Rocket Range in 1962. He had served as Minister of Industry in the Norwegian government between 1971 and 1972, and was head of the board of Statoil when he was chairing the satellite remote sensing committee. The committee report, delivered about eight months after the committee's birth, presented a long list of arguments about why it was important that the Norwegian government intensify its efforts to facilitate satellite-based remote sensing. Satellites could enable surveillance of Norway's vast, offshore Exclusive Economic Zone and of marine and atmospheric pollution, assist in the mapping of sea ice and the locations of natural resources at sea and ashore, and they could improve weather forecasts and facilitate scientific progress in oceanography, meteorology, biology, and archaeology. The commission concluded that "Norwegian interests in remote sensing and remote analysis by means of satellites are considered substantial. This is due in part to the fact that we are responsible for vast offshore territories, some of which are located far from our airports. Extensive and variegated industrial activities are carried out in these areas (fisheries and oil- and gas excavation), which crucially depend on environmental conditions and which may have impacts on these conditions."[58]

Johannessen repeated this quote in his application to the Dean of the Faculty of Mathematics and Natural Sciences, showing how Lied's committee had prepared the stage by establishing remote sensing as of great "national interest" before Johannessen started formulating the application to set up the Nansen Remote Sensing Center.[59] Indeed, it is also possible that OECD's efforts to spread the discourse of useful science to industries and governments had smoothed the way, but there are no traces of such influence in Johannessen's application.

Third, the ground was well prepared with respect to Johannessen's argument that the oceanographic community in Bergen had built up significant competence in satellite-based remote sensing. Lied's committee had also identified the oceanographic community in Bergen as having established "close and fruitful" collaboration with US organizations in applying data from US satellites for maritime purposes and ice studies.[60] Johannessen himself had played an active role in forging this collaboration, but the remote sensing community in Bergen would not have materialized unless a variety of individuals and institutions had joined forces with Johannessen and the local team of "emerging" remote sensing experts.

In 1977, Johannessen had been invited by the NASA Goddard Space Flight Center in Washington, D.C., to participate in a team assembled to validate data from a satellite scheduled to be launched in 1978: Nimbus 7.[61] Launched in November, during the spring of 1979 "Team Johannessen" and participants in the Norwegian Coastal Current Project, as well as some other scientists and institutions, took part in a joint Norwegian-USA remote sensing experiment off the Norwegian coast. Among the US institutions participating in the project were the US Air Force, the Naval Research Laboratory in Washington, D.C., and the Coastal Studies Institute at Louisiana State University.[62] This project – called Norwegian Remote Sensing Experiment (NORSEX 79) – had two explicit goals: to "build up national expertise in some remote sensing methods within the marine sciences," and to "evaluat[e] and appl[y] these methods on important oceanographic phenomena in the Norwegian Coastal Waters, Norwegian Sea, Barents Sea and the Polar Ocean."[63]

While the Norwegian Coastal Current Project ended in 1980, the analysis of the NORSEX 79 experiment culminated with a research article in *Science* in 1983.[64] The efforts to validate and use data from satellites in oceanographic research continued in two new projects: the Marginal Ice Zone Experiment (MIZEX) and the Norwegian Maritime Remote Sensing Experiment (NORMARSEN). The former was a "mega-science" project lasting from 1983 to 1987, and consisted of several field experiments led by Johannessen. The 1984 experiment was the largest ever conducted in a marginal ice zone, the area where the

Polar Ice meets the Polar Sea. It lasted for two months and engaged an interdisciplinary team of more than 200 scientists from 11 countries, with seven ships, four helicopters, eight aircraft, and four satellite systems at their disposal.[65] The results were published as several reports in *Science*. It was through these projects and this collaboration with a variety of individuals and institutions that the oceanographic community in Bergen had been able to build up significant competence on satellite-based remote sensing.

The final argument used by Johannessen was that the Nansen Remote Sensing Center would enable collaboration between scientists and experts with a variety of skillsets, as was required to realize the potential represented by satellite-based research. Here, too, the stage was prepared in advance. As we have seen above, in the mid-1970s the Norwegian government called for more interdisciplinary and problem-oriented research to address practical societal problems. In the early 1980s, many scholars and administrators at the University of Bergen were engaged in setting up independent centers outside the formal framework of the university. The first proposal to set up an independent research center in Bergen came from scholars engaged in Middle East and Islamic studies, in 1980. According to Forland, the main inspiration was that scholars at the University of Lund, Sweden, were preparing a similar institution. The first center actually established in Bergen in this period was the Center for Humanistic Women's Studies, established in 1984. The Center for Middle East and Islamic Studies was established in 1988. In other words, the Nansen Remote Sensing Center was one of several interdisciplinary centers established in Bergen around the same time.[66]

Why focus on this preparatory work? The point is that much of the established literature has identified a very limited number of events in the 1960s and 1970s as the triggers of reforms and changes in the ways of organizing and funding science from the 1970s onward. Gibbons and colleagues emphasized changes in economic conditions. Forman and others have focused on student protests, the "revolt of the client," and activities within OECD. These events and activities have been considered triggers for a new "state of mind," or new attitudes towards science and universities, which in turn have been portrayed as the main underlying cause for subsequent changes throughout the "integration phase" or the "age of postmodernity." Above, we have extended the list of events and activities contributing, or that may have contributed, to the materialization of the Nansen Remote Sensing Center in Bergen.[67]

Chasing external funding

The Nansen Remote Sensing Center was formally established on November 28 1986, and officially opened on June 16 1987. The non-profit foundation has depended, and still depends, strongly on external funding to exist.[68] Johannessen had identified some niches to which the center could contribute with important insights, such as shipping, fishing, fish-farming, and oil exploration and production. In 1989, the Nansen Remote Sensing Center made efforts to intervene in another emerging niche: environmental research. In 1990, the center even added "environmental" to its name, and the Nansen *Environmental* and Remote Sensing Center came into being. In retrospect, Johannessen suggests that the decision was pragmatic: dependent on external funding, the center needed "more legs to stand on."[69]

The annual report for 1991, an extended account marking the center's fifth anniversary, gives a vivid picture of the vast range of activities and sources of funding enabling the Nansen Center to exist as an independent, nonprofit institution. One project focused on validation and use of satellite data, another studied physical processes of the ice edge of importance to future petroleum activity in the Barents Sea, and yet another project aimed to develop methods for "real-time ice monitoring" to provide ice warning services for fishing vessels and other users. There were also several projects addressing topics related to climate change. One aimed to contribute to the development of global ocean models and carbon cycle models, and to determine the sensitivity of various physical processes to anthropogenic emissions, and another project investigated whether it was possible to inject CO_2 into the ocean and thus reduce emissions to the atmosphere. Yet another project aimed to develop a carbon cycle model for the North Atlantic. The list of funders of various projects included oil companies like Statoil, Hydro and Saga Petroleum, the Norwegian State Pollution Agency, the Norwegian Ministry of the Environment, the Norwegian Space Center, the European Space Agency, the Norwegian Research Council for Science and the Humanities (NAVF), and the U.S. Office of Naval Research.[70]

The annual report for 1991 also shows that the Nansen Center had great ambitions and expectations regarding environmental research. In fact, the main vision of the center was presented as making "a significant contribution to the understanding of regional and global environmental problems through research and development."[71] A year later, the center saw "a substantial growth potential in climate-related projects of theoretical as well as experimental nature." While all climate-related projects presented in the 1991 account were funded by Norwegian companies and agencies, the Nansen Center was preparing to take part in two research projects funded by the European Community.[72]

Apparently, climate and environmental research did not turn out to be as lucrative as expected. In 1993, the Chairman of the Board complained that "in spite of both domestic and international environmental awareness, grants are scarce. Intensified marketing is thus required, directed in particular towards the European Union, the European Space Agency and other international sources of funding."[73] Similar complaints were articulated during subsequent years.[74]

In spite of these disappointments, the Nansen Environmental and Remote Sensing Center managed to expand, both economically and physically. In 1991, the center received funding amounting to 12.7 million NOK. In 1997, the budget had increased to 18.9 million NOK, and in 2007 to 37.2 million NOK. Several new sister institutions have been established in other countries. In 1992, the Nansen International Environmental and Remote Sensing Center was set up in St. Petersburg, Russia. In 1998, the Nansen Environmental Research Center India was set up in Kerala, and in 2003, the Nansen-Zhu International Research Center was set up in collaboration with Institute of Atmospheric Physics at the Chinese Academy of Sciences, Beijing. In 2010 and 2012, two new Nansen Centers were established in South Africa and Bangladesh, respectively.[75] With these centers, many international projects started in the Arctic and Indian Oceans, including studies of teleconnections between the Arctic and the Asian monsoon system. Thus, the Nansen Center has contributed to involving Bergen in international research. The center also expanded at home in 2000, when it became a partner in the "Bjerknes Collaboration" and co-founder of the Bjerknes Centre for Climate Research. At present, the International Nansen Group has a staff of more than 200, including PhD students.[76]

The Bjerknes Centre for Climate Research

Around 1997, the rumor that the Norwegian Minister of Church Affairs, Education and Research was planning to introduce a Center of Excellence (CoE) arrangement reached scientists at the Nansen Environmental and Remote Sensing Center. Through this new invention, Norwegian scientific communities could gain access to a generous, and relatively long-term, source of funding from the Norwegian government.[77] Scientists at the Nansen Center convened a joint meeting to discuss how to proceed in order to prepare an application for CoE status for a climate research center in Bergen. According to Helge Drange, a mathematician and climate modeler employed at the Nansen Center at the time, three young scientists at the meeting "took, or were granted" a mandate to formulate a "vision" for climate collaboration in Bergen. These were marine geologist Eystein Jansen, meteorologist Nils Gunnar Kvamstø and Helge Drange.[78]

The three scientists did not merely formulate a vision for climate collaboration. During the next two years, they struggled to forge a collaboration among the Nansen Remote Sensing Center, the Institute of Marine Research, and the University of Bergen. After three tough rounds of negotiations, the three institutions signed a formal agreement in August 2000. Eystein Jansen was appointed director of the Bjerknes Centre, the institution set up to co-ordinate and manage the collaboration among the signatories.[79] In 2002, the Norwegian Research Council announced that an application from Bergen had passed through the needle's eye. The new center's continued existence was secured, and Bergen had received another geophysical research center outside the formal structure of the university.[80] How did the new institution come into being?

Jansen started preparing a CoE application to the Norwegian Research Council, in which he presented the objective of the Bjerknes Centre for Climate Research as "to become an acknowledged center for conducting cutting-edge climate research and training to further our knowledge of past, present, and future climate change with emphasis on Northwest Europe, the Northern Seas, the North Atlantic and the Arctic Regions."[81]

The application contained several arguments explaining why the new center deserved CoE status. First, the research conducted at the center addressed "one of the key environmental and socio-economic challenges facing the world today." Yet due to the "intrinsic complexity" of the climate system, long-term research efforts were required if climate scientists were to understand and predict future changes with reasonable accuracy. Jansen referred to a report recently published by the Intergovernmental Panel on Climate Change (IPCC) to support this view: "The 3rd assessment report of the IPCC clearly states the need for a long-standing, wide approach research to reduce uncertainties and replenish shortfalls in our current knowledge." Gaining a better understanding of these key processes was "key to many aspects of economy, politics, welfare, and cultural-social affairs."[82]

Second, Bergen's physical location was important. Again, Jansen could lean on IPCC: "The nearness to the Arctic is important, particularly as the Arctic region, as summarized in the 3rd IPCC assessment report, is the region where future global warming is expected to be most extensive." The Arctic region was important because several "key climate processes of global or hemispheric significance occur in the region," and because paleoclimatic evidence indicated that the North Atlantic region had been particularly subject to abrupt climate changes in the past.[83]

Lastly, the scientific community in Bergen had built up a "combination of expertise" which was well designed to address many of the open questions and uncertainties regarding human impacts on the future climate. The Bjerknes Collaboration consisted of scientists studying past abrupt changes from oceanic and terrestrial palaeoclimatic archives, others studying ocean dynamics, and yet others specializing on modeling. A particularly important asset was the recently developed Bergen Climate Model. This coupled global ocean-ice-atmosphere model would serve as a "hub" for many of the activities at the Bjerknes Centre.[84] We will look more closely at the production of this model in chapter 9.

The efforts to forge collaborations among the three institutions in Bergen, and the efforts to convince the Norwegian Research Council that the scientific community in Bergen was deserving of a Center of Excellence, were important to the materialization of a new research center in Bergen in the early 2000s. Yet, again the ground was well prepared in advance. First, the Norwegian government's decision to introduce a Center of Excellence arrangement was the outcome of deliberations and investigations conducted by other people at other sites. The Norwegian government presented the idea to the parliament in a White Paper entitled "Research at a crossroad" in 1999.[85] Identifying several other sites in which the arrangement had already been implemented: "An increasing number of countries prioritize excellence in research by introducing particular arrangements. A frequently used term is 'centers of excellence.' Such arrangements have been introduced in Australia, Austria, Japan, Canada, Korea, Denmark, and Finland, and are due to be implemented in the Netherlands."

The government presented the arrangement as a means to "develop more scientists and scientific communities at a high international level." It pointed out that evaluations had shown that such arrangements often turned out to be very successful. Indeed, they tended to perform "better than originally expected."[86] The government's White Paper was an attempt to interest the Norwegian parliamentarians in the Center of Excellence arrangement. The document itself was the culmination of a lot of preparatory work, and further efforts following the presentation to the parliament contributed to setting the stage for a new, independent research center in Bergen in the early 2000s.

Second, Jansen's proposition that the new center would help address one of the key "environmental and socio-economic challenges facing the world today" may indeed have appeared convincing to the people engaged in evaluating the CoE applications forwarded to the

Norwegian Research Council in 2002. Yet 15 years earlier this statement would probably have appeared disputable or grossly exaggerated. According to historian Spencer R. Weart, a meeting in Villach, Austria, in 1985 was the first occasion when a group of climate scientists "went beyond the usual call for more research to take a more activist stance" on global warming.[87] However, when the World Commission on Environment and Development presented the report *Our Common Future* in 1987, the question of global warming was still but one concern among many. The report covered the topic over a few pages in a chapter addressing concerns about increasing demands for energy.[88]

Things changed during subsequent years. In 1988, the World Meteorological Organization (WMO) and the United Nations Environment Programme (UNEP) established the Intergovernmental Panel on Climate Change (IPCC), which produced two assessment reports during the 1990s and a third report in 2001. In 1992, the UN followed up the World Commission's report by convening the United Nations Conference on Environment and Development. One outcome was the formulation of a United Nations Framework Convention on Climate Change. According to historian Paul N. Edwards, these events implied that "[a]n era of global climate politics had dawned(...)."[89]

In Norway, the issue of global warming became a central political issue in 1989, when the government presented a White Paper discussing how to follow up the policy advice formulated in *Our Common Future*. The topic of global warming received a more prominent position in this document than in *Our Common Future*. It was one of the first topics addressed in several chapters of the White Paper.[90] The government also set up an Inter-Ministerial Climate Group, tasked with formulating advice on climate policies and preparing participation in international negotiations on global climate agreements.[91] We will discuss this in more detail in the next chapter. The purpose here is to identify some of the domestic and international sites in which people had been busy preparing the ground for the Bjerknes Centre by "lifting up" the climate issue.

Lastly, the proposition that the scientists in the Bjerknes Collaboration were particularly well prepared to address many of the open questions regarding climate change was probably not hard to accept in 2002. The scientists at the Nansen Remote Sensing Center had entered the niche of climate research in 1989, and had argued that oceanography was indispensable for climate research to succeed in gaining a better understanding of the climate system and to produce realistic scenarios of future climate change.[92] During the 1990s, both meteorologists and oceanographers in Bergen were busy building numerical models to calculate the world and the climate of the future.

In 1997, the future partners of the Bjerknes Collaboration joined forces with three Oslo-based institutions (Norwegian Meteorological Institute, the University of Oslo, and the Norwegian Institute for Air Research (NILU)) in a project named RegClim (Regional Climate Development under Global Warming), the first coordinated project on natural scientific research on climate change funded by the Norwegian Research Council. Its main objective was to estimate climate changes in Norway and the surrounding seas with more precision than provided by global climate models.[93] During the early phase of this project – which lasted for nearly 10 years – some of the scientists in Bergen started to construct a model connecting processes in the sea and the atmosphere.[94] The outcome was the Bergen Climate Model, the tool presented in the CoE application as a crucial asset for the scientists in the Bjerknes Collaboration. The Bergen Climate Model was the only model produced in a Nordic country used by the IPCC in preparation of the fourth assessment report on climate change, published in 2007.[95] Obviously, nobody knew in 2002 that the model would become *that* successful. Still, it shows that local efforts to build expertise on climate research contributed to the establishment of the Bjerknes Centre for Climate Research in Bergen.

Adding events and sites

In this chapter, we have investigated some of the work conducted by people at different sites, which contributed to changing the modes of organizing and funding geophysical research in Bergen from the mid-1970s to the early 2000s. We have also discussed findings by other scholars, and have argued that much of this literature is too reductionist because it relies on a relatively narrow set of events and sites to explain the emergence of new phenomena. Some are even busy *subtracting* events in order to grant more importance and influence to others.

In this chapter, we have aimed to *add* events and sites contributing to the emergence of new ways of organizing and conducting research in Bergen. We do not deny that student protests and discourse production within OECD had implications for later events, but their importance appears exaggerated when we draw attention to the many other sites where people were busy advocating changes in the modes of organizing and conducting research. We have shown that scientists in Bergen advocated interdisciplinary collaboration in ocean studies and in satellite-based research to achieve further progress in their disciplines, alongside arguments about possible uses for the insights produced. We have also shown that people in the Norwegian government, the Norwegian Space Agency, Tenneco Oil, Rieber Shipping, the World Commission on Environment and Development, and at meetings in Villach, Austria, and Rio de Janeiro contributed to preparing the stage for changes occurring in Bergen.

We have not been concerned with evaluating these changes. Some commentators fear and despise them, others applaud them, and yet others appear ambivalent. Forman belongs to the first group, and he does not hesitate to speak on behalf of most professors when criticizing the turn to postmodernity:

"[T]hose within our institutions of higher learning who wish them restructured and repurposed to comport with the present historical moment are generally to be found only in its topmost administrative ranks and bottommost academic ranks. Professors understand that such restructuring and repurposing would bring complete intellectual chaos – not to mention the elimination of almost every gratifying feature of the professorial profession."[96]

Gibbons et al. appear to be more ambivalent. They do not portray the onset of mode 2 science as particularly troubling, but they do articulate concerns about possible problematic consequences. One such concern is that the procedures for evaluating the quality of scientific findings have become less transparent.[97] Yet others point to positive effects. The prime mover and former leader of the Nansen Environmental and Remote Sensing Center, Ola M. Johannessen, speaks warmly about a new "push" that brings science forward.[98] In a similar tone, the former leader of the Bjerknes Centre, Eystein Jansen, speaks of a "vitalization" of geophysical research in Bergen after the Bjerknes Centre for Climate Research was established.[99] The various "insider" appeals for reforms during the course of events also indicate that local scientists welcomed some of these changes when they appeared.

9
Geophysics in the age
of climate change

In the late 1980s, questions of global warming and anthropogenic impacts on climate became hot topics in Norway and elsewhere. As the Norwegian government started preparing the follow-up to the UN report *Our Common Future* (1987), geophysicists in Bergen soon started directing their attention to questions and problems related to climate change and/or anthropogenic impacts on climate. Climate-related research has become an increasingly important source of funding for geophysical research in Bergen, and it has brought about changes in the content and conduct of scientific inquiries. Furthermore, it has granted Bergen geophysics a high standing in the international climate research community.[1] Finally, the turn to climate research has brought geophysicists in Bergen onto the stage of public and political controversy. Some geophysicists have become active participants in public debates on a variety of topics related to climate change, including the quality of scientific findings, and what political strategies to pursue to gain more knowledge and combat anthropogenic global warming.[2]

In this chapter, we investigate how meteorology and oceanography in Bergen became part of the emerging interdisciplinary field of climate research, and how the entry into this field has brought about changes in research questions and research strategies. Finally, we ask whether the entry into this interdisciplinary field has come with a downside: Do the recurring climate disputes and the slow progress in climate policy formation indicate that people have lost faith in the capacity of geophysicists and other climate scientists to calculate the world?

Global warming becomes a matter of concern

In 1938, the engineer Guy Stewart Callendar proposed that human beings' burning of fossil fuels could have an impact on the planet's climate. Yet nearly half a century passed before scientists started to voice concerns in a concerted manner. As shown by the historian Yngve Nilsen, in the early 1980s several Norwegian meteorologists participated in a committee set up by the Norwegian government to formulate advice on science policy regarding climate research. The immediate context was a conference convened by the World Meteorological Organization (WMO) in 1979. This World Climate Conference had concluded that climate change – natural and man-made – required more attention. The Norwegian committee produced a report entitled "Impacts in Norway of possible climate changes." According to Nilsen, this committee focused more on the possible positive consequences of global warming than on potential problems and concerns. Still, the committee concluded that research aiming to increase the understanding about climate change deserved "very high priority."[3]

To be sure, not everyone expected only positive effects. In 1984, the meteorologist Øystein Hov encouraged the Norwegian government and Norwegian geophysicists to take warnings about global warming seriously. In a feature article in *Aftenposten*, Hov cited a warning articulated by the US National Research Council: "In our calm assessment we may be overlooking things that should alarm us." It was time, Hov argued, that Norwegian scientists made this warning "our own."[4]

The topic of anthropogenic global warming also received some attention at the Geophysical Institute in Bergen in the early 1980s. In 1983, Swedish meteorologist Bert Bolin gave a guest lecture at the Geophysical Institute entitled "A study of the carbon cycle and how the atmosphere's content of carbon dioxide may come to increase due to future use of fossil fuels."[5] Bolin was one of the prime movers in the revitalization of climate studies, and especially studies of connections between anthropogenic emissions and climate change. He had been one of the key speakers at WMO's World Climate Conference in 1979, and later became the first leader of the Intergovernmental Panel on Climate Change (IPCC).[6] A survey of the Geophysical Institute's Annual Reports suggests, however, that Bolin's talk did not trigger an immediate re-orientation or revitalization of climate studies at the institute. The first explicit reference to concerns about climate change as justification for ongoing research appears in the Annual Report of 1988.[7] We will return to this event below.

As mentioned in Chapter 8, one historian has identified a meeting in Villach, Austria, in 1985 as the moment when some climate scientists went beyond the usual call for more research to take a more "activist stance."[8] Two years later, the World Commission on Environment and

Development – set up by the UN's General Assembly in 1983 – brought the topic to a broader audience with the report *Our Common Future*. The commission's mandate was to identify the main challenges facing humankind, and to elaborate on possible strategies for coping with them.[9] The commission identified anthropogenic climate change as a serious matter of concern, and pointed explicitly to the warnings articulated at the Villach meeting in 1985: "After reviewing the latest evidence on the greenhouse effect in October 1985 at a meeting in Villach, Austria (...), scientists from 29 industrialized and developing countries concluded that climate change must be considered a 'plausible and serious probability.'"[10]

The Commission also embraced the Villach scientists' proposal of a "four-track strategy" to approach the issue. First, it was important to improve the monitoring and assessment of the evolving phenomena. Second, it was necessary to increase the research efforts to improve knowledge about the origins, mechanisms, and effects of these phenomena. Third, it was important to negotiate "internationally agreed policies" for the reduction of climate gases, and lastly, to adopt strategies for coping with the effects of climate change.[11]

The World Commission's report contributed to placing the issue of global warming on the agendas of a variety of governments and other institutions. The United Nations General Assembly committed all member governments and UN agencies to consider the World Commission's report and recommendations. As shown by the historian Iris Borowy, the UN mobilized an immense apparatus to circulate the report to governments and other agencies. A Centre for Our Common Future opened in Geneva in 1988, its only task to spread the report's message.[12] About the same time, UNEP and WMO set up IPCC, whose mandate was to review the scientific literature on climate change, on social and economic impacts of climate change, and on possible response strategies.[13]

The Norwegian government had a particularly good reason to attend to the recommendations formulated in *Our Common Future*: the Chair of the World Commission on Environment and Development, Gro Harlem Brundtland, was also Norway's Prime Minister when the UN started disseminating the report to governments and non-governmental organizations. In fact, the report often goes by the name "the Brundtland Report." The Brundtland government imme-diately started to elaborate on how it would follow up the report. It forwarded the report to more than 600 private and public institutions in Norway, and encouraged them to comment on its content and for-mulate proposals for adequate policies.[14] It also started to prepare a White Paper to the Norwegian parliament. Completed in 1989, the White Paper *Environment and Development. Norway's follow-up of the World Commission's report* granted the issue of global warming a more prominent position than did *Our Common Future*.[15] The government also set up an Inter-Ministerial Climate Group with the mandate to formulate advice on climate policy.[16] The Inter-Ministerial Climate Group soon started encouraging scientific communities in Norway to provide reports on various topics relevant to policy formation on the issue of global warming.

Entering the niche of interdisciplinary climate research

When the Nansen Remote Sensing Center came into being in 1986, global warming was not on the research agenda. Ola M. Johannessen did not mention the topic in the application forwarded to the Dean of the Faculty of Mathematics and Natural Sciences. Nor did the scientists at the Geophysical Institute pay much attention to the topic during the early and mid-1980s. Things changed around 1989. In the 1988 Annual Report, written sometime after March 1989, the author points out that "[d]eep-water formation in the Greenland Sea is a relevant topic due to the increasing concentration of CO_2 in the atmosphere."[17] The context was a presentation of Johannessen's activities in 1988. Johannessen had coordinated the Marginal Ice Zone Experiment (MIZEX) between 1980 and 1988, and from October 1 1988 the project continued as Seasonal Ice-Zone Experiment (SIZEX). The explicit purpose of SIZEX was to study "the sea ice and how it interacts with the sea and the atmo-sphere (...)."[18] This had been an object of inquiry in some of the former projects, like NORSEX and MIZEX, as well. Indeed, the phenomenon of deep-water formation had been an object of inquiry for Norwegian oceanographers, including oceanographers in Bergen, for much longer. Fridtjof Nansen had already published pioneering work on the topic in 1906.[19] Håkon Mosby had studied deep-water formation in Arctic and Antarctic waters between the 1930s and 1960s, and his successor Arne Foldvik had continued the studies between the 1960s and 1980s.[20] Ola M. Johannessen also continued in this tradition. Deep-water formation had received attention because the oceanographers considered it an "engine" in the global ocean circulation, a key to understanding the physical movements of the sea.[21] The novelty in 1989 was that this old object of inquiry became relevant in a new way: it became a key to understanding anthropogenic global warming.

By March 1989, Johannessen had already, on several occasions, referred to concerns about global warming as part of the motivation for conducting the SIZEX experiments.[22] On February 7 1989, the Norwegian News Agency (NTB) presented a story about the SIZEX experiment entitled "The greenhouse effect challenges Arctic scientists." Johannessen explained that the amount of carbon dioxide in the atmosphere had increased significantly since the middle of the previous century. The greenhouse effect, he warned, could lead to a warmer climate, increased sea level, and dramatic changes in precipitation patterns.[23] The newspaper *Nordlys* presented a more extensive story about the project titled "CO_2 in the sea" the very same day. About two weeks later, the newspaper *Aftenposten* reported on the SIZEX project in a story entitled "Wants to know why the Polar Ice is melting."[24]

It was probably not difficult to convince scientists and politicians alike that oceanography deserved to be included in the interdisciplinary field of climate research. During 1990, the Inter-Ministerial Climate Group was fully engaged in collecting reports on global warming from various scientific and administrative institutions in Norway. The group received about 40 reports. One of them was written by a group of scientists, mainly meteorologists, affiliated with the Norwegian Institute of Air Science (NILU). "The greenhouse effect and the climate development" contained an open invitation to oceanographers to join forces in the study of global warming.[25] The group argued that prediction of future climate change required close collaboration between meteorologists and oceanographers. The path forward was to build "coupled atmospheric and ocean circulation models," and collaboration between the disciplines was essential to reach this goal.[26]

The NILU group completed its report in April 1990, and scientists at the Nansen Center quickly responded to the invitation. In December 1990, they forwarded their own report to the Inter-Ministerial Climate Group. Contributors to the report – "The ocean's impact upon the atmospheric CO_2-budget and global climate change" (1990) – were oceanographer Peter M. Haugan, marine geologist Eystein Jansen, oceanographer Ola M. Johannessen, and marine biologist Ulf Lie, who presented it as a supplement to the NILU report.[27] They questioned the findings of a recent report by the IPCC that had proposed that global average temperatures would increase by 0.3 degrees Celsius per decade in the next century. Haugan and colleagues argued that model studies recently conducted in Hamburg challenged this conclusion. The scientists in Hamburg had coupled two ocean models to an atmospheric model, and the experiments that followed suggested that global warming would be slower than proposed by IPCC, at least in the short term. In the longer term, however – specified as about 50 years ahead – the simulations came closer to IPCC's findings, due mainly to a complete melting of the ice in Arctic waters during summer and during gradually longer periods of the year.[28]

A crucial question was whether the increase in atmospheric temperature in the Arctic region could trigger changes in the carbon cycle. CO_2 was an important greenhouse gas, which circulated between the atmosphere and the sea. It was absorbed at the sea surface, and brought down to deeper layers by deep-water formation and some other mechanisms. The sea also released CO_2 to the atmosphere, but available evidence suggested that the sea absorbed more CO_2 than it released. Yet it was possible that this relationship could be reversed, and that the sea would eventually become a net exporter of CO_2. A consequence was that CO_2 eventually would accumulate in the atmosphere at a greater rate than expected based on current data.[29]

The authors presented a frightening scenario. The interactions between the atmosphere and the sea were a "very sensitive system" and increases in the greenhouse effect could potentially trigger changes comparable to events during the Younger Dryas, about 10,000 years ago, when the previous Ice Age ended.[30] This was a *worst-case* scenario. The main message was that the various processes in the sea were poorly represented in existing climate models.[31] Changes in these processes represented the single most important source of uncertainty in contemporary attempts to estimate how the planet would respond to continued emissions of carbon dioxide to the atmosphere.[32] If the problem of anthropogenic global warming was to be understood, it was decisive to study the sea.

The scientists also formulated concrete proposals on domestic science policy. The report had established that processes in Arctic waters played a crucial role in the global climate system, and the authors argued that Norwegian scientific communities had "world class" competence in such processes. Thus they were well positioned to contribute to reducing uncertainties regarding future climate change. The international community of scientists, the authors added, already expected Norway to play a crucial role in this area.

Yet reducing uncertainties was not all that was at stake. The authors provided some pragmatic reasons why Norway should prioritize climate research. They saw a "unique opportunity" for Norway to strengthen its position in international climate research. The path forward was to build on established competence and experience, which meant investigating and modeling processes in Arctic waters.[33]

The authors also included some policy advice that seems to involve a concern about the *location* of the climate research expertise in Norway. The NILU scientists had proposed that it was important to build up *one* climate research community in a small country like Norway. It would be rational to "unite the forces" of various competences and to equip this community of scientists with adequate computing power.[34] The NILU group did not specify where this community should be located, but a qualified guess is that the group had the capital, Oslo, in mind. The Bergen scientists may indeed have seen this proposal as a threat to their own opportunities of playing a central role in the climate research community emerging in Norway. That may be the reason why they suggested that Norway should *not* try to be competitive with regard to computing power. It would be very costly for Norway to try to be in the front line of computer development. It would be more cost effective to focus on process studies and competence building, and to conduct model experiments with available computer power. Norway already had two super-computers capable of running "heavy" model experiments. One was located in Trondheim, and the other at the IBM Bergen Scientific Centre in Bergen.[35]

At the Geophysical Institute's Department of Meteorology, the turn to climate research came with two new employees in 1989 and 1990. The first was Øystein Hov, one of the "early warners" about global warming in Norway. He had worked for several years at the Norwegian Institute of Air Research (NILU) in Oslo before obtaining a professorship at the Geophysical Institute, where he stayed until 1997, when he returned to NILU. He served as the institute's leader starting in 1990.[36] Hov studied the transportation of chemical compounds in the atmosphere. The other new employee was Sigbjørn Grønås, who replaced Hilding Sundqvist, as the latter accepted a professorship at the University of Stockholm. Sundqvist had been engaged in developing computer models for application in numerical weather prediction (NWP), and Grønås's work continued along these lines. Yet he quickly developed an interest in climate-related research. Shortly after taking office, he started a colloquium in which scientists could meet and discuss questions regarding global warming. The first IPCC report on global warming was published the same year Grønås took office, and this report was one of the first documents discussed at his colloquium.[37]

Thus, by late 1990, three different groups of geophysicists in Bergen were about to take the initial steps into the emerging field of interdisciplinary climate research. One group focused on chemical meteorology, and another on physical meteorology. Finally, the oceanographers had established that their discipline was indispensable for reaching the goal of making realistic predictions of future climate change. New questions had been formulated: Could the stability of the climate system be disturbed? Would increased emissions of CO_2 lead to a reduction of the ocean's capacity to absorb CO_2 from the atmosphere? Was it likely that abrupt climate change could occur at some point in the future? The path forward was to build coupled atmospheric and ocean circulation models to enable realistic predictions of future events. This required collaboration between disciplines.

Building and coupling models

In the early 1990s, however, the different groups were working separately. Scientists at the Nansen Environmental and Remote Sensing Center set out to build models of ocean circulation, especially CO_2 circulation in the sea. At the Geophysical Institute, the physical meteorologists set out to build models of atmospheric circulation. Hov's group of chemical meteorologists used atmospheric models developed by physical meteorologists to study the atmospheric circulation of chemical compounds. This approach to climate research was discontinued in Bergen after Hov returned to Oslo in 1997. Concomitantly, scientists from the two other groups started working more closely together on the task of coupling models of the atmosphere and the sea.[38]

In 1990, the Nansen Center decided to employ a PhD candidate, mathematician Helge Drange, as part of its strategy to engage in climate research, and he set out to build a carbon cycle model for the North Atlantic. Drange started with a global ocean circulation model, known as MICOM, developed by a research team in Miami; it was a "weather forecasting" model for the sea. Drange's PhD project involved adding some carbon chemistry and simple biology to this general model, to describe how the sea absorbed CO_2 from the atmosphere, how it circulated in the sea, and how it returned to the atmosphere. Working on the MICOM model for about four years, Drange gained valuable experience with global ocean circulation modeling.[39]

At the Geophysical Institute, meteorologist Nils Gunnar Kvamstø was one of the first scientists to engage in the task of designing a numerical model to simulate climate change. After completing a PhD thesis on questions related to weather forecasting, in 1993, he obtained a postdoctoral position in an international climate research project known as the Norwegian Ocean and Climate Project (NOClim), a large project consisting of scientific institutions from several Nordic countries funded by the Nordic Council of Ministers. Its purpose was to build a global, atmospheric climate model. According to Kvamstø, the NOClim project marked the beginning of global climate modeling in the Nordic region. He recalls that the NOClim group started out with a French atmospheric model, called ARPEGE, and set out to make it more detailed and realistic. The group gained invaluable experience with climate modeling, but the project turned out to be too ambitious. "If that project had been carried out today," Kvamstø notes today, "I would probably be out of business."[40]

In 1997, Drange and Kvamstø became participants in RegClim (Regional Climate Development under Global Warming), an interdisciplinary project with participants from the Norwegian Institute of Marine Research, the Geophysical Institute in Bergen, the Nansen Environmental and Remote Sensing Center, and three institutions in Oslo.[41] Its main objective was to "regionalize" a global climate model by adding more details on a regional level limited to Norway and the surrounding sea territories than was permitted by the global models.[42] According to Kvamstø, the collaboration between the RegClim partners was difficult. Members disagreed on methodological choices and approaches, leading Drange and Kvamstø to work on the side to connect the two models on which they had gained competence during the previous years: ARPEGE and MICOM. With funding from the University of Bergen, they put together a small team to work on coupling of the two models.[43] The outcome of this work was the Bergen Climate Model (BCM), which was an important asset when the climate scientists affiliated with the Bjerknes Centre for Climate Research applied for Center of Excellence status in 2001.

The Bjerknes Centre for Climate Research was formed in 2000, and the ball started rolling. In 2003, the center obtained generous funding from the government as a Center of Excellence, and the same year six scientists published an important paper describing and evaluating the new, coupled, Bergen Climate Model. The paper presented the results of tests conducted to establish whether the model was capable of producing simulations of past climate changes that corresponded with available observational data.[44] The authors concluded that "[t]he model indeed captures the main features of the observed climate, and in particular the simulation of radiation, clouds, and freshwater fluxes is well produced." They also evaluated the simulations of processes in the sea as "realistic," and suggested that the Bergen Climate Model would form the basis for many forthcoming studies of "the global climate system in general, and the regional climate system in particular."[45]

Reaching up, reaching out

The Bergen Climate Model became an important tool to study the effects of continued emissions of CO_2 and other greenhouse gases, but it was also a ticket to becoming relevant within the international climate science community. In 2003, the Bjerknes Centre announced that it had been invited to perform a number of model simulations and scenario predictions on behalf of IPCC as part of the preparation for the next assessment report on climate change, due to be issued in 2007. IPCC had also invited the Bjerknes Centre director, geologist Eystein Jansen, to be Coordinating Lead Author of a chapter on paleoclimate in the report accounting for the physical basis of climate change.[46]

Providing analyses for IPCC became an important part of the center's activity in subsequent years. In late June and early July 2006, the Bjerknes Centre hosted a "Lead Author" meeting of IPCC's Working Group I, responsible for the report on the "Physical Science Basis" regarding climate change. The 170 participants gathered at Solstrand Hotel outside Bergen to prepare the report's final version. Although mainly a closed meeting, the organizers had prepared a detailed media plan including a press briefing and an opportunity for journalists to ask questions, which resulted in extensive media coverage.[47]

IPCC Working Group I's report on the physical science basis for climate change was published in February 2007, with important contributions from Bergen's climate science community. The Bjerknes Centre contributed with one co-ordinating lead author, one lead author, and four contributing authors. Equally important, it had contributed global climate scenario simulations from the Bergen Climate Model, one out of four European centers, and the only center in the Nordic countries to perform such simulations.[48]

The Bjerknes Centre was beginning to climb up in the international climate research community. As early as 2006, an international evaluation committee established that the Bjerknes Centre was "on the forefront of development in their field," that it was on the way to become "one of the leading centres worldwide," and that it had "an excellent reputation at the national and international levels."[49] On the day the physical science basis report was released in 2007, the newspaper *Bergens Tidende* published an interview with Bjerknes Centre Director Eystein Jansen. The journalist suggested that Jansen was "perhaps more powerful than any Norwegian politician" at the moment. He had been part of a process through which the international scientific community had managed to formulate more certain conclusions than in earlier reports: "The estimates are the same as in the previous report. But we are more certain. And it is more likely that it will be warmer than indicated by our prognoses, than that it will be colder," Jansen argued.[50]

The Bjerknes Centre scientists also worked systematically to bring research results regarding anthropogenic climate change to broader audiences. The center's annual reports regularly commented on its employees' outreach activities. "An important mission of the BCCR is to enhance the public awareness and understanding of key processes involved in the climate system and the potential consequences of climate change," we learn in the 2004 Annual Report. Bjerknes scientists had published 32 popular articles, held eight invited lectures and contributed to more than 36 entries in the mass media.[51] The 2006 report announced a "very positive" development in the media exposure, and a "breakthrough with national media." From 2005 to 2006, the number of Norwegian media reports increased from 89 to 162 entries. Increased visibility strengthened the recognition of the center's expertise in the media, and the center was used "progressively more often as an authoritative source for climate issues in the media," the report established.[52]

In 2006 and 2007, the scientists received substantial assistance in their effort to raise awareness about climate change. First, the former US presidential candidate, Al Gore, attracted much attention to the issue with his book and documentary movie "An Inconvenient Truth." In 2007, the climate issue gained further publicity when the Norwegian Nobel Peace Prize Committee decided to award the Peace Prize to Al Gore and the Intergovernmental Panel for Climate Change (IPCC) for their efforts to produce and disseminate knowledge about anthropogenic climate change and the measures to counteract it. The Nobel Committee explicitly sought to "contribute to a sharper focus on the processes and decisions that appear to be necessary to protect the world's future climate, and thereby to reduce the threat to the security of mankind."[53] In the Bjerknes Centre's 2007 Annual Report, Jansen claimed that "2007 was the year when climate change was acknowledged nationally and globally as the main challenge facing humanity."[54] The rise to prominence of climate research in Bergen was accompanied by a rise to prominence of the issue of anthropogenic climate change.

Can CO_2 be drowned in the sea?

The strategy of focusing on the development of models, and particularly the efforts to construct a coupled atmosphere-ocean model, turned out to be a success. Yet building models to predict climate change was not the only strategy pursued when the Bergen oceanographers entered the field of climate research around 1990. Drange's goal of developing a model to simulate the transport of CO_2 in the sea was accompanied by another question shortly after he started on his PhD project: Could CO_2 be drowned in the sea by means of shallow-water injection? Drange and his supervisor at the Nansen Center, Peter Mosby Haugan, set out to investigate a possible contribution to the task of reducing anthropogenic emissions of CO_2 to the atmosphere. Their solution differed from another form of CO_2 storage with which the Norwegian oil company Statoil started experimenting about the same time: storing CO_2 in geological formations beneath the seabed.[55] The solution of injecting CO_2 directly into the sea was not entirely new when Drange and Haugan started investigating this particular option in the early 1990s. The Italian physicist Cesare Marchetti had published a paper on deep-water storage of carbon dioxide in 1977, and some other scientists had followed up on his line of inquiry.[56] A common characteristic of these studies was that the CO_2 would have to be injected into the sea at great depths, or in shallow waters in particular locations where sinking currents would transport the injected CO_2 to great depths. Marchetti's 1977 paper discussed the latter option, and he used the Strait of Gibraltar as an example of a site in which injected CO_2 would be transported to great depths in the Atlantic Ocean by means of naturally existing currents. The problem with Marchetti's solution was that it depended on very specific topographical characteristics. The problem with the other solution – direct injection at great depths – was that it was expensive, demanding considerable energy. The novelty of the hypothesis proposed by Drange and others at the Nansen Center was that gas injected into shallow waters of about 200–400 meters would possibly form "self-generating" sinking currents, as CO_2-enriched water was heavier than water with less CO_2. Perhaps the sinking current would be created *as* the CO_2 was injected. If this worked, shallow-water injection could be applicable even in areas without strong sinking currents of the kind existing off Gibraltar. This mechanism could potentially enable CO_2 injection in Norwegian waters, where oil and gas platforms emitted CO_2 to the atmosphere – and at lower costs than in the case of direct injection into deep waters.[57]

The Nansen Center scientists carried out preliminary investigations funded by Statoil in 1991, and produced a report for the Norwegian Pollution Control Authority the same year.[58] The question became an integral part of Drange's PhD project. The double ambition of his PhD project is reflected in the thesis title: "An isopycnic coordinate carbon cycle model of the North Atlantic; and the possibility of disposing of fossil fuel CO_2 in the ocean."[59] While the thesis was defended in 1994, Drange and his supervisor, Peter Mosby Haugan, had managed to draw quite a bit of attention to this hypothesis with a publication in *Nature* in 1992. This paper – entitled "Sequestration of CO_2 in the deep ocean by shallow injection" – helped spread the idea to other scientists as well as to a wider public. On June 4 1992, a few days after the publication in *Nature*, the newspaper *Bergens Tidende* reported on a "Bergen CO sensation." In this news story, Drange articulated his hope that their method could be applicable to the North Sea petroleum industry as well as to industrial emission sites worldwide located near the sea. Pollution from power plants running on coal, oil, and gas represented about one-third of the world's total CO_2 emissions, and Drange proposed that the method had potential to reduce CO_2 emissions on that scale.[60] The newspaper *Aftenposten* covered the story the following day, including an interview with Olav Kårstad, a representative from Statoil's research center in Trondheim. Kårstad reported that Statoil had contributed to the funding of the project from the beginning, and that the company considered this research an important and interesting part of the company's engagement in CO_2 research. He was excited about the scientists' work, and would attend to their research with great interest in the future.[61]

Haugan, Drange and other scientists at the Nansen Center continued to elaborate on the topic during the 1990s.[62] Two interrelated problems required further study. The first regarded the question of acidification of the sea as an outcome of CO_2 injection. Haugan and Drange discussed this question in a 1996 article. They recognized that acidification could have adverse effects on marine life. However, the main impact of shallow injection would be felt in deeper layers of the sea. As most of the biological production occurred in the upper layers of the sea, including that part which was most important for commercial fisheries, it was not obvious that acidification would amount to a serious problem.[63]

The other problem regarded the need to test the theory in practice by means of field experiments. All Norwegian experiments regarding shallow-water injection had been computer simulations. In 1992, Norwegian representatives to the International Energy Agency (IEA) had managed to place Drange and Haugan's idea on the agenda of this international forum.[64] However, a few years lapsed before concrete plans for field experiments materialized. Important progress was made in Kyoto in 1997, in connection with the conference convened to follow up the UN Framework Convention on Climate Change signed in Rio in 1992. Representatives of the governments of Norway, Japan, and the United States set up a consortium with the mandate to facilitate theoretical and experimental work on ocean storage of CO_2. The governments of Canada and Australia, as well as the Swiss/Swedish company Asea Brown Boveri (ABB), soon joined the group. Another important step was a decision by the European Community to fund a large project on Global Storage of Anthropogenic Carbon (GOSAC) under its Environment and Climate Program. The project ran from 1998 to 2001. Seven European modeling groups contributed to the project, and one of the groups consisted of scientists from the Nansen Environmental and Remote Sensing Center.[65]

The consortium established in 1997 started planning a three-step experiment to be executed at a facility in Hawaii during the summer of 2001. Due to fierce opposition from local indigenous groups and environmental organizations, however, the consortium never obtained the necessary permissions to initiate the experiments.[66] In a feature article published in *Bergens Tidende* in September 2001, Drange and his Nansen Center colleague Guttorm Alendal reported that the project was "postponed indefinitely" due to this opposition.[67]

The consortium formulated a plan B: to conduct the experiments off the Norwegian coast, west of the town of Kristiansund. Consortium participants forwarded an application containing a detailed experimental plan and impact assessments to the Norwegian government. The Norwegian State Pollution Control Agency gave permission to conduct the experiment in February 2002, but shortly thereafter, the environmental organizations Greenpeace and World Wildlife Fund forwarded complaints to the government. The Minister of the Environment, Børge Brende, decided that the application required more thorough consideration before it was accepted.[68] In a feature article published in *Aftenposten* on July 3 2002, Drange criticized Brende for halting the project. The experiment involved injecting 5.4 tons of CO_2 into the water. This amount, Drange argued, equaled 20 seconds of emissions from Norwegian cars. He pointed out that the world's oceans contained 50 times more CO_2 than the atmosphere, and that they naturally absorbed one-third of global emissions. Its capacity to store CO_2 was much greater than this, he argued, but "still the Minister of the Environment (...) considers an experiment of injecting 5.4 tons of CO_2 too dangerous."[69]

A few days later, a political advisor at the Ministry of the Environment, Bjørn Skaar, addressed Drange's "misunderstanding." The Ministry had not decided whether the project should be permitted; it had merely decided that the government had to evaluate whether the project conformed with Norwegian pollution legislation. It was unclear, he argued, whether the emissions would harm living organisms in the sea. Furthermore, a thorough treatment of the issue would allow other affected parties to articulate their views. This was a "democratic right," and, in addition, it would help illuminate the issue.[70]

The next day, the State Pollution Authority again announced that it considered the environmental impacts to be acceptable.[71] Yet in August, the Minister of the Environment once more intervened and turned down the application.[72]

Three weeks later, Nansen Center employee Guttorm Alendal criticized Børge Brende for this decision in an opinion piece in *Aftenposten*. Alendal accused the Minister of the Environment of halting an experiment in which the Ministry of the Environment had been both an initiator and funder. He observed that Brende had referred to international agreements regulating pollution of international waters as justification for the decision. However, Alendal suspected that Brende had listened too much to representatives from environmental organizations like Greenpeace and World Wildlife Fund. He also indicated that the minister had turned down the project for "tactical reasons": the decision to turn down the project was made four days prior to the opening of the United Nation's World Summit on Sustainable Development in Johannesburg.[73]

In March 2003, Peter Mosby Haugan published a letter in *Aftenposten*, arguing that Brende had to reconsider his former decision. One argument was that the decision could harm Norwegian international research collaboration, particularly with the partners making investments and preparations for the experiment prior to the minister's denial. Haugan also pointed out that the problems related to the handling of CO_2 required clarification; even the IPCC had recently decided to produce a special report on capture and storage of CO_2, in the sea as well as in geological formations. Active obstruction of research aiming to provide relevant information on these matters, Haugan argued, could not be an acceptable policy.[74]

Brende never reconsidered his decision, and the international consortium never carried out any CO_2 sequestration experiment off the Norwegian coast. In 2003, Haugan became engaged in an experiment on a smaller scale off the coast of San Francisco.[75] While not identical to the experiments planned in Hawaii and off the Norwegian coast, there were similarities. The objective was to investigate whether it was possible to store CO_2 in local "lakes" on strategic sites at the sea bottom. Haugan was also invited to contribute as lead author to IPCC's special report on carbon capture and storage, published in 2005. Entitled "Carbon Dioxide Capture and Storage" (2005), the report was a formal part of the work of IPCC's Working Group III, elaborating on strategies for combating climate change, and consequently a part of the knowledge base for IPCC's Fourth Assessment Report, for which it received the Nobel Peace Prize in 2007.[76] As Haugan was one of the lead authors of the report, he was one of the scientists who received a personal Nobel Peace Prize certificate for this contribution. The IPCC's 2005 special report on CO_2 capture and storage seems to mark the end of serious consideration of CO_2 sequestration as a means to combat anthropogenic global warming. Perhaps another climate related concern – the one regarding increasing acidification of the sea – has contributed to this outcome.

Where to locate offshore windmills

In 2009, Bergen geophysicists started exploring another solution to the concern about anthropogenic climate change: the construction of offshore wind farms. The immediate context was that in 2007, the Research Council of Norway announced an invitation for scientific communities to apply for status and funding as "Centers for Environment-friendly Energy Research" (FME).[77] In June 2008, representatives from various research institutions and industries in the Bergen area convened to discuss the possibility of forming a consortium to prepare an application. Some of the institutions represented at the meeting, including the research institutions at the University of Bergen, Uni Research, and Christian Michelsen Research, and the companies BKK and Norwind, decided to work together toward this end. According to a former research director at Uni Research, Svein Winther, the initial plan was to focus on the area where Bergen had "international recognition" and "unique competence": geophysics. Thus, the center application was to focus on "the wind field and the dynamic interaction on different scales (...) between the wind field on one hand and the ocean, each turbine, and the complete wind farm, on the other hand."[78]

The list of topics expanded during negotiations with other potential industry partners, such as Statoil and Statkraft, which requested a stronger technological profile, and the University of Agder, University of Stavanger, and Aalborg University were invited to join the consortium to strengthen its competence in areas like marine operations, mechanical systems, maintenance, and communication systems. One concern was that a research cluster in Trondheim – which was also preparing an FME application – had already developed strong competence in these areas. Representatives from the Bergen-based consortium and the cluster in Trondheim even negotiated about forming a common application. However, these negotiations came to naught. Christian Michelsen Research coordinated the preparation of the FME application forwarded to the Research Council of Norway, and in February 2009 the Norwegian Centre for Offshore Wind Energy (NORCOWE) obtained FME status. The new center officially opened in October of that year.[79]

An explicit goal for NORCOWE was to make offshore wind farms competitive suppliers of energy. It was crucial to reduce costs and optimize the use of resources in the "whole value chain" for offshore wind energy.[80] To Bergen geophysicists, involvement in NORCOWE involved new research questions and new challenges regarding data collection. In the broadest sense, oceanographers and meteorologists started investigating the optimal placement of wind farms and the optimal placement of individual windmills within each wind farm. More specifically, this involved studying processes in the ocean for potential locations for wind farms, and in the atmospheric space above the surface – known as the marine atmospheric boundary layer (MABL).

Measuring winds in the MABL up to 200–300 meters, the area of relevance to offshore windmills, required technological innovation. Meteorologist Joachim Reuder at the Geophysical Institute became the leader of a group exploring the possibility of measuring wind speeds by means of a device known as "lidar" (Light Detection and Ranging). After initial trials at Stavanger Airport in 2013, some lidars were placed atop an onshore wind turbine in the Netherlands. Using laser beams, the lidars collected data on the wind flowing into the windmill, and the turbulence, or "wake," produced behind it. Collecting data about the wind wakes was particularly important because of the implications for locating other windmills in a large, offshore wind farm. The data collected by the lidars were compared to data from more conventional measurement devices, and the scientists concluded that the new device was capable of providing valuable and reliable data. In spring 2015, the scientists set out to test the device offshore. Collaborating with several German research institutions, they installed their measuring devices at the FINO1 research platform, located close to an existing wind farm in the German Bight, where they studied how turbulence from the windmills in the front of the wind farm influenced the productivity of windmills located farther behind. Yet their most important ambition was to develop methods for studying turbulence. According to Reuder, the experiments in the German Bight were "by far the greatest accomplishment of the Norcowe centre. The measurement campaign was unique and the most extensive campaign of its kind undertaken so far. We were able to test several new methods specifically targeting the characterization of turbulence. The results are promising, but we must take a closer look before we can say anything more than that."[81]

The Bergen oceanographers contributed to NORCOWE by addressing three main topics. They measured surface waves, currents, temperature, and salinity in the water column to define the characteristics and variability of the sea at various sites, particularly sites of interest as location for offshore wind farms. Next, they conducted investigations to increase the understanding of dynamics of the upper ocean, and lastly, they investigated how the presence of large offshore wind farms could influence ocean circulation, surface wave fields, and ecosystems.[82] All these topics required investigation of the interface between the sea and the atmosphere, or, to put it in the scientists' own terms, between the lower part of the MABL and the oceanic mixed layer (OML). As had the meteorologists, the oceanographers set out to develop and deploy a variety of measurement devices to collect data on the phenomena investigated.[83]

The funding of the NORCOWE center lasted for eight years, and the project formally ended on March 31 2017. Toward the end of the project, partners from NORCOWE joined forces with representatives from the research cluster in Trondheim – who had been involved in another FME funded offshore wind project called NOWITECH (Norwegian Research Centre for Offshore Wind Technology) – and applied for another round of FME funding for a new offshore wind center (COWIND), but the Research Council of Norway did not approve this application. In NORCOWE's Final Report (May 2017), the partners announced that NORCOWE would continue as a research network, and indicated that they would continue efforts to apply for new project funding from the Research Council of Norway, the EU program Horizon 2020, and industries.[84] Even if the prospects for continued engagement in large-scale offshore wind projects appear to be uncertain, Bergen geophysicists will probably engage in research on climate- and energy-related topics for years to come. In 2016, the Geophysical Institute established a new professorship in offshore wind, and NORCOWE participant Finn Gunnar Nielsen was hired on September 1. The institute has also started a master's program on energy, in which Nielsen and his colleagues will introduce students to offshore wind and other forms of renewable energy production.[85]

Remaining on top?

In the early 1990s, oceanographers, meteorologists, biologists, mathematicians, and geologists started to address the puzzle about connections between anthropogenic influences on the planet's climate. About a decade later, the creators of the Bjerknes Centre for Climate Research articulated their ambition that the center become "a leading international centre for climate research" and "the key provider of top quality knowledge on climate change to stakeholders, i.e. policy makers, industry, and the general public."[86] It appears reasonable to conclude that the goal was reached by 2007. Has it maintained its position? There are many indications that the center, and thus the Bergen climate science community, is still going strong. By the time the Center of Excellence period ended in 2012, the government had already come up with a new grant securing the funding of core activities until 2021. Formally, the new grant was reserved for a separate body named the

Centre for Climate Dynamics, but it was organized as an integral part of the Bjerknes Centre for Climate Research.[87] The Bjerknes Centre has continued to expand with regard to funding. In the first year as a Center of Excellence, the Bjerknes Centre's budget amounted to 63.1 million NOK. In 2007 it had increased to about 75 million NOK, and in 2016 it amounted to 174 million NOK. An even better indicator of the center's international standing is the fact that an increasing proportion of the funding stems from the European Commission – an arena in which the center has to compete with other institutions for grants for scientific projects. In 2007, eight percent of the funding came from the EU, and in 2016 it was 19 percent. Regarding scientific publications, in 2016 the center delivered an "all-time high" of more than 200 papers, most of them published in prestigious international journals. In addition, a new modeling tool applied and partly developed at the center – the Norwegian Earth System Model – was among the most used and cited models in the world.[88]

A crisis of credibility?

The entry into the interdisciplinary field of climate research turned out to be a successful move for the geophysicists in Bergen. They entered the new stage just as the climate issue started to change character from one environmental concern among many others to the "main challenge facing humanity." However, has the entry into this new field come with a downside? Do the recurring climate disputes and the slow progress in climate policy formation indicate that people have lost faith in the capacity of geophysicists and other climate scientists to calculate the world?

Several scholars seem to consider a "decline in respect of science" a crucial characteristic of the current age. In chapter 8, Paul Forman spoke of a turn from modernity to postmodernity in the late 1960s or early 1970s, a turn involving an abrupt devaluation of science and many of the values and virtues associated with it. Some historians investigating the history of climate research have painted a similar picture. In his book *The discovery of global warming* (2008), Spencer

R. Weart concludes that the discovery of global warming was a "social product," a "limited consensus of judgements arising in countless discussions among thousands of experts." At the same time, he criticizes social scientists and scholars in the interdisciplinary field of science studies for drawing the conclusion that global warming was "*nothing but* a social construction – more like a myth invented by a community than a fact like a rock that you could hold in your hand." Weart found it reasonable to conclude with the IPCC that it was "*very likely* that serious global warming, caused by human actions, is coming in our own lifetimes."[89] The unambiguous message is that it is time to stop questioning the facts presented by the climate scientists, and to move on the next stage of finding appropriate measures to combat global warming. An implicit premise in the argument is that widespread doubts about the climate scientists' findings constitute an important obstacle to moving on to "the next stage."

Paul N. Edwards draws similar conclusions in the book *A vast machine. Computer models, climate data, and the politics of global warming* (2010/2013). Like Weart, Edwards launches an attack on social scientists and other scholars associated with relativism and social constructivism. At an early stage, he argues, scholars in this tradition formulated a fruitful attack on problematic views about science.[90] Yet, he continues, the movement has gone too far: "Science became little more than ideology and group think, with which any belief at all might come to count as knowledge." The result was a "corrosive suspicion of all scientific knowledge." Again, echoing Weart, Edwards ensures the reader that it is time to put skepticism aside. "It is now virtually certain that CO_2 concentrations will reach 550 ppm (the doubling point) sometime in the middle of this century."[91]

Both historians seem to assume that a serious skepticism about climate scientists' capability to calculate the world has taken hold of broad segments of the public, and that this skepticism is an important cause for the slow progress on climate policy formation. The argument corresponds well to Forman's postmodernity thesis. Yet there are at least two good reasons to question this diagnosis. First, as we have seen in this chapter, the issue of global warming has risen from one concern among many others in the late 1980s to one of the main challenges facing humanity and even world peace. That was the status when the Nobel Committee granted the Peace Prize to IPCC and Al Gore in 2007, and it seems reasonable to conclude that the climate issue still holds this privileged position among the various topics of concern circulating in the media and the sites of policy formation. A clear indication can be found on the UN homepage, where it accounts for its efforts to "promote sustainable development." Development involves lifting the poor out of poverty. Achieving this in a sustainable manner means combating climate change:

"Close to 40 per cent of the population of the developing world lived in extreme poverty only two decades ago. Since then, the world has halved extreme poverty, with the UN's Millennium Development Goals greatly contributing to this progress (...) At the same time, as climate change poses a growing challenge to the world's development objectives, the UN supported negotiations to adopt a meaningful and universal global climate agreement in 2015. The UN is also working to develop a financing for development framework to ensure that both the sustainable development agenda and climate action are properly resourced."[92]

The issue of sustainable development has become, and remains, an issue about fighting poverty and fighting climate change. The other issues discussed in *Our Common Future* in 1987, like population growth, increasing dependence on non-renewable resources, food production and food security, conservation of ecosystems, and urbanization are further down the list. Thus, perhaps a more accurate diagnosis is that the facts produced by geophysicists and other climate scientists have never moved policy makers and the public as much as during the most recent decades.

Another reason to question the diagnosis presented by Weart and Edwards is that many controversies regarding global warming are controversies about how to approach the problem of anthropogenic global warming – and not about whether it is real. When politicians and others argue about which means or strategies to adopt to combat climate change, then the existence of the problem is an undisputed premise in the controversy. As shown by historian Kristin Asdal, Norwegian climate controversies during the late 1980s and throughout the

1990s were often controversies about what were the most effective or desirable means to combat global warming. In the early phase, politicians, experts, and various interest groups were engaged in controversies about whether taxes and duties should be introduced to reduce domestic emissions of CO_2. Later, the focus gradually shifted to another means: an internationally negotiated quota regime.[93] Asdal pointed out that economists – not natural scientists – were the main providers of knowledge in these controversies. Importantly, she also observed that the question about the "reality" of global warming was not a topic in these controversies: "The facts about the greenhouse effect, established by natural scientists, appear to have been accepted as an undisputed premise. It was not a part of the discussions or controversies among the various Ministries."[94]

Asdal's analysis is limited to Norway, and to a limited period of time. However, the members of the UN Framework Convention on Climate Change, signed in Rio in 1992 and entering into force in 1994, have met at least once every year since 1995 to try to negotiate agreements on how to combat climate change.[95] Those of us who take the climate scientists' warnings seriously and who would like to see firm political action can only conclude that progress has been slow. Yet it seems premature to ascribe this slow progress to a widespread skepticism about the scientists' capacity to calculate the world. Controversies about which means and strategies to choose may be difficult to close, as the alternatives on the menu may have implications for the distribution of benefits and burdens among countries, industry sectors, and other stakeholders. The stakes are high. Yet the fact that these controversies appear difficult to settle is not necessarily an indication that the participants in the disputes consider the climate scientists' facts mere fictions.

10
Reflections

The occasion for writing this book is the 100th anniversary of the Geophysical Institute in Bergen, an institution built on research traditions that go back even farther in time. Bergen has been a center for a specific way of understanding the world, a geophysical worldview based on the idea that through precise observations in the field and insights from physics, one could use calculations to see into the future. These ideas have spread through international networks, and some of the insights have had a significant impact on geophysics elsewhere, in particular in physical oceanography and meteorology. Bergen is celebrated as the site at which weather forecasting found a scientific basis around 1920, and it was a center for NATO's civil oceanographic research in the 1960s. Since around 1990, the city's geophysical community has worked to become a leading center for climate research, and its remote sensing center has spawned sister institutions around the world.

A 100th anniversary offers a chance to reflect on the origins and development of the activities that are still going on. In this chapter we will tease out some of the long lines in the history of geophysics as seen from Bergen. We will show how many of the same issues have been raised by generation after generation of researchers and how many of the practices remain the same, but also how new questions have appeared or reappeared in new ways. Geophysics in Bergen has focused on the parts of the planet in perpetual motion, and the research has focused on the atmosphere, the oceans, and, for a time, the geomagnetic field and cosmic rays.

Our choice of putting the shared geophysical goal of calculating the world at the center has allowed us to show how meteorology and physical oceanography have shared origins, differentiated into disciplinary specialties with limited contact, and reconnected under the umbrella of climate studies in the 1990s. However, before discussing the historical lines, we will discuss how the geophysical community has understood its own past, focusing on the Bergen school of meteorology.

The Bergen school remembering itself

Geophysicists in Bergen have long a tradition of writing and celebrating their history, especially the Bergen school of meteorology. Efforts to situate the weather forecasting unit in Vilhelm Bjerknes's attic in Bergen as an epoch-making event in the history of meteorology began around the Second World War, when first the Geophysical Institute in Leipzig, and then the Weather Forecasting Office in Bergen, celebrated their 25th anniversaries.[1] In June 1958, a Scandinavian-American congress on air mass and frontal theory attracted some 120 meteorologists, celebrating both the 40th anniversary of the Bergen school and the 30th anniversary of the institute building.[2] Four years later, in October 1962, the 100th anniversary of the birth of Vilhelm Bjerknes was celebrated with visits from Tor Bergeron and Edward Lorenz from MIT, and with a special edition of the journal he had helped found, *Geofysiske publikasjoner*.[3] These and later events, such as a new anthology for the 50th anniversary of the forecasting unit,[4] the international symposium celebrating the 75th anniversary of the Bergen school front cyclone model in 1994,[5] an institutional history written for the 90th anniversary,[6] and various events for the 100th anniversary of the Geophysical Institute in 2017, including this book, reflect how the past is kept alive and relevant.

The Bergen school meteorologists produced different and competing historical narratives explaining their past. The most famous account, and by far the most reproduced, focuses on the biography of Vilhelm Bjerknes, and positions the Bergen school as the beginning of scientific weather forecasting. The story was first told by Bjerknes in 1938 during a celebration of the 25th anniversary of the Geophysical Institute in Leipzig, and again at the 25th anniversary of the weather forecasting unit in Bergen in 1943.[7] In the first presentation, Bjerknes framed weather forecasting as being as old as man's ability for thought and observations, but riddled with all kinds of superstitions. From around 1600 onward, new instruments like Galileo's thermometer heralded the advent of a more quantitative description of the atmosphere. Next came the development of the telegraph network, and the astronomer Urbain Le Verrier's report that the storm that wreaked havoc on the French Navy during the Crimean War in 1854 could have been predicted, if only the weather observations along the network had been put on the same map. This led to a host of national weather services. However, Bjerknes argued, by the turn of the century, the initial wave of optimism had long waned. Meteorologists observed the weather maps, but did not understand them:

"The pioneers of meteorology, who had started out with such great dreams, began to feel discouraged. There was no hope for the progress of weather forecasting. They allowed the weather service, which had already been set up, to run on mechanically, serving day-to-day needs. And their institutes, specially established for forecasting the weather, turned their scientific interests towards a more placid branch of meteorology, namely climatology."[8]

Climatology was, to Bjerknes, a science of a lower order. Instead of investigating the *causes* of weather, climatologists were looking for patterns in the *results* of the weather. In his own narrative, it is Bjerknes's attempt to "attack the forecast problem as a purely scientific one" through the use of physics that saved weather forecasting from intellectual stagnation.[9]

Instead of building on the work of earlier meteorologists, Bjerknes's narrative about the birth of the Bergen school started with his own father's investigations into "action at a distance." While in Göttingen in 1856, the mathematician Carl Anton Bjerknes had asked: "[I]f two bodies move in a liquid, will they not then, through the liquid as intervening medium, mutually affect each other's movements? And will not an observer who sees the bodies but not the liquid, believe that he is witnessing action at a distance?" [10] A decade later, Carl Anton Bjerknes arrived at equations that describe how two bodies moving in a liquid mutually affect each other's movements, and in 1881, together with his then 19-year-old son, Vilhelm, he demonstrated an instrument that reproduced the effect to great praise at the Paris International Electric Exhibition.[11] The displays showed how two harmoniously pulsating balls submerged in a fluid acted upon one another as though they were electrically charged, attracting or repelling each other. The main purpose of the experiment was to create an analog, not just for how material objects interact at a distance, but for electromagnetism.

To Bjerknes, the circulation theorem he first presented at a lecture in Stockholm in 1897 was a direct continuation of his father's work.[12] Instead of being a metaphor for action at a distance, the intervening medium was now interesting in and of itself: by using thermo- and hydrodynamics, one could calculate the movement of fluids in the real world, one step at a time. Earlier physical equations to describe the circulation of fluids, in particular Hermann von Helmholtz's theorem of vorticity conservation and William Kelvin's theorem on the conservation of circulation, had applied to ideal fluids with no viscosity, density, or friction.[13] This meant that circulation and vortex motions were eternal and could neither come into being nor perish – which is exactly what characterizes motion in the atmosphere. Instead, in 1904, Bjerknes suggested that the theorem should consist of seven basic equations, and laid out a program for scientific, or exact, weather forecasting. The principle was simple: "A system's future state can be calculated if I know its state at a certain time and those laws according to which its state changes."[14]

Through the annual grants Bjerknes received from the Carnegie Institution in Washington to pursue his weather forecasting program, which were renewed until the Second World War, he hired assistants, many of whom became leading figures in Norwegian geophysics.[15] Progress was made in Leipzig, but was interrupted by war: One by one, his students and assistants were called up for war service, and five of his ten doctoral students perished. After moving to Bergen with his assistants, he convinced Norwegian Prime Minister Gunnar Knudsen to fund the tightly knit weather observation network along the coast of western Norway to benefit farmers at harvest time:

"I had been clear and firm in my demands. I understood that weather maps at the time were far too summary in the information they gave. I requested that the number of telegraphic weather stations in Southern Norway be increased tenfold, from 9 to 90. Then I hoped, as I expressed it, that the observations would show us *the weather's face*, alluding to the portraits found in newspapers made up solely of dots: ten dots yield no physiognomy, but ten thousand can give the characteristic wrinkles and lines by which a face is known. I was extremely anxious to see if this tenfold increase in the number of stations would be enough to let us see the weather's face."[16]

In Bjerknes's narrative, the dense observation network, the perspective of seeing the atmosphere through the lens of physics, and the intense labor of his young assistants created the Bergen school. And in the process weather forecasting, for the first time, found a scientific basis. But this was not the only historical narrative created by the Bergen school.

In his widely distributed textbook *Introduction to Meteorology* (1941), Sverre Petterssen aligned Bjerknes' biographical approach with the history of science in general. Meteorology, like all other sciences, went through three distinct phases, Petterssen argued. The first phase was characterized by accidental observations and attempts to give a natural explanation of phenomena. For meteorology, this phase began in ancient Greece, when Aristotle published *Meteorology* (350 BC), which among other topics discussed the formation of clouds, rainbows, and ocean currents, as well as how continents change over geological time. The goal of Aristotle's treatise was to find the motions common to the four elements of air and water, earth and fire, and thus to give natural explanations for natural phenomena ranging from rainbows to earthquakes. Such studies eventually awoke sufficient interest for systematic observations to begin bringing science from the first accidental phase to a second, descriptive, phase. In order for science to advance even further on its road to becoming exact, one needed a theoretical breakthrough, where the laws *behind* the phenomena were described: the laws of nature.

"The interest thus created usually results in an organized program of observations, followed by a systematic working up of observations, more or less along statistical lines. Eventually, the theorist tries to extract from the observations and the empirical rules [of] the laws of nature that govern the phenomena, and this marks the transition from the descriptive to the exact epoch in the development of the science in question."[17]

Starting his narrative in antiquity allowed Petterssen to align meteorology with the notion that all sciences are essentially the same, and that they all go through the same linear progress. This perspective concentrates on great men and their contributions to observation, explanation, and prediction: Hippocrates wrote the first treatises on medical climatology around 400 BC, and fifty years later, Aristotle wrote *Meteorology* based on his own observations.[18] The next stage, the era of systematic observations, began when Leonardo da Vinci constructed an improved weather vane and a mechanical moisture indicator in 1500, and when Galileo Galilei made the first instruments for measuring temperature in 1597.[19] Like Bjerknes, Petterssen also mentioned Galilei's student Evangelista Torricelli, who in 1643 built the first mercury barometer, an instrument that measures atmospheric pressure. In a letter written in 1644, Torricelli beautifully summed up the results of his research as follows: "We live submerged at the bottom of an ocean of the element air, which by unquestioned experiments is known to have weight."[20] Finally, the perspective allowed Petterssen to present Vilhelm Bjerknes and the Bergen school as the vanguard of the exact phase, the pinnacle of science.

When Carl Ludvig Godske, in a popular textbook in 1956, presented the history of meteorology to a wider audience, he made subtle but fundamental changes to Petterssen's phases. Instead of Petterssen's "accidental" phase, Godske began with a "descriptive" stage, characterized by collecting numerical observations and the systematic mapping of natural phenomena. Then followed an "analytical or diagnostic" phase, in which science searched for laws of nature to explain how phenomena behaved. For weather forecasting, the development of the telegraph network was a necessary condition for this phase to occur. The third phase was characterized by the ability to make predictions, in particular the Bergen school. However, unlike Petterssen's "exact phase," Godske did not tie the predictions to the use of hydrodynamic methods. He also added a fourth and final phase, "applied science." Instead of merely making predictions, this phase was characterized by the predictive powers being used to stage interventions in nature: "It is in particular triumphs at this stage that justify the proud saying: *knowledge is power.*"[21]

Petterssen stressed that weather forecasting was simply "physics, mathematics, mechanics, and chemistry applied to the atmosphere," and that it had to wait for progress in these sciences in order to progress to the exact stage. Thanks to the Bergen school, however, this was changing: "[I]n view of the rapid advances in recent years, it seems safe to say that the science of meteorology is now in a state of transition from the descriptive to the exact state, although the road to complete exactitude may be long and winding."[22] Godske, on the other hand, prefaced his retrospective by stating that meteorology was in its adolescence, and that it would probably never achieve the status of a "classical" science. Rather than seeing the Bergen school of meteorology as a revolution, Godske saw it as rediscovering new insights: in the 1830s, the German meteorologist Heinrich Wilhelm Dove had discovered that weather was connected to warm and cold air flows. Further, before committing suicide in 1865, the British meteorologist and captain of *HMS Beagle* for Charles Darwin's circumnavigation, Robert FitzRoy, had identified both air masses and the weather fronts: "Yes, he discovered much of what created the Bergen school's renown."[23] While Godske also quoted Vilhelm Bjerknes's own words at length, his role in the birth of the Bergen school was as a team leader, and not a practitioner: "J. Bjerknes, H. Solberg and T. Bergeron constituted the triumvirate of the Bergen school. (...) As good comrades, they worked together, animated like scientific communism."[24] The Bergen school was a practical approach to analyzing weather maps, while the theoretical work was continued only after Vilhelm Bjerknes had left for Oslo. The theoretical Oslo school was continued by his last assistant, Einar Høiland.

What Petterssen and Godske's narratives had in common, in addition to focusing mainly on the internal logic of scientific activities, was that they created a hierarchy where some approaches to meteorological research were considered more advanced than others. Godske's study of minimum temperatures in the Bergen valley, for instance, would in Petterssen's framework be categorized as belonging to the second, descriptive phase, because of its reliance on statistics and maps. This was less advanced and therefore inferior to the hydrodynamic methods Petterssen pursued, which made predictions based on the laws behind the phenomena. In Godske's framework, however, the hierarchy was reversed: not only did the study of minimum temperatures make predictions, but the knowledge about how cold streams would form and flow laid the foundation for intervention, for instance through building small stone walls to divert the cold air away from crops. This belonged to the "applied" stage, and was therefore more advanced than forecasting that could merely predict.

Finally, to Tor Bergeron, the problem to be explained in the history of meteorology was not progress, but the lack thereof. Focusing exclusively on weather forecasting, he argued: "No other dynamic-thermodynamic problems are subject to such a daily and world-wide idle talk and professional attack, and yet within no comparable field are progress and success so modest."[25] The explanation, Bergeron argued, was that forecasting depends on meteorological observations, tools to communicate and make use of the observations, and models of the atmospheric structure to interpret the observations. In order to make progress all these factors had to be linked, which was seldom the case. Rather, "[it] seems to be a rule in our science: Progress is impeded by want of meteorological knowledge on the part of the theoreticians and by a too poor mathematical training of weather-men."[26] When progress did occur, it was not thanks to epoch-making breakthroughs by great men, but by research schools that succeeded in combining observations, tools, and methods in novel ways. This would eventually meet a dead end, a block, where the research school's philosophy became dogma, and it became impossible to recognize observations or facts that did not fit the perspective.[27]

According to Bergeron, the secret to the Bergen school's success was new observations, new tools, and new models combined in new ways. The dense network of surface observations set up in 1918 had made it possible to see details that had hitherto not been possible. Next, the insistence on composite weather maps where *all* observations were utilized in the analysis, rather than making one map for each variable, had turned the weather map into a new kind of tool. In addition, the construction of a three-dimensional model of the life cycle of extratropical cyclones would not have been possible without access to upper-air data from Bjerknes's time in Leipzig. Finally, the sheer lack of experience by the main investigators, Halvor Solberg, Jacob Bjerknes and Bergeron, was a strong advantage: "[T]he three just-mentioned investigators were not beforehand overburdened with meteorological knowledge and therefore were unbiased and open to new ideas."[28]

From Bergeron's perspective, the Bergen school of meteorology was part of a complex continuum, not a solitary revolution bringing science from one phase to another. First, it built on the work at Vilhelm Bjerknes's Leipzig school. Second, it existed in parallel with other research schools elsewhere. In particular, Bergeron highlighted work done in Vienna at more or less the same time, which used a different form of physics. The Bergen school had used Lagrangian physics, which can be compared to describing the flow in a river from the perspective of a floating balloon. The Vienna school had used the physics of Euler, which can be compared to describing the flow of a river from a fixed observation point, or, from their position in the Alps, studying the atmosphere from a fixed mountain observatory.

Finally, Bergeron argued that the Bergen school, like other research schools, eventually reached a dead end, a "block." First, it did not contain the dynamic and thermodynamic principles necessary to calculate the steering and development of weather systems over periods longer than 48 hours. Second, the ideal model was of little assistance when different weather systems interacted, or a cyclone hit an obstacle, such as a mountain. Third, when electronic computers became available after the Second World War, the equations developed by Bjerknes were too complex to be useful. Instead, yet another research school carried the torch: the Chicago school. Under the spiritual guidance of Carl-Gustaf Rossby, this research school introduced simplified general models of large-scale atmospheric movements, including planetary waves in the outer atmosphere (Rossby waves), the jet stream, and methods for mathematical calculations using computers. Instead of using Bjerknes's thermo- and hydrodynamics, the starting point was Rossby's wave formula, treating the atmosphere as having different layers, and calculating the motion in one layer at the time. Instead of seven equations, where the results of each equation were used in the solving of the others, the Chicago school made it possible to use a one-equation system.[29]

The competing narratives were part of ongoing debate about meteorological hierarchies, in which interpretations of the past and the nature of scientific progress were a battleground. To Bjerknes, the future's weather was the only great problem in meteorology. By following in his father's footsteps, and with help from competent and enthusiastic assistants, he had personally heralded a new era of scientific weather forecasting. Petterssen, on the other hand, aligned the progress of meteorology with phases all sciences go through. It was the hydrodynamic methods and hunt for mechanisms, rather than the individuals involved, that had brought meteorology to the "exact" phase. Godske was worried that weather forecasting was overshadowing other problems in meteorology, and consequently, rather than focusing on hydrodynamics, his historical narrative emphasized prediction as a step on the way to applying of meteorological knowledge to changing the behavior of nature. This placed his own research at the pinnacle of science, more advanced than merely being able to predict tomorrow's weather. Finally, Bergeron, whose main interest, after discovering the occluded front, was the formation of precipitation in clouds, argued against the idea of scientific phases, and reflected his experience of different researchers in different places having different interests and building on each other's work.

The historical narratives focusing on a single discipline ultimately served to sever the contact between meteorology and oceanography. From Mohn on the *Vøringen* expedition, to Bjerknes's circulation theorem initially explicitly aimed at both the atmosphere and the oceans, to Helland-Hansen and Nansen's climate study in 1917, and the joint expeditions in the 1920s, meteorology and oceanography had developed hand in hand as specialties within the same geophysical worldview. Yet all narratives about the history of the Bergen school focused solely on meteorology. Since the contemporary links between the atmosphere and the ocean were severed, so were the connections in the past.

Striking continuities

Rather than focusing only on what in hindsight turned out to have been successful, we have in this book strived to reflect what has been seen as important in light of its contemporary context – including efforts that in hindsight may appear like failures or blind alleys. We have shown that the geophysical community finds its roots in expeditions, polar exploration, concerns about fisheries and agriculture, and an ambition of making predictions through treating the world as a giant math puzzle. The orchestrated efforts aimed at making numerical observations of nature, developing physical equations, and through this making predictions through calculations began with Henrik Mohn and the *Vøringen* expedition in the 1870s. However, Mohn failed to build a school with followers. Instead, it was Vilhelm Bjerknes's circulation theorem that today is considered the genesis of the unique strand of geophysics that developed in Bergen. His vision, in essence, was to combine the state of the atmosphere or the oceans at a specific point in time with sufficient insights into the laws of nature. This made it possible, at least in theory, to calculate a new state a short time into the future – which again could be used to make new calculations. Being able to see into the future then only depended on being able to make the calculations faster than nature changed. This belief that the world could be calculated is the running theme in the history of geophysics seen from Bergen. When geophysicists today bring their tools to the field to gather observations, and then bring them back home to do calculations, they are fundamentally doing the same thing as Henrik Mohn did while surrounded by gentlemen scientists on the steamship *Vøringen* in the 1870s.

Over time there have been different opinions about how and what to observe, what and how to calculate, and which parts and at what scales one should study nature. The very first publication from the Geophysical Institute, a climate study by Bjørn Helland-Hansen and Fridtjof Nansen, is a useful illustration: initially their goal was to make seasonal predictions based on surface temperatures of the oceans farther south in the Gulf Stream. The ocean moves more slowly than the atmosphere, and it was thus common sense that the water masses could influence temperatures on land at some point in the future. However, when it turned out that the surface temperatures at sea were mainly decided by the atmosphere and wind direction, not the other way around, the two proceeded to assemble a geophysical world-view that saw the oceans, the atmosphere, the magnetic field, and energy from the sun as integrated and interrelated pieces of the same puzzle. This integrated perspective reached a peak when Ernst Calwagen in 1922 published his unbroken temperature curve from an altitude of

1000 meters to 1200 meters below the ocean surface. Arguably, Karl Falch Wasserfall's seasonal forecasts from the early 1930s, based on comparing periodic variations in the magnetic field and surface temperatures, were an expression of the same integrated perspective.

As we showed in the third chapter of the book, Vilhelm Bjerknes saw the efforts of assembling an integrated geophysical worldview as premature. Rather than looking for patterns and correlations, he argued that geophysicists should investigate the mechanisms "behind" the different physical phenomena; only then could science be exact. In hindsight, this line of reasoning was the most influential. But this focus on mechanisms also contributed to the different disciplines growing apart. Rather than an interesting (but complex and complicated) continuity, the surface of the ocean became a barrier between disciplines. Meteorology concerned itself with the atmosphere; oceanographers were interested in what went on below the surface. There were notable exceptions, such as Harald Ulrik Sverdrup's investigations of the interactions between ice and the atmosphere on the *Maud* expedition and Svalbard; the weather ships combining weather observations with oceanographic measurements producing some of the world's longest consistent time series from the deep ocean; attempts at joint oceanographic/meteorological sessions under the auspices of the International Union of Geodesy and Geophysics; and a small involvement in a Danish project in the Kattegat aimed at investigating ocean-atmosphere interaction in the 1970s. But these were the exceptions, rather than the rule.

Another reason why the disciplines grew apart is found in how they related to the field. Meteorologists relied on setting up observation networks, and spent most of their time analyzing observations done by others. Oceanographers, on the other hand, identified strongly with the practice of going into the field to gather observations. When Carl Ludvig Godske bid weather forecasting and the Bergen school of meteorology farewell and instead focused on weather in landscapes, he shared the oceanographers' emphasis on fieldwork. Still, Godske's landscapes were on land, not at sea. Generally praised for his outreach and teaching, his science has been seen as relatively lackluster and unfocused. His focus on climate research came about at a point in time when this was frowned upon by the weather forecasters as backwards. Yet by bringing the first computer to Bergen, and by arguing that when time scales or geographical scales changed then methods for observation and analysis must also change, he did clear the way for later developments. Starting in the 1980s and 1990s, climate studies have reunited geophysics, requiring vast observations from a variety of sources, time series, supercomputers, and different experts working together. In Bergen, the coupled Bergen Climate Model developed in the 1990s and later used by the International Panel on Climate Change (IPCC) symbolizes how the disciplines have reunited under the umbrella of climate research.

Bergen's influence as a global center for geophysical research has also changed over time. Around the First World War, Bergen was a world capital first for marine science, and then for weather forecasting. However, beginning as early as the 1930s, several of the leading researchers left for new positions elsewhere, in particular the United States. While the Second World War in the United States led to strong links between geophysicists and the military that continued into the Cold War era, in Bergen the war was the beginning of a stagnant period that lasted for 20 years. Although in 1946 the Geophysical Institute was at the center for the new University of Bergen's first building project, the priority was to build new departments – not to strengthen an Institute that was already understood to be a world leader. It was not until 1960 that the combination of the University gearing up for mass education, military funding, and Guro Gjellestad's redefinition of Section C resulted in the renewed expansion of geophysics. Although oceanographer and institute head Håkon Mosby was central in organizing postwar oceanography internationally, the main view is that in a growing field, Bergen became less important. Financial muscle and the ability to attract leading and ambitious minds and have them collaborate, or the lack thereof, were important reasons why the center of gravity shifted elsewhere.

Turning field experiences into numerical observations requires instruments, and the historical trajectory here has been one of professionalization and standardization. Until the 1960s, oceanographers often went on fieldwork with instruments they had developed themselves or in close collaboration with staff technicians. While the Nansen bottle was established as a global standard for taking water samples at fixed points while registering the precise in situ temperature, there were numerous processes they could not measure. Developing instruments for direct current measurements has been another sustained ambition in the history of geophysics in Bergen. This reached a climax with the NATO buoy project, which in the early 1960s led to the development of a world-leading current meter. However, after producing some forty units, in 1966 the instrument gave birth to a commercial company, Aanderaa Instruments. Similar companies were established elsewhere around the same time. Since the 1970s, instruments have mainly been bought off the shelf while only maintenance and adjustments are done in-house. This has produced more and more standardized observations, while the colorful instrument makers in the Institute's workshops have become increasingly redundant.

The current meter illustrates the postwar ambition of putting the planet under constant surveillance, producing detailed time series that require computers to analyze. This went hand in hand with tools for calculations that have become faster and cheaper, and have greatly increased capacities. In addition to doing more complex calculations, the process of getting the observations from the instruments to the computer has been automated. After the Second World War, Godske introduced first the punch card machines, and in 1958, EMMA and Norway's first commercial computer. This soon grew into the University of Bergen's IT department. In the 1980s, when desktop office computers started to proliferate, the IBM center in Bergen provided supercomputers for modeling. In the mid-1990s, the expertise gained resulted in the Bergen Climate Model, BCM. This was a "coupled model" that included both the atmosphere and the oceans.

Despite the importance of geophysicists in Norwegian nation building, personified in particular by Fridtjof Nansen, who in Norway is celebrated as playing a major role in helping the country gain its independence from Sweden in 1905, the research community in Bergen was for a long time financed locally. Over the entrance to the main offices on the first floor of the Geophysical Institute's building (Geofysen) a plaque reads "Erected by the citizens of Bergen." Construction of the building was made possible by private donations, mainly from the town's wealthy business and shipping elite. During his first five years in Bergen, Vilhelm Bjerknes received his salary through private donations to the Bergen Museum, and it seems likely that the weather forecasting unit was set up to prove its practical utility to the benefactors. While private funding provided independence and autonomy, it also made the research community susceptible to shifts in the economy. By 1920, the economy in Bergen had crashed. After the opening of Geofysen in 1928, followed two years later by the establishment of the Chr. Michelsen Institute based on an inheritance from 1925, it would be half a century before the next privately funded expansion took place: the Nansen Remote Sensing Center, established in 1986 to specialize in the application of satellite-based data in studies of the sea and the sea ice in Arctic waters on the basis of external research funding. Meanwhile, the government had become the dominant source of funding through the Norwegian Research Councils set up in the aftermath of the Second World War, as well as through state funding to the University. The Geophysical Commission, established in 1917, which prior to the war coordinated the geophysical budget proposals to the government, slowly lost its relevance, and was discontinued in 1999 after more than a decade of inactivity.[30] While it seems clear that geophysicists over time have lost much influence over what research to fund, this issue needs further research.[31]

Nature is also acting

Another issue we wish we could have explored more deeply is how the geophysicists have observed nature changing over time.[32] From 1900 to 1920, the Norwegian Sea experienced a period of cooling, with lower air and sea temperatures than at any later point in time. From 1920 to 1960 the ocean got warmer and saltier. Over a five-year period, the average ocean temperatures increased by 2° Celsius, and the salinity in the Faroe-Shetland channel increased to such a great extent that the oceanographers wondered, for a period, if the explanation was merely faulty instruments. Next, increased southward transport of ice and freshwater by the East Greenland Current in the late 1960s led to "the great salinity anomaly" that could be traced for fourteen years (1968 to 1982). Finally, the trend from 1970s to the present is that the abyss has also become warmer. The temperature in the deep ocean, once believed to be stable, and even used to calibrate instruments, has increased by 0.1° C. The temperature increase has gone hand in hand with freshwater from rainfall, rivers and melting ice and glaciers that now penetrates more deeply than ever before.[33]

Nowhere on the planet have climatic changes had a greater impact than in the Arctic. Geophysicists explain this by "polar amplification," a phenomenon whereby global temperature increases are magnified near the poles. In the north, Arctic amplification has led the polar ice cap to decrease by about 40 percent between 1979 and 2017. The proportion of thick multiyear ice has decreased even more rapidly, with a dramatic reduction in the ice volume.

In 2004, oceanographers in Bergen suggested that by the end of the 21st century, the Arctic might experience ice-free summers for the first time in human history if the CO_2 concentration doubles.[34] Later model projections have shown that the Arctic might become nearly ice-free as early as 2020.[35] However, the extent of the melting depends on different emission scenarios. Human release of climate gases is estimated to explain approximately 60 percent of the observed rate of decline.[36] The other main factor is natural variation, in particular the Atlantic Multidecadal Oscillation (AMO), which is believed to now be entering a cooling period that will last for some three decades, before heating up again around 2050.[37] Understanding and researching these and similar phenomena, as well as testing and improving the models, requires long time series.

The warming of the Arctic will have significant effects on local ecosystems, fisheries, shipping, petroleum exploration and production, tourism, and naval operations in the area, as well as on weather patterns and the climate on land, but the exact implications are unknown. Beyond expectations of more extreme weather events, ranging from periods of extremely cold weather to heavier rainfall and more frequent and violent storms, the effects of the expansion of ice-free areas are unknown. An important source of uncertainty are the effects of different feedback mechanisms, including storms in the area which will break up the ice cover, causing more rapid melting in the summer. What effect a smaller reflective ice cover will have on the absorption of energy from the sun, and what impact the melting glaciers on Greenland will have on the ocean currents, are other unknowns.

Through field observations and theoretical calculations, geophysicists have identified a number of mechanisms, and continue to do so. Around the turn of the 20th century, Vagn Walfrid Ekman explained the dead water phenomenon and the Ekman spiral. Helland-Hansen and Nansen observed eddies and, for the first time, deep water formation that plays a key part in the global circulation system.[38] In the 1960s and 1970s, geophysicists from Bergen produced new insights into bottom water formation in the Weddell Sea.[39] In the 1990s, they observed for the first time dense 100-meter-wide plumes inside a mesoscale chimney in the Northern Greenland Sea, a phenomenon that until then had only been theorized.[40] Plumes in deep water formation areas consist of distinct water masses passing through other water masses, in this case dense water sinking to more than five hundred meters into the depths, which may be important in creating the bottom water that spreads across all the world's oceans. Researchers continue monitoring the melting Arctic, and work to improve the models for projections, investigating in particular what impact the declining sea ice will have on ocean circulation and extreme weather both in the Arctic and farther south.

Motivations and publications

One of the questions we asked when conducting the interviews for this book was which scientific virtues have motivated the geophysicists in their careers. Four answers dominate: producing new scientific insights, experiencing the insights being applied, educating the next generation, and being recognized for their efforts. A recurring metaphor has been to present science as a brick wall where each slab builds on those below, and makes it possible for others to build further. Hence, a motivation is to find a "loose-fitting brick," and improve it to bolster the scientific structure. Seeing the insights being used in decision making is also a strong motivator. This can range from fishermen deciding to stay ashore to avoid a coming storm that the weather forecasters have predicted to deciding the positioning of windmills to increase their electricity yield. Third, several geophysicists see their role as taking part in a collective quest that spans generations, highlighting educating the next generation.

Yet more common are virtues that put the individual at the center. Having a high h-index, a numerical measurement of publications and citations, is worn as a badge of honor. Many see this as tangible proof that the research is read and referred to by colleagues, and has an impact. Likewise, prizes and honorary positions, memberships in science academies, or prominent positions in institutions and international organizations are proof of high standing in the scientific community. The most exclusive honor is probably the Royal Norwegian Order of Saint Olav, which only seven geophysicists from Bergen have received: Fridtjof Nansen, Bjørn Helland-Hansen, Harald Ulrik Sverdrup, Jonas Ekman Fjeldstad, Håkon Mosby, Yngvar Gjessing, and Ola M. Johannessen.

In the last half century, the collective aspects of science have become increasingly apparent in scientific publications, and both the proportion of coauthored publications and the number of authors per paper has increased dramatically. In the 1960s, 21 percent of the publications were coauthored, with an average of two authors per coauthored publication. In the 1970s, the proportion of coauthorship had increased to 36 percent, and these had on average 2.5 authors. From 1980 to 1988, the proportion was 57 percent and the number of authors 3.6 per publication. Between 2000 and 2010, 71 percent were coauthored, with an average of five authors per publication. In our latest complete publication record from 2010 to 2015, 84 percent of all publications were coauthored, with an average of 6.8 authors per publication. Also the "production volume" has increased: geophysicists in Bergen have published more titles in the last twenty years than in the previous eighty. Huge tomes, decades in the making, have completely disappeared, displaced by shorter journal papers in international journals; the community has expanded markedly, and it seems that today, recruitment is based more on strong publication records than on field practice.

Publication practices have changed both in genre and in the ways in which scientific arguments are made. Yearbook publications, books, and measurement series have virtually disappeared. The reports that dominated from the 1960s to the 1980s now make up less than ten percent of the publications.[41] Since 2000, the proportion of research published in international journals has increased from 30 to 70 percent, and four out of five publications are now written in English and aimed at the international science community. Unlike the first publications that put emphasis on procedure and methods, today's publications put results at center stage. Furthermore, it seems that the use of specialized terminology has increased, in particular in leading geophysical journals with clear limitations on the number of words allowed. While the classical works in oceanography and meteorology, *The Norwegian Sea* (1909) and the original Bergen school paper (1918), were accessible to most readers, there is today a clear divide between publications aimed at colleagues and those aimed at the general public.

Making the invisible visible

While geophysical insights are integral to today's society, there are major differences in how detectable the different applications are. Some aspects are taken for granted to the extent that they are virtually invisible. When we get onboard a plane, we take for granted not only that the pilot is informed about the weather conditions underway, but that the forecasts are accurate and trustworthy. If extreme weather events or floods are about to cause threats to life and property, we expect advance warning. We take for granted that when we order goods from China, they will arrive safely in our ports. Neither the geophysical infrastructure – ranging from satellites and radiosondes, to weather stations on land and buoys at sea, to gliders and other instruments in the deep oceans – nor the supercomputers that calculate the predictions receive much public attention. Only when mistakes are made, such as when we experience rainfall that has not been forecasted, does the invisible become visible.

Climate science, on the other hand, has for decades been under nearly constant public scrutiny and debate, especially when it comes to humanity's impact on the global climate. Although the notion that increased concentrations of CO_2 in the atmosphere can lead to increased temperatures can be traced back to Vilhelm Bjerknes's colleague, Svante Arrhenius, in Stockholm in 1896,[42] it was only in the 1970s that geophysicists began to sound the alarm. In Norway, *Our Common Future* (1987) was what put man-made climate change on the agenda. The following decade, the geophysical community in Bergen was the first in the country to succeed in creating a coupled climate model that again saw the ocean and the atmosphere as a whole. This went hand in hand with attempts to develop technical solutions to pump CO_2 out of the atmosphere and into the deep oceans, but this line of research was later abandoned. Now, the focus is instead on alternative energy sources and the consequences of man-made climate change. Through prolonged effort in making both observations and calculations, geophysicists have made the issue of climate change visible.

The Bjerknes Centre for Climate Research, established in 2000, is the latest institutional addition to the geophysical community in Bergen. Between 2003 and 2012 it received status and funding as a national Center of Excellence, and in 2007 it was the only Nordic research center to contribute computer simulations to the International Panel on Climate Change's fourth assessment report. The main goal of the center is to understand and quantify the climate system in increasing detail, focusing in particular on northern Europe and the polar regions.

In 2017, the Bjerknes Centre for Climate Research moved into the newly refurbished west wing at the Geophysical Institute, which now houses about half the scientists attached to the center. The collaboration now engages 247 people: 127 scientists, 41 postdoctoral researchers, 58 PhD candidates, and 21 technicians and administrators.[43] The climate research community has representatives from 34 countries, and 153 people (62 percent) are non-Norwegians. It is one of the largest climate research units in Europe, and its researchers come from the

Geophysical Institute, the Institute of Marine Research, UNI Research, and the Nansen Environmental and Remote Sensing Center.

In its 100th year, the Geophysical Institute remains a cornerstone for the geophysical research community in Bergen. When visiting the building, it is impossible not to notice its history, whether glancing at the busts and paintings on the walls, when rushing past the exhibition of instruments outside the cafeteria, or when sitting down to eat underneath framed historical weather maps. Bergen is one of the places where the belief that it is possible to calculate the future through observations and calculations based on insights from physics was born. While the instruments for observations, the theoretical framework, and the tools and methods used for calculations have all changed through time, this fundamental goal of calculating the world remains the same. After a hundred years, it can be said that this undertaking has been successful to the extent that it has become almost invisible. Many of the issues that geophysicists continue to research, climate change in particular, are also invisible to the naked eye. This does not make these topics less important. It is our hope that this book has contributed to rendering the invisible visible.

Appendix

Bibliography

**Ahlmann, Hans W:son and Bjørn Hel-
land-Hansen.** "Sambandet mellan konti-
nentala nivåförändringar, Norsk-hafvets
oceanografi och de pleistocena inland-
sisarna omkring detta haf." *Geologisk
Förenings Förhandlingar.* Vol. 40, no. 4.
1918: 783–792.

Amundsen, Roald. *The South Pole. An Account
of the Norwegian Antarctic Expedition in
the "Fram" 1910–1912, Vol. 2.* Translated by
A.G. Chater. John Murray, London. 1912.

Andersen, Peter. "An application of the
H. Johnsen weather types as discriminant
functions for the Bergen precipitation."
*Årbok for Universitetet i Bergen. Mat.-
Naturv. Serie.* No. 8. 1969: 1–20.

Andersen, Peter. "First order empirical probabilities of transition of the H. Johansen
weather types, 1904–1957." *Årbok for
Universitetet i Bergen. Mat.-Naturv. Serie.*
No. 2. 1972: 1–20

Andersen, Peter. "On the synoptic-scale distribution of cloudiness over southern Norway
in relation to prevailing H. Johansen
weather types." *Årbok for Universitetet i Bergen. Mat.-Naturv. Serie.* No. 5. 1973a: 1–22.

Andersen, Peter. "The distribution of monthly
precipitation in southern Norway in
relation to prevailing H. Johansen weather
types." *Årbok for Universitetet i Bergen.
Mat.-Naturv. Serie.* No. 1. 1973b: 1–20.

Andersen, Peter. "Surface winds in southern Norway in relation to prevailing

H. Johansen weather types." *Meteorologiske
Annaler.* Vol. 6, no. 14. 1975: 377–399.

Annual Reports for the Geophysical Institute,
1917–1988. Until 1946, the Annual Reports
were printed in the Bergen Museum
yearbooks, afterwards in the University of
Bergen yearbooks.

Arrhenius, Svante. "On the Influence of Carbonic Acid in the Air upon the Temperature
of the Ground." *The London, Edinburgh and
Dublin Philosophical Magazine and Journal
of Science.* Series 5, Volume 41. April 1896:
237–276.

Arrhenius, Svante August. *Lehrbuch der
kosmischen Physik.* Zweiter Teil. Hirzel,
Leipzig. 1903.

Asdal, Kristin. *Politikkens natur – Naturens
politikk.* Universitetsforlaget, Oslo. 2011.

Auerbach, Max. "Bericht über die Expedition
des 'Armauer Hansen'." *Verhandlungen des
Naturwissenschaftlichen Vereins in Karls-
ruhe.* Vol. 26. 1914: 3–54.

Baker, James. "Ocean Instruments and Experimental Design." In: Warren 1981: 396–433.

Barnes, Harold. *Oceanography and Marine
Biology. A book of techniques.* Ruskin
House, London. 1959.

Ben-David, Joseph. *Fundamental research
and the universities. Some comments on
international differences.* Organization for
Economic Co-operation and Development,
Paris. 1968.

Bengtson, Lennart. "Some historical aspects of extra-tropical cyclone research." In: Grønås and Shapiro (eds.) 1994.

Benson, Keith and Helen M. Rozwadowski (eds.). *EXTREMES: Oceanography's Adventures at the Poles.* Science History Publications / Watson Publishing International. Sagamore Beach, USA. 2007.

Benum, Edgar. "Et nytt forskningspolitisk regime? Grunnforskningen, OECD og Norge 1965–1972." *Historisk tidsskrift.* Vol. 86, no. 4. Universitetsforlaget. 2007: 551–574.

Bergeron, Tor. "Methods in Scientific Weather Analysis and Forecasting: An Outline in the History of Ideas and Hints at a Program." In: Bolin 1959: 440–474.

Bergeron, Tor. "The Bergen school." *Geofysiske Publikasjoner. Special issue: In memory of Vilhelm Bjerknes on the 100th anniversary of his birth.* Vol. 24. 1962: 16–21.

Bergeron, Tor. "Synoptic meteorlogy: A historical review." *Pure and Applied Geophysics.* Vol. 119. 1981: 443–473.

Bergeron, Tor; Olaf Devik and Carl Ludvig Godske. "In memory of Vilhelm Bjerknes on the 100th anniversary of his birth." *Geofysiske Publikasjoner.* Vol. 24. 1962: 1–37.

Berntsen, Drude. "The pioneer era in Norwegian scientific computing (1948–1962)." In: Bubenko 2003: 23–32.

Bigelow, Frank. "Storms, Storm Tracks and Weather Forecasting." *Weather Bureau Bulletin No. 20.* U.S. Department of Agriculture / Weather Bureau, Washington, D.C. 1897.

Bjerknes Centre for Climate Research. Annual Reports 2003–2016. https://bjerknes.uib.no/artikler/arsrapporter

Bjerknes Centre for Climate Research. CoE application to the Research Council of Norway (Application deadline January 11, 2002). Unpublished document. Copy borrowed from Nils Gunnar Kvamstø.

Bjerknes, Ernst. *Med ski, velosiped og skissebok. Minner fra åtti-årenes Norge.* Oslo. 1944.

Bjerknes, Jacob. "Über die Fortbewegung der Konvergenz- und Divergenzlinien." *Meteorologische Zeitschrift.* 1917: 345–349.

Bjerknes, Jacob. "On the structure of moving cyclones." *Geofysiske Publikationer.* Vol. 1, no. 2. 1921: 1–8.

Bjerknes, Jacob. "Diagnostic and Prognostic Application of Mountain Observations." *Geofysiske Publikasjoner.* Vol. 3, no. 6. 1926: 3–38.

Bjerknes, Jacob. "Aktuelle forskningsoppgaver i meteorologien." *Universitetets radioforedrag.* Aschehoug, Oslo. 1930.

Bjerknes, Jacob. "Exploration de quelques perturbations atmosphériques à l'aide de sondages rapprochés dans le temps." *Geofysiske Publikasjoner,* Vol 9, no. 9. 1933: 3–54.

Bjerknes, Jacob. "Die Theorie der aussertropischen Zyklonenbildung." *Meteorologische Zeitschrift.* Vol. 54. 1937: 460–466.

Bjerknes, Jacob. and Halvor Solberg. "Meteorological Conditions for the formation of rain." *Geofysiske Publikasjoner.* Vol. 2, no. 3. 1923: 3–61.

Bjerknes, Jacob and Carl Ludvig Godske. "On the Theory of Cyclone Formation at Extra-Tropical Fronts." *Astrophysica Norvegica.* Vol. 1, no. 6. Oslo. 1936: 199–235.

Bjerknes, Jacob and Erik Palmén. "Aerologische Analyse einer Zyklone. Case 1: March 28–30, 1928." *Beiträge zur Physik der freien Atmosphäre.* Band 21, heft 1. 1933.

Bjerknes, Jacob and Erik Palmén. "Investigations of Selected European Cyclones by Means of Serial Ascents. Case 3: December 20–31, 1930." *Geofysiske Publikasjoner.* Vol. 11, no. 4. 1935: 3–15.

Bjerknes, Jacob and Erik Palmén. "Investigations of selected European cyclones by means of serial ascents. Case 4: February 15–17, 1935." *Geofysiske Publikasjoner.* Vol. 12, no. 3. 1937: 5–62.

Bjerknes, Jacob and Halvor Solberg. "Life Cycle of Cyclones and the Polar Front Theory of Atmospheric Circulation." *Geofysiske Publikasjoner.* Vol. 3, no. 1. 1926: 3–18.

Bjerknes, Vilhelm. "Über die Dämpfung schneller electrischer Schwingungen." *Annalen der Physik und Chemie*. Vol. 44. 1891a: 74–79.

Bjerknes, Vilhelm. "Über die Erscheinung der multiplen Resonanz elektrischer Wellen." *Annalen der Physik und Chemie*. Vol. 44. 1891b: 92–101.

Bjerknes, Vilhelm. "Über den Zeitlichen Verlauf der Schwingungen im primären Hertz'schen Leiter." *Annalen der Physik und Chemie*. Vol. 44. 1891c: 513–526.

Bjerknes, Vilhelm. "Die Resonanzerscheinung und das Absorptionsvermögen der Metalle für die Energie electrischer Wellen." *Annalen der Physik und Chemie*. Vol. 47. 1892: 69–76.

Bjerknes, Vilhelm. "Verschiedene Formen der multiplen Resonanz." *Annalen der Physik und Chemie*. Vol. 55. 1895a: 58–63.

Bjerknes, Vilhelm. "Über electrische Resonanz." *Annalen der Physik und Chemie*. Vol. 55. 1895b: 121–169.

Bjerknes, Vilhelm. "Über einen hydrodynamischen Fundamentalsatz und seine Anwendung besonders auf die Mechanik der Atmosphäre und des Weltmeeres." *Kongl. Svenska Vetenskaps-Akadademiens Handlingar*. Vol. 31, no. 4. 1898: 1–35.

Bjerknes, Vilhelm. "Les actions hydrodynamiques à distance d'après la théorie de C.A. Bjerknes." *Rapports présentés au Congrès International de Physique réuni à Paris en 1900*. Paris. 1900: 251–276.

Bjerknes, Vilhelm. *Vorlesungen über hydrodynamische Fernkräfte nach C.A. Bjerknes' Theorie*. Leipzig. Vol. 1, 1900; Vol. 2, 1902.

Bjerknes, Vilhelm. "Das Problem der Wettervorhersage, betrachtet vom Standpunkte der Mechanik und der Physik." *Meteorologische Zeitschrift*. Vol. 21. 1904: 1–7.

Bjerknes, Vilhelm. *Fields of Force. A course of lectures in mathematical physics delivered December 1 to 23, 1905*. Columbia University Press. 1906.

Bjerknes, Vilhelm. "Preface." In: Nansen 1906: 1–2.

Bjerknes, Vilhelm. "Veirforutsigelse: Foredrag ved Geofysikermøtet i Gøteborg 28. august 1918." *Naturen*. 1919: 321–246.

Bjerknes, Vilhelm. "Om forutsigelse av regn. Foredrag ved Videnskapsselskapets aarsmøte i Kristiania 3dje mai 1919." *Naturen*. 1920: 321–346.

Bjerknes, Vilhelm. "On the dynamics of the circular vortex: with applications to the atmosphere and atmospheric vortex and wave motions." *Geofysiske publikasjoner*. Vol. 2, no. 4. 1921: 1–88.

Bjerknes, Vilhelm. "On quasi-static wavemotion in barotropic fluid strata." *Geofysiske Publikasjoner*. Vol. 3, no. 2. 1926: 3–24.

Bjerknes, Vilhelm. "Leipzig – Bergen: jubilee address on the 25th anniversary of the Geophysical Institute of the University of Leipzig (1938)." Translated by Lisa Shields. *Met Éireann*. Historical Note No. 2. 1997 [1938]: 1–17.

Bjerknes, Vilhelm and Johan Wilhelm Sandström. "Über die Darstellung des hydrographischen Beobachtungsmaterial durch Schnitte, die als Grundlage der theoretischen Diskussion der Meerescirkulation und ihrer Ursachen dienen können." *Kungl. Vetenskaps- och Vitterhets-Samhället i Göteborgs Handlingar*. Vol. 4. 1901.

Bjerknes, Vilhelm and Johan Wilhelm Sandström. *Dynamic Meteorology and Hydrography. 1. Statistics*. Carnegie Institution of Washington, Washington, D.C. 1910.

Bjerknes, Vilhelm; Jacob Bjerknes; Halvor Solberg and Tor Bergeron. *Physikalische Hydrodynamik mit Anwendung auf die dynamische Meteorologie*. Springer-Verlag, Berlin/Heidelberg. 1933.

Bjerknes, Vilhelm; Theodor Hesselberg; Olaf Devik and Johan Wilhelm Sandström. *Dynamic Meteorology and Hydrography. 2. Kinematics*. Carnegie Institution of Washington, Washington, D.C. 1911.

Bjerknes, Vilhelm; Theodore Hesselberg; Carl Ludvig Godske and Finn Spinnangr. *Vervarslinga på Vestlandet 25 år: festskrift utgitt i anledning 25-års jubileet 1. juli 1943*. John Griegs Boktrykkeri, Bergen. 1944.

Bock, Ortwin and Karen B. Helle. *Fridtjof Nansen and the Neuron*. Bodoni forlag, Bergen. 2016.

Bolin, Bert (ed.). *The Atmosphere and the Sea in Motion: Scientific Contributions to the Rossby Memorial Volume*. Rockefeller Institute Press, New York. 1959.

Bolin, Bert (ed.). *The greenhouse effect, climatic change and ecosystems*. Wiley. Chichester. 1986.

Borge, Aagot. *Diary*. Gunnar Ellingsen's private archive.

Broch, Harald Beyer. "'Du må ha troen for å lykkes.' Tro og praksis blant nordnorske kystfiskere i det 21. Århundre." *Tidsskrift for kulturforskning*. No. 1, 2016: 73–90.

Brunt, David. *Physical and Dynamical Meteorology*. Cambridge University Press. 2011 [1939].

Bubenko Jr., Janis; John Impagliazzo and Arne Sølvberg (eds.). *History of Nordic Computing: Conference proceedings*. IFI-PACIT, vol. 174. Springer, USA. 2003.

Byers, Horace B. "Carl-Gustaf Arvid Rossby, 1898–1957. A Biographical Memoir." *National Academy of Sciences Biographical Memoir*. National Academy of Sciences, Washington, D.C. 1960: 246–270.

Bø, Reidar; Arne Foldvik and Thor Kvinge. "Retrieval of Current Meters and Study of Thermohaline Convection in the Weddel Sea." *Antarctic Journal*. 1973: 147–148.

Bøyum, Gunnvald. "The energy exchange between sea and atmosphere at ocean weather station M, I and A." *Geofysiske Publikasjoner*. Vol. XXVI, no. 7. 1965: 1–19.

Carsey, Frank D. (ed.). Microwave Remote Sensing of Sea Ice. *Geophysical Monograph Series*. Vol. 68. 1992.

Charney, Jule. "On the Scale of Atmospheric Motions." *Geofysiske Publikasjoner*. Vol. 17, no. 2. 1948.

Charney, Jule; Ragnar Fjørtoft and John von Neumann. "Numerical integration of the barotropic vorticity equation." *Tellus*. Vol. 2, no. 4. 1950: 237–254.

Coen, Deborah R. "Scaling Down: The 'Austrian' Climate Between Empire and Republic." In: Fleming et al. 2006: 115–140.

Collett, Jon Peter. *Making Sense of Space. The History of Norwegian Space Activities*. 1995.

Committee on Oceanography. *Oceanography 1960 to 1970: 12. Marine sciences in the United States 1958*. National Academy of Sciences/National Research Council, Washington, D.C. 1959.

Conway, Eric M. "Drowning in data: Satellite oceanography and information overload in the Earth Sciences." *Historical Studies in the Physical and Biological Sciences*. Vol. 37, no. 1. 2006: 127–151.

Dahl, Odd and Thor Ramberg. *Med muldyr og kano gjennem tropisk Sydamerika*. Gyldendal Norsk Forlag, Oslo. 1927.

Dahl, Odd. "The Capability of the Aanderaa Recording and Telemetering Instrument." *Progress in Oceanography*. Vol. 5. 1969: 103–106.

Dahl, Odd. *Trollmann og rundbrenner*. Gyldendal Norsk Forlag, Oslo. 1981: 153

Dahl, Per F. *From Nuclear Transmutation to Nuclear Fission, 1932–1939*. Institute of Physics Publishing, Bristol and Philadelphia. 2002.

Dahle, E.A. and O. Kjærland. "The capsizing of M/S Helland-Hansen. The investigation and recommendations for preventing similar accidents." *Norwegian Maritime Research*. Vol. 8, no. 3. 1980: 2–13.

Damas, Désiré. *La croisière atlantique de l'Armauer Hansen, mai-juin 1922*. H. Vaillant-Carmanne, Liège. 1922.

Dannevig, Petter. *Norsk Geofysisk Forening 1917–1991*. Oslo. 1991.

Day, Deborah. "A History of 'The Oceans: Their Physics, Chemistry and General Biology.'" *Scripps Institution of Oceanography Archives*. 2003.

De Figueiredo, Mark Anthony. *The Hawaii Carbon Dioxide Ocean Sequestration Field Experiment: A Case Study in Public Perceptions and Institutional Effectiveness*. Master Thesis. Massachusetts Institute of Technology. Technology and Policy

Program, Massachusetts Institute of Technology. Dept. of Civil and Environmental Engineering. MIT. June 2003. http://hdl.handle.net/1721.1/16929

Deacon, George E.R. "H. Mosby. The Waters of the Atlantic Antarctic Ocean..." *ICES Journal of Marine Science*. Vol. 10, no. 3. 1935: 326–7.

Deacon, Margaret; Tony Rice and Colin Summerhayes (eds.). *Understanding the Oceans: A Century of Ocean Exploration*. Routledge, London/New York. 2003 [2001].

Defant, Albert; Gerhard Neumann; Bert Schröder and Georg Wüst. "Aus den wissenschaftlichen Ergebnissen der Internationalen Golfstrom-Unternehmung 1938." *Annalen der Hydrographie und Maritimen Meteorologie*. Hamburg. Published over five issues, 1940–1941.

Defant, Albert. "Die absolute Topographie des physikalischen Meeresniveaus und die Druckflaten sowie die Wassenbewegung im Atlantischen Ozean." *Wissenschaftliche Ergebnisse der deutschen atlantischen Expedition auf dem Forschungs- und Vermessungsschiff "Meteor", 1925–1927*. De Gruyter, Berlin. Vol. 6. 1941: 191–260.

Den interdepartementale klimagruppen. "Drivhuseffekten. Virkninger og tiltak. Rapport fra Den interdepartementale klimagruppen." Miljøverndepartementet, Oslo. 1991.

Den norske kyststrøm. "Some preliminary results from a synoptic experiment in the Norwegian Coastal Current (SEX 75)." Geofysisk institutt, Universitetet i Bergen. 1976.

Den norske kyststrøm. "Currents and hydrography in the Norwegian Coastal Current off Utsira during JONSDAP-76." Geofysisk institutt, Universitetet i Bergen. 1977.

Den norske kyststrøm. "Remote Sensing Experiment in the Norwegian Coastal Waters: Spring 1979." Geofysisk Institutt, Universitetet i Bergen. 1979.

Devik, Olaf. "Vilhelm Bjerknes: Family background. Education, student of Hertz." *Geofysiske Publikasjoner. Special issue: In memory of Vilhelm Bjerknes on the 100th anniversary of his birth*. Vol. XXIV. 1962: 7–10.

Devik, Olaf and Peter Thrane. *Norsk Geofysisk forening 1917–1967*. Geofysisk Forening, Oslo. 1967.

Dickson, Bob and Svein Østerhus. "One hundred years in the Norwegian Sea." *Norsk Geografisk Tidsskrift*. Vol. 61, no. 2. 2007: 56–75.

Doel, Ronald E. "Constituting the Postwar Earth Sciences: The Military's Influence on the Environmental Sciences in the USA after 1945." *Social Studies of Science*. No. 5. 2003: 645–666.

Drange, Helge. "An isopycnic coordinate carbon cycle model for the North Atlantic; and the possibility of disposing of fossil fuel CO_2 in the ocean." PhD thesis. Geophysical Institute, University of Bergen. 1994.

Drange, Helge (ed.). *The Nordic Seas: An Integrated Perspective*. Geophysics Monograph Series, Vol. 158. American Geophysical Union, Washington, D.C. 2005.

Drivenes, Einar Arne and Harald Dag Jølle (eds.). *Norsk polarhistorie, bind 1: Ekspedisjonene*. Gyldendal Norsk Forlag, Oslo. 2004a.

Drivenes, Einar-Arne and Harald Dag Jølle (eds.). *Norsk polarhistorie, bind 2: Vitenskapene*. Gyldendal Norsk Forlag, Oslo. 2004b.

Dubois, John L.; Robert P. Multhauf and Charles A. Ziegler. *The Invention and Development of the Radiosonde with A Catalog of Upper-Atmospheric Telemetering Probes in the National Museum of American History, Smithsonian Institution*. Smithsonian Institution Press, Washington, D.C. 2002.

Døssand, Atle; Arnljot Løseth and Åsa Elstad. "Ekspansjon i eksportfiskeria 1720–1880." *Norges fiskeri- og kysthistorie* Vol. III. 2014.

Edwards, Paul N. *A Vast Machine. Computer Models, Climate Data, and the Politics of Global Warming*. MIT Press, USA. 2010/2013.

Egeland, Alv and William J. Burke. *Carl Størmer, Auroral Pioneer.* Springer Astrophysics and Space Science Library, Vol. 393. Springer-Verlag, Berlin/Heidelberg. 2013.

Eggvin, Jens. "The Movements of a Cold Water Front. Temperature Variations along the Norwegian Coast based on Surface Thermograph Records." *Report on Norwegian Fishery and Marine Investigations.* Vol. VI, no. 5. 1940: 1–153.

Ehrmann, Werner and Manfred Wendisch (eds.). *Geophysics and Meteorology at the University of Leipzig. On the Occasion of the 100th Anniversary of the Foundation of the Geophysical Institute in 1913.* Leipziger Universitätsverlag GmbH, Leipzig. 2013: 14.

Ekman, Vagn Walfrid. "On the influence of the earth's rotation on ocean-currents." *Arkiv för Matematik, Astronomi och Fysik.* Bd. 2, no. 11. Almqvist & Viksells Boktryckeri, Uppsala. 1905: 1–52.

Ekman, Vagn Walfrid. "On Dead-Water: Being a description of the so-called phenomenon often hindering the headway and navigation of ships in Norwegian Fjords and elsewhere, and an experimental investigation of its causes etc." In: Nansen 1906: 1–152.

Ekman, Vagn Walfrid. "Die Zusammendrückbarkeit des Meerwassers nebst einigen Werten für Wasser und Quecksilber." *Conseil permanent international pour l'exploration de la mer.* No. 43. 1908: 1–47.

Ekman, Vagn Walfrid. "On a new Repeating Current-meter." *ICES: Publications de circonstance.* No. 91, Copenhagen. 1926: 1–32.

Ekman, Vagn Walfrid. "Fridjof Nansen. 10/10 1861–13/5 1930." *Svensk Geografisk Årsbok 1930.* Sydsvenska geografiska sällskapet, Lund. 1931: 170.

Ekman, Vagn Walfrid and B. Helland-Hansen. "Measurement of Ocean Currents: Experiments in the North Atlantic." *Kungliga Fysiografiska Sällskapets i Lund förhandlingar.* Bd. 1, no. 1. 1931. [offprint]

Eliassen, Arnt; Duncan C. Blanchard and Tor Bergeron. "The Life and Science of Tor Bergeron." *Bulletin of the American Meteorological Society.* Vol. 39, no. 4. April 1978: 387–392.

Eliassen, Arnt. "Vilhelm Bjerknes and his students." *Annual Review of Fluid Mechanics.* Vol. 14. 1982: 1–11.

Eliassen, Arnt. "Vilhelm Bjerknes' early studies of atmospheric motions and their connection with the cyclone model of the Bergen school." In: Grønås and Shapiro 1994: 2–12.

Eliassen, Arnt. "Jacob Aall Bonnevie Bjerknes, 1897–1975. A Biographical Memoir." *National Academy of Sciences Biographical Memoir.* National Academies Press, Washington, D.C. 1995: 3–21.

Ellingsen, Gunnar. "Instrumentutvikling med NATO-bistand." In: Hovland 2007: 102–115.

Ellingsen, Gunnar. *Varme havstrømmmer og kald krig. 'Bergensstrømmåleren' og vitenskapen om havstrømmer fra 1870-årene til 1960-årene.* PhD thesis, University of Bergen. 2013.

Ellingsen, Gunnar. "'I just gave the right kind of maps to the right young men...' About narratives and beginnings in the history of meteorology." *Manuscript,* submitted in 2017.

Enebakk, Vidar. "Hansteen's magnetometer and the origin of the magnetic crusade." *British Journal for the History of Science.* Vol. 47, no. 4. 2014: 587–608.

Faugli, Per Einar; Arne H. Erlandsen and Olianne Eikenæs (eds.). "Inngrep i vassdrag; konsekvenser og tiltak – en kunnskapsoppsummering." *NVE Publikasjon Nr. 13.* 1993.

Fjeldstad, Jonas Ekman. "Foreløbig beretning om resultatene av Mauds reise langs Sibirien." *Naturen.* 1923a. [Offprint]

Fjeldstad, Jonas Ekman. "Litt om tidevandet i Nordishavet." *Naturen.* 1923b. [Offprint]

Fjeldstad, Jonas Ekman. "Contribution to the dynamics of free progressive tidal waves." *Norwegian North Polar Expedition (1918–1925). Scientific results, Vol. 4, Oceanography.* Geofysisk Institutt Special Reports no. 3, Bergen. 1929.

Fjeldstad, Jonas Ekman. "Interne wellen." *Geofysiske publikasjoner.* Vol. 10, no. 6. 1935: 3–35.

Fjeldstad, Jonas Ekman. "Internal waves of tidal origin." *Geofysiske publikasjoner.* Vol. 25, no. 5a & 5b. 1964.

Fjørtoft, Ragnar. "Graphical integration of the barotropic vorticity equation." *Videnskaps-akademiets institutt for vær- og klimaforskning: Rapporter og oversikts-artikler 1. juli 1951–30. juni 1952.* Report no. 6. 1952.

Fleming, James R. *Weathering the Storm. Sverre Petterssen, the D-Day Forecast, and the Rise of Modern Meteorology.* University of Chicago Press/American Meteorological Society. 2001.

Fleming, James R. "Sverre Petterssen, the Bergen School, and the Forecasts for D-Day." *History of Meteorology.* Vol. 1, no. 1. 2004: 75–83.

Fleming, James R.; Vladimir Jankovic and Deborah Coen (eds.). *Intimate Universality. Local and Global Themes in the History of Weather and Climate.* Science History Publications, Sagamore Beach. 2006.

Fleming, James R. *Inventing Atmospheric Science. Bjerknes, Rossby, Wexler and the Foundations of Modern Meteorology.* MIT Press, Cambridge, Massachusetts. 2016.

Foldvik, Arne; Tor Gammelsrød and Tor Tørresen. "Hydrographic observations from the Weddell Sea during the Norwegian Antarctic Research Expedition 1976/77." *Polar Research.* Vol. 3, no. 2. 1985a: 177–193.

Foldvik, Arne; Tor Gammelsrød and Tor Tørresen. "Physical oceanography studies in the Weddell Sea during the Norwegian Antarctic Research Expedition 1978/79." *Polar Research.* Vol. 3, no. 2. 1985b: 195–207.

Foldvik, Arne; Tor Gammelsrød and Tor Tørresen. "Circulation and water masses on the southern Weddell Sea shelf." In: Jacobs 1985: 5–20.

Foldvik, Arne et al. "Ice shelf water overflow and bottom water formation in the southern Weddell Sea." *Journal of Geophysical Research.* Vol. 109. 2004.

Forland, Astrid and Anders Haaland (eds.). *Universitetet i Bergens historie. Bind. 1.* Universitetet i Bergen/Bergens Museum, Bergen. 1996a.

Forland, Astrid and Edgar Hovland (eds.) *Forskningsformidling gjennom 75 år. Selskapet til Vitenskapenes Fremme 1927–2002.* Bergen Museums Skrifter nr. 13. Bergen Museum/Universitetet i Bergen. 2002.

Forman, Paul. "On the Historical Forms of Knowledge Production and Curation: Modernity Entailed Disciplinarity, Postmodernity Entails Antidisciplinarity." *Osiris.* Vol. 27. 2012: 56–97.

Frankel, Henry R. *The Continental Drift Controversy, Vol. I–IV.* Cambridge University Press, Cambridge. 2012.

Freeland, Howard J. (ed.). *The origins of the Institute of Ocean Sciences.* Institute of Ocean Sciences. 2003: 1–11.

Friedman, Robert Marc. "Nobel Physics Prize in perspective." *Nature.* Vol. 292, August 27, 1981: 793–798.

Friedman, Robert Marc. *Appropriating the weather: Vilhelm Bjerknes and the construction of a modern meteorology.* Cornell University Press, Ithaca, NY. 1989.

Friedman, Robert Marc. *The Expeditions of Harald Ulrik Sverdrup: Contexts for Shaping and Ocean Science.* Scripps Institute of Oceanography Publication Series, University of California, San Diego. 1994.

Friedman, Robert Marc. "Civilization and National Honour. The Rise of Norwegian Geophysical and Cosmic Science." In: Collett 1995: 28–37.

Friedman, Robert Marc. "Polar Dreams and California Sardines: Harald Ulrik Sverdrup and the study of ocean circulation prior to World War II." In: Deacon et al. 2003 [2001]: 158–172.

Friedman, Robert Marc. "Nansenism." In: Drivenes and Jølle 2004b: 107–173.

Frouin, Robert; Catherine Gautier; Kristina B. Katsaros and Richard J. Lind. "A Comparison of Satellite and Empirical Formula Techniques for Estimating Insolation over the Oceans." *Journal of Applied Meteorology*. Vol. 27, no. 9. 1988: 1016–1023.

Fujiwhara, Sakuhei. "The Men who had a Deep Insight into Nature – The birth of the Norwegian School." *Astronomy and Meteorology* (Japanese popular science magazine), August 1949. [Translated offprint]

Fulsås, Narve. "En æressag for vor nation." In: Drivenes and Jølle 2004a: 173–209.

Furevik, Tore; Mats Bentsen; Helge Drange; Ina T. Kindem; Nils Gunnar Kvamstø and Asgeir Sorteberg. "Description and evaluation of the Bergen Climate Model: ARPEGE coupled with MICOM." *Climate Dynamics*. No. 21. 2003: 27–51. http://bora.uib.no/bitstream/handle/1956/424/BCCR-furevik.pdf

Gade, Herman G. "Some hydrographic observations of the inner Oslofjord during 1959." *Hvalrådets skrifter*. No. 46. 1963: 1–62.

Gade, Herman G. "Hydrografi. Oslofjorden og dens forurensningsproblemer. 1. Undersøkelser 1962–1965." *Norsk Institutt for Vannforskning, Delrapport 12*. Oslo. 1967: 1–104.

Gade, Herman G. "Melting of Ice in Sea Water: A Primitive Model with Application to the Antarctic Ice Shelf and Icebergs." *Journal of Physical Oceanography*. Vol. 9, no. 1. 1979: 189–198.

Gade, Herman G. "Features of fjord and ocean interaction." In: Hurdle 1986: 183–189.

Gade, Herman G. "When ice melts in sea water: A review." *Atmosphere-Ocean*. Vol. 31, no. 1. 1993: 139–165.

Gammelsrød, Tor; Svein Østerhus and Øystein Godøy. "Decadal variations of ocean climate in the Norwegian Sea observed at Ocean Station 'Mike'" (66°N 2°E). *ICES Marine Science Symposia: Hydrobiological Variability in the ICES Area, 1980–1989*. Vol. 195. 1992: 68–75.

Geison, Gerald L. "Scientific Change, Emerging Specialties and Research Schools." *History of Science*. Vol. 19, no. 3. 1981: 20–40.

Geison, Gerald L. and Frederic L. Holmes (eds.). *Research schools: Historical Reappraisals*. *Osiris*, special issue. Vol. 8. 1993.

Gibbons, Michael; Camille Limoges; Helga Nowotny; Simon Schwartzman; Peter Scott and Martin Trow. *The new production of knowledge. The dynamics of science and research in contemporary societies*. Sage. Los Angeles, CA. 1994.

Gjessing, Yngvar Trygvassonn. "Local climates and growth climates of the Sognefjord region. The ratiation climate." *Meteorologiske annaler*. Vol. 5, no. 10. 1969.

Godske, Carl Ludvig. "Die Störungen des Zirkularen Wirbels Einer Homogen-Inkompressiblen Flüssigkeit." *University Observatory, Oslo, Publication No. 11*. Avhandlinger utgitt av Det Norske Vitenskaps-Akademi I Oslo. 1. Matematisk-naturvitenskapelig klasse, no. 1. 1934.

Godske, Carl Ludvig. "On the Minimum Temperatures in the Bergen Valley." *Bergen Museums Årbok: Naturvitenskapelig Rekke*. No. 11. John Grieg, Bergen. 1944. [Offprint]

Godske, Carl Ludvig. "Det vitenskapelige arbeid ved Vervarslinga på Vestlandet." In: Bjerknes et al. 1944: 47–56.

Godske, Carl Ludvig. "Naturforskningen og samfunnet." In: Godske 1946: 9–26.

Godske, Carl Ludvig. (ed.). *Forskning og Framsteg*. J.W. Eides Forlag, Bergen. 1946.

Godske, Carl Ludvig. "Representativeness, a fundamental problem in meteorology." Union Geodesique et Geophysique International/Association de Meteorologie. *Reunion d'Oslo. Programme et Resume des Memoires*. Los Angeles. 1948: 21.

Godske, Carl Ludvig. *Forelesning i statistikk til anvendt matematikk bifag*. Studentsamskibnaden i Bergen. 1952.

Godske, Carl Ludvig. *Hvordan blir været?* J.W. Cappelens forlag, Oslo. 1956.

Godske, Carl Ludvig. "Oddvar Bjørgum in Memoriam." *Matematisk Tidsskrift*. Vol. 10. 1962: 5–7.

Godske, Carl Ludvig. *Statistics of Meteorological Variables. Final report*. US Air Force contract no. AF 61 (052)-416. University of Bergen, Norway. 1965.

Godske, Carl Ludvig. "The Future of meteorological data analysis." *WMO Technical Note No. 100.* 1969: 52–63.

Godske, Carl Ludvig. *Oppdagerferden.* Forskning og friluftsliv, Bergen. 1971.

Godske, Carl Ludvig; Peter Andersen; Thor Jakobsson and Harald Johansen. *Further Studies of Statistical Meteorology II. Final Report 1- Jan. 1967–31. Dec. 1968.* US Air Force contract no. F61052 67 C 0023. University of Bergen, Norway. 1969.

Godske, Carl Ludvig; Tor Bergeron; Jacob Bjerknes and Robert C. Bundgaard. *Dynamical meteorology and weather forecasting.* American Meteorological Society, Boston. 1957.

Goksøyr, Jostein. "De ikke-biologiske realfagene." In: Roll-Hansen et al. 1996: 127–244.

Golmen, Lars G. and Svein Østerhus. "Ny standard for utrekning av salinitet." *Naturen.* No. 4. 1983: 135–137.

Grønås, Sigbjørn. "Vilhelm Bjerknes' vision for scientific weather prediction." In: Drange 2005: 357–366.

Grønås, Sigbjørn and Magne Lystad. "Meteorologi: Moderne værvarsling – Fra Vilhelm Bjerknes' visjon fra 1904 til i dag." In: Johnsen 2017: 129–150.

Grønås, Sigbjørn and Melvyn A. Shapiro (eds). *The Life Cycles of Extratropical Cyclones.* Vol. 1. Alma Mater Forlag, Bergen. 1994.

Guldberg, Cato Maximilan and Henrik Mohn. *Études sur les mouvements de l'atmosphère.* A.W. Brøgger, Kristiania. 1876/1880.

Haaland, Anders. "Bergen Museums historie 1825–1945." In: Forland and Haaland 1996: 9–186.

Habermas, Jürgen. *Legitimation crisis.* Heinemann. London. 1976.

Hamblin, Jacob Darwin. *Oceanographers and the cold war: disciples of marine science.* University of Washington Press, Seattle. 2005.

Hamblin, Jacob Darwin. "Seeing the Ocean in the Shadow of Bergen Values." *Isis.* Vol. 105. 2014: 352–363.

Han, Ki Won. *The Rise of Oceanography in the United States, 1900–1940.* PhD dissertation in history, University of California, Berkeley. 2010.

Harper, Kristine C. "The Scandinavian Tag-Team: Providers of atmospheric reality to numerical weather prediction efforts in the United States (1948–1955)." *History of Meteorology.* Vol. 1, no. 1. 2004: 84–91.

Harper, Kristine C. *Weather and Climate: Decade by Decade.* Facts On File, New York. 2007.

Harper, Kristine. *Weather by the Numbers: The Genesis of Modern Meteorology.* MIT Press, Cambridge, MA. 2008.

Harris, Rollin A. "Some indications of land in the vicinity of the North Pole." *National Geographic.* 1904: 255–61.

Harris, Rollin A. *Arctic tides.* U.S. Coast and Geodetic Service, Government Printing Office, Washington. 1911.

Hartvedt, Gunnar Hagen (ed.). *Bergen Byleksikon.* Kunnskapsforlaget, Oslo. 1999 [1994].

Haugan, Peter M.; Eystein Jansen; Ola M. Johannessen and Ulf Lie. *Havets innvirkning på det atmosfæriske CO_2-budsjettet og den globale klimautviklingen. Rapport til Miljøverndepartementet som ledd i den interdepartementale klimautredningen.* Nansen senter for miljø og fjernmåling, Bergen. 1990.

Haugan, Peter M. and Helge Drange. "Sequestration of CO_2 in the deep ocean by shallow injection." *Nature.* Vol. 357, no NN. 28 May, 1992: 318–320.

Haugan, Peter M. and Helge Drange. Effects of CO_2 on the ocean environment. *Energy Conversion and Management.* Vol. 37, issues 6–8, June-August 1996: 1019–1022.

Haugan, Peter M. "On the production and use of scientific knowledge about ocean sequestration." *Greenhouse Gas Control Technologies proceedings of the 6th International Conference on Greenhouse Gas Control Technologies,* 1–4 October 2002, Kyoto, Japan. 2003: 719–724.

Haugan, Peter M. "Bygningens Historie." 2008. http://www.uib.no/gfi/57708/bygningens-historie

Haugan, Peter. "Oseanografi." In: Johnsen 2017: 151–171.

Helland-Hansen, Bjørn. "Fysisk oseanografi" ["Physical Oceanography"]. In: Murray and Hjort 1912: 210–306.

Helland-Hansen, Bjørn. "Det første Atlanterhavstogt." *Naturen.* October, 1913a. [Offprint]

Helland-Hansen, Bjørn. *Armauer Hansen's første togt. Festskrift utgit i anledning av Aars og Voss' skoles 50-aars jubilæum.* Kristiania, 1913b.

Helland-Hansen, Bjørn. "Eine Untersuchungsfahrt im Antlantischen Ozean mit dem Motorschiff: 'Armauer Hansen' im Sommer 1913." *Internationale Revue der gesamten Hydrobiologie und Hydrographie.* Vol. 7, no. 1. 1914: 61–83.

Helland-Hansen, Bjørn. "Nogen hydrografiske metoder." *Forhandlinger ved de Skandinaviske naturforskeres 16. møte i Kristiania den 10.-15. juli 1916.* Kristiania. 1918: 357–9.

Helland-Hansen, Bjørn. "Det Geofysiske Institut i Bergen." In: *Bergens Museums Årbok 1924–25.* John Griegs Boktrykkeri, Bergen. 1925. [Offprint]

Helland-Hansen, Bjørn. "Nybygningen for Det Geofysiske Institutt." *Bergens Museums Årsberetning 1928–1929.* John Griegs Boktrykkeri, Bergen. 1929: 3–24. [Offprint]

Helland-Hansen, Bjørn. "New Investigations in the Norwegian Sea." International Association of Physical Oceanography, Procès-Verbaux. No. 2, 1937: 101–103.

Helland-Hansen, Bjørn. "International Association of Physical Oceanography, Washington Assembly, September 1939. Presidental Address: The international survey of the Gulf Stream area." *Association d'Océanographie Physique, Procès-Verbaux.* No. 3. 1940: 52–67.

Helland-Hansen, Bjørn and August Brinkmann. "Bergens Museums Biologiske stasjon." *Bergens Museum Årbok, 1924–25.* John Griegs Boktrykkeri, Bergen. 1925. [Offprint].

Helland-Hansen, Bjørn and Fridtjof Nansen. *The Norwegian Sea: its physical oceanography based upon the Norwegian Researches, 1900–1904.* Report on the Norwegian Fishery and Marine Investigations. Vol. 2, no. 2. Det Mallingske Bogtrykkeri, Kristiania. 1909.

Helland-Hansen, Bjørn and Fridtjof Nansen. "Oceanography." In: Amundsen 1912: 404–438.

Helland-Hansen, Bjørn and Fridtjof Nansen. "Temperatur-Schwankungen des Nordatlantischen Ozeans und in der Atmosphäre. Einleitende Studien über die Ursachen der klimatologischen Schwankungen." *Videnskapsselskapets skrifter, matematisk-naturvidenskapelig klasse.* No. 9, Kristiania, 1917: 1–341.

Helland-Hansen, Bjørn and Fridtjof Nansen. *Temperature Variations on the North Atlantic Ocean and in the Atmosphere. Introductory studies on the cause of climatological variations.* Smithsonian Miscellaneous Collections. Vol. 70, no. 4. 1920a: 1–456.

Helland-Hansen, Bjørn and Fridtjof Nansen. "Klimavekslinger og deres aarsaker." *Naturen.* 1920b: 12–28, 101–116, 347–361.

Helland-Hansen, Bjørn and Fridtjof Nansen. "The Eastern North Atlantic." *Geofysiske Publikasjoner.* Vol. IV, no. 2. 1927: 1–71. Also published as *The Cruises of the 'Armauer Hansen.'* Vol. I, no. 1. 1927.

Helsvig, Kim G. *Universitetet i Oslo 1975–2011. Mot en ny samfunnskontrakt?* Vol. 6 of Universtitetet i Oslo 1811–2011. Unipub, Oslo. 2011.

Hesselberg, Theodor. "Ernst G. Calwagen [Obituary]." *La Météorologie. Revue mensuelle de Météorologie et de physique du globe.* Tome III. 1927: 15–17.

Heuss, Theodor. *Anthon Dohrn. A Life for Science.* Springer, New York/Berlin. 1991 [1940].

Hognestad, Per T. *Norske Havforskeres Forening gjennom 50 år.* Norske Havforskeres Forening/Trykkeriet, Høgskolen i Bodø. 1999.

Hogstad, Reidun Dagny. "Vêrskipsstasjon M 1948–1998." *Unpublished manuscript.* Provided by Svein Østerhus. 1998.

Hoppe, Gunnar and Valter Schytt. "Memorial to Hans W. Ahlmann 1889–1974." *Geological Society of America Memorials*. Vol. 6. 1975: 1–5.

Hove, Joakim and Peter M. Haugan. "Dynamics of a CO_2-seawater interface in the deep ocean." *Journal of Marine Research*. No. 63. 2005: 563–577.

Hovland, Edgar (ed.). *I vinden. Geofysisk institutt 90 år.* Fagbokforlaget, Bergen. 2007.

Hufford, Gary L. and James M. Seabrooke. "Oceanography of the Weddell Sea in 1969 (IWSOE)." *United States Coast Gard Oceanographic Report No. 31.* CG 373–31. 1970.

Huntington, Elleworth. "The sun and the weather: New light on their relation." *Geographical Review*. Vol. 5, no. 6. 1918: 483–491.

Hurdle, Burton G. (ed.). *The Nordic Seas.* Springer, New York. 1986.

IPCC. Homepage. https://www.ipcc.ch/organization/organization_history.shtml

Iversen, Trond. "Norklima: RegClim spesial." *Klima*. No. 2. 2008: 34–47.

Jacobs, Stanley S. (ed.). *Oceanology of the Antarctic Continental Shelf.* Antarctic Research Series, Vol. 43, American Geophysical Union, Washington, D.C. 1985.

Jewell, Ralph. "Tor Bergeron's First Year in the Bergen School." *Pure and Applied Geophysics*. Vol. 119. 1980/81.

Jewell, Ralph. "The Meteorological Judgment of Vilhelm Bjerknes." *Social Research*. Vol. 51, no. 3. 1984: 783–807.

Jewell, Ralph. "Vilhelm Bjerknes: How the Bergen school came into existence" (translation). In: Shapiro and Grønås 1994.

Johannessen, Ola M. "Undersøkelse av Oslofjordens forurensning 1962–1966, Underbilag D3–1: Strømundersøkelser i Drøbaksundet mai-juni 1963." *Oslofjordprosjektet*. Norsk Institutt for Vannforskning, Blindern. 1965.

Johannessen, Ola M. "Some current measurements in the Drøbak Sound, the narrow entrance to the Oslofjord." *Hvalrådets skrifter*. No. 52. Oslo. 1968: 3–38.

Johannessen, Ola M. "Nansen senter for fjernmåling i Bergen." Application to Dean of Faculty of Natural Sciences and Mathematics. Unpublished document. Copy borrowed from Ola M. Johannessen. 1985.

Johannessen, Ola M.; William J. Campbell; Robert Shuchman; Stein Sandven; Per Gloersen; Johnny A. Johannessen; Edward G.J. Osberger and Peter M. Haugan. "Microwave Study Programs of Air-Ice-Ocean Interactive Processes in the Seasonal Ice Zone of the Greenland and Barents Seas." In: Carsey 1992: 261–289.

Johannessen, Ola M.; Lennart Bengtsson; Martin W. Miles; Svetlana I. Kuzmina; Vladimir A. Semenov; Genrikh V. Alekseev; Andrei P. Nagurnyi; Victor F. Zakharov; Leonid P. Bobylev; Lasse H. Pettersson, Klaus Hasselmann and Howard P. Cattle. "Arctic climate change: Observed and modelled temperature and sea-ice variability." *Tellus A*. Vol. NNN, no. 4. August 2004.

Johannessen, Ola M.; Kjetil Lygre and Tor Eldevik. "Convective Chimneys and Plumes in the Northern Greenland Sea." In: Drange 2005: 251–272.

Johannessen, Ola M.; Svetlana I. Kuzmina; Leonid P. Bobylev and Martin W. Miles. "Surface air temperature variability and trends in the Arctic: new amplification assessment and regionalization." *Tellus A*. Vol. NNN, 28234. 2016.

Johannessen, Ola M. (ed.). *Arctic Sea Ice: Past, Present and Future.* Springer. 2018, in press.

Johansen, Hjalmar. *Selv-Anden paa 86,14'. Optegnelser fra den norske polarfærd 1893–96.* H. Aschehoug & Cos Forlag, Kristiania. 1896.

Johansen, Harald. *Litt om hyppighetsfordeling av sirkulasjonstyper i Vest-Norge.* Publikationer Fra Det Danske Meteorologiske Institut. No. 17. Copenhagen. 1964.

Johansen, Karl Egil. *Norges Fiskeri- og kysthistorie Bind III: En næring i omforming, 1880–1970.* Fagbokforlaget, Bergen. 2014.

Johnsen, Magnar Gullikstad (ed.). *Norsk Geofysisk Forening 100 år. En samling artikler i anledning foreningens 100-års-jubileum i 2017.* Norsk Geofysisk Forening/Nordisk Trykk. 2017.

Johnson, John William and Robert L. Wiegler. *Investigation of Current Measurement in Estuarine and Coastal Waters.* California State Water Pollution Control Board Publication no. 19. Berkeley/Sacramento. 1959.

Jølle, Harald Dag. "Polar prestisje og vitenskapelig ære. Fridtjof Nansens kamp for havforskning ved Universitetet." *Historisk Tidsskrift.* Vol. 88, no. 4. 2009: 611–637.

Jølle, Harald Dag. *Nansen: Oppdageren, bind 1.* Gyldendal, Oslo. 2011.

Jølle, Harald D. and Kari A. Myklebust. "Fridtjof Nansen og Aeroarctic-selskapet 1924–1930." *Historisk Tidsskrift.* Vol. 93, no. 3. 2014: 445–469.

Kalvaag, Hilde K. "Stjerne-Guro." *Hubro.* No. 4, 2008: 34–35.

Kohler, Robert E. *Landscapes and Labscapes. Exploring the Lab-Field Border in Biology.* 2002.

Krick, Irving P. and Roscoe Fleming. *Sun, Sea and Sky: Weather in our world and in our lives.* J.B. Lippincott, Philadelphia. 1954.

Krige, John. "NATO and the Strengthening of Western Science in the Post-Sputnik Era." *Minerva.* Vol. 38, no. 1. 2000: 81–108.

Krige, John. *American Hegemony and the Postwar Reconstruction of Science in Europe.* MIT Press, Cambridge MA/London. 2006.

Kristiansen, Thorleif Aass. *Meteorologi på reise. Veivalg og impulser i Arnt Eliassen og Ragnar Fjørtofts forskerkarrierer.* PhD thesis, University of Bergen, Bergen. 2017.

Krogness, Ole Andreas. "De magnetiske stormers betydning i meteorologien." *Naturen.* 1917: 63–81, 104–120, 185–192, 236–246.

Krogness, Ole Andreas. "Rapport om rustdannelse i vannledningsrørene under Bergen." *Report to Bergen municipality*, 1930a.

Krogness, Ole Andreas. "Rapport om vagabonderende strømme i det underjordiske rør- og kabelnett i Bergen." *Report to Bergen municipality.* 1930b.

Krogness, Ole Andreas. "Jordmagnetismen og dens forbindelse med nordlys og vær." *Naturen.* 1933: 266–277; 308–317; 339–347; 361–366.

Kutzbach, Gisela. *The Thermal Theory of Cyclones. A History of Meteorological Thought in the Nineteenth Century.* American Meteorological Society, Historical Monograph Series, Boston Massachusetts. 1979.

Kvinge, Thor; Odd H. Sælen and Frank Cleveland. "Samling av gamle instrumenter." *Reports in Meteorology and Oceanography.* No. 1. Geofysisk Institutt, Universitetet i Bergen. 2005.

Larsen, Reidulv and Steinar Berger. *Nordlysobservatoriet – historie og erindringer.* Universitetet i Tromsø, Tromsø. 2000.

Latour, Bruno. *Reassembling the social. An introduction to actor-network theory.* Oxford University Press, UK. 2005.

Liljequist, Gösta H. "Tor Bergeron. A Biography." *Pure and Applied Geophysics.* Vol. 119, no. 3. 1981: 409–442.

Ling, Arthur R. and Thomas H. Pope. "Tornöe's Optical Method of Determining Alchol and Extract in Beer." *Journal of the Federated Institutes of Brewing.* Vol. 7, no. 2. 1901: 170–181.

Livingstone, David. *Putting Science in its Place.* Chicago. 2003.

Lorenz, Edward N. "Deterministic Nonperiodic Flow." *Journal of the Atmospheric Sciences.* Vol. 20, no. 2. 1963: 130–141.

Lund, O.N.; Helmer Dahl and Otto Ottesen (eds.). *Leksikon for maskinister.* Aktietrykkeriet i Trondhjem/Det norske maskinistforbund, Trondheim. 1951: 213–217.

Manne, Rolf. "De norske saltverker – ein fiasko med positive etterverknader" [The Norwegian salt works – a failure with positive repercussions]. *Osterøy i soge og samtid. Sogeskrift for Osterøy.* 2009: 103–113.

Marchetti, Cesare. "On geoengineering and the CO_2 problem." *Climatic Change.* Vol. 1, no. 1. 1977: 59–68.

Meleshko, Valentin and Tatyana Pavlova. "Projections of sea ice in the Arctic for the

21st century." In: Johannessen (ed.). 2018: Chapter 10.

Metz, Bert; Ogunlade Davidson; Heleen C. de Coninck; Manuela Loos and Leo A. Meyer. *IPCC Special Report on Carbon Dioxide Capture and Storage.* Prepared by Working Group III of the Intergovernmental Panel on Climate Change. Cambridge University Press, Cambridge, United Kingdom and New York, USA. 2005.

Midttun, Lars. "Jens Konrad Eggvin." *Fiskets Gang.* No. 9. 1989: 24.

Mieghem, Jacques Van. "L'œuvre scientifique de Jules Jaumotte, Directeur de l'Institut Royal Météorologique de Belgique." *Ciel et Terre.* Vol. 57. 1941: 97–117.

Miles, Travis et al. "Glider observations of the Dotson Ice Shelf outflow." *Deep Sea Research Part II: Topical Studies in Oceanography.* Vol. 123, no. 1. 2016: 16–29.

Mills, Eric L. "Pacific Waters and the P. O. G. The origin of physical oceanography on the west coast of Canada." In: Freeland 2003: 1–11.

Mills, Eric L. "Mathematics in Neptune's Garden: Making the Physics of the Sea Quantitative, 1876–1900." In: Rozwadowski and van Keuren 2004: 39–64.

Mills, Eric L. "From Discovery to discovery: the hydrology of the Southern Ocean, 1885–1937." *Archives of Natural History.* Vol. 32, no. 3. 2005: 246–264.

Mills, Eric L. *The Fluid Envelope of Our Planet. How the Study of Ocean Currents Became a Science.* University of Toronto Press, Toronto. 2009.

Mills, Eric L. *Biological Oceanography: An Early History, 1870–1960.* University of Toronto Press, Toronto. 2012.

Misund, Ole A.; Dag W. Aksnes; Hanne H. Christensen and Thro B. Arlov. "A Norwegian pillar in Svalbard: The development of the University Centre in Svalbard (UNIS)." *Polar Record.* Vol. 53, No. 3. May 2017: 233–244.

Mohn, Henrik. *Det Norske Meteorologiske Instituts Storm-Atlas.* Bentzen, Christiania. 1870.

Mohn, Henrik. "Norges Klima." In: Schübeler 1879: 14–25.

Mohn, Henrik. "Die norwegische Nordmeer-Expedition." *Petermanns Mittheilungen.* Supplement no. 63. Gotha. 1880.

Mohn, Henrik. *The Norwegian North-Atlantic Expedition 1876–1878: X. Meteorology.* Grøndahl & Søn, Kristiania. 1883.

Mohn, Henrik. "Die Strömungen des europäischen Nordmeeres." *Petermanns Mittheilungen.* Supplement no. 79. Gotha. 1885.

Mohn, Henrik. *The Norwegian North-Atlantic Expedition 1876–1878: XVIII. The North Ocean, its depths, temperature and circulation.* Grøndahl & Søns Bogtrykkeri, Kristiania. 1887.

Mohr, Wilhelm. *Mine livserindringer.* Eget forlag, Bergen. 1969.

Monomer, Mark. *Mapping the Storm: A Look Back at How Meteorologists Learned to Map and Predict the Weather.* University of Chicago Press, Chicago. 1999.

Montgomery, Raymond B. "Observations of vertical humidity distribution above the ocean surface and their relation to evaporation." *Massachusetts Institute of Technology Meteorological Papers.* Vol. 7, no. 4. 1940.

Morgan, Judith. "Ellen Revelle Eckis: A Scripps Legacy." *SIO Reference Series No. 99-12.* 1999: 34–46.

Mork, Martin. "Jonas Ekman Fjeldstad: Minnetale i det norske vitenskaps-akademi." *Det norske vitenskaps-akademis årbok.* Oslo. 1986. [Offprint]

Mork, Martin. "Glimt fra 70-åras oseanografi på Geofysen. Noen personlige betraktninger." In: Utaaker 1999: 77–78.

Mosby, Håkon. "Sunshine and radiation." *Norwegian North Polar Expedition (1918–1925) Scientific results.* Vol. 1a. Meteorology. Geofysisk Institutt Special Reports no. 7, Bergen. 1933.

Mosby, Håkon. "The Sea-Surface and the Air." *Scientific Results of the Norwegian Antarctic Expeditions 1927–28.* Vol. 1, no. 10. Det Norske Videnskabsakadem, Oslo. 1933.

Mosby, Håkon. "The Waters of the Atlantic Antarctic Ocean." *Scientific Results of the Norwegian Antarctic Expeditions 1927–28.* Vol. 2, no. 11. Det Norske Videnskabs-akademi, Oslo. 1934.

Mosby, Håkon. "Verdunstung und Strahlung auf dem Meere." *Annalen der Hydrographie und Maritimen Meteorologie.* Vol. 64. 1936: 281–286.

Mosby, Håkon. "Oceanografi." In: Lund et al. 1951: 377.

Mosby, Håkon. "Havstrømmer." In: Lund et al. 1951: 213–217.

Mosby, Håkon. "Oberflächenströmungen in der Meerenge bei Tromsö." *Archiv für Meteorologie, Geophysik und Bioklimatologie,* Serie A. Vol. 7, no. 1. 1954: 378–384.

Mosby, Håkon. "Bjørn Helland Hansen, 1877–1957." *ICES Journal of Marine Sciences.* Vol. 23, no. 3. 1958: 321–323.

Mosby, Håkon. "M/S Helland-Hansen." *Naturen.* 1958: 402–412.

Mosby, Håkon. "Deep water in the Norwegian Sea." *Geofysiske Publikasjoner.* Vol. XXI, no. 3. 1959: 1–62.

Mosby, Håkon. "International Organizations in the Marine Sciences." *ICSU Review.* Vol. 1. 1959: 179–183.

Mosby, Håkon. "Water, Salt and Heat Balance of the North Polar Sea and of the Norwegian Sea." *Geofysiske Publikasjoner.* Vol. XXIV, no. 11. 1962: 289–313.

Mosby, Håkon. *Fridtjof Nansen Minneforelesningen III: Bunnvannsdannelse i havet.* Det norske videnskaps-akademi i Oslo. Universitetsforlaget, Oslo. 1967. Offprint.

Mosby, Håkon. "Atlantic Water in the Norwegian Sea." *Geofysiske Publikasjoner.* Vol. 28, no. 1. 1970: 1–60.

Mosby, Håkon. *Quo vadis universitas?* Universitetsforlaget, Oslo. 1971.

Mosby, Håkon. "En oversikt over norsk oseanografi fram til den 2. verdenskrig." In: Sakshaug et al. 1976: 209–323.

Mosby, Håkon. "Historical foreword." In: Hurdle 1986: vi–ix.

Mosby, Olav. *Diary.* Peter M. Haugan's archive.

Munk, Walter and Deborah Day. "Harald U. Sverdrup and the War Years." *Oceanography.* Vol. 15, no. 4. 2002: 7–29.

Murray, Sir John and Johan Hjort. *Atlanterhavet. Fra overflaten til havdypets mørke.* H. Aschehough & Co, Kristiania. 1912.

Murray, Sir John and Johan Hjort (eds.). *The Depths of the Ocean: A general account of the modern science of oceanography based largely on the scientific researches of the Norwegian steamer Michael Sars in the North Atlantic.* Macmillan, London. 1912.

Nansen Environmental and Remote Sensing Center. Annual reports 1991–2016. https://www.nersc.no/about/annual-reports

Nansen Environmental and Remote Sensing Center. Homepage. https://www.nersc.no/about/nansen-group

Nansen, Fridtjof. *På ski over Grønland/The First Crossing of Greenland.* H. Aschehoug & Co, Oslo 1890.

Nansen, Fridtjof. *Fram over Polhavet. Den norske polarfærd 1893–1896/Farthest North. Being the Record of a Voyage of Exploration of the Ship "Fram" 1893–96 and of a Fifteen Months' Sleigh Journey by Dr. Nansen and Lieut. Johansen.* H. Aschehoug & Co Forlag, Kristiania. 1897.

Nansen, Fridtjof (ed.). *The Norwegian North Polar Expedition, 1893–1896. Scientific Results.* Vol. V, no. XV. A.W. Brøgger, Kristiania. 1906.

Nebeker, Fredrik. *Calculating the Weather: Meteorology in the 20th Century.* International Geophysics Series, vol. 60. Academic Press, San Diego/London. 1995.

Nicholls, Keith W. and Svein Østerhus. "Interannual variability and ventilation timescales in the ocean cavity beneath Filchner-Ronne Ice Shelf, Antarctica." *Journal of Geophysical Research.* Vol. 109. 2004.

Nierenberg, William A. "Harald Ulrik Sverdrup, 1888–1957." *National Academy of Sciences, Biographical Memoir.* National Academies Press, Washington, D.C. 1996: 337–375.

Nilsen, Yngve and Magnus Vollset. *Vinden dreier. Meteorologiens historie i Norge*. Scandinavian Academic Press, Oslo. 2016.

Niskin, Shale J. "A Reversing-thermometer Mechanism for Attachment to Oceanographic Devices." *Limnology and Oceanography*. Vol. 9, no. 4. 1964: 591–594.

Nobel Foundation. https://www.nobelprize.org/nobel_prizes/peace/laureates/2007/press.html

Norcowe. Annual reports 2009–2017. http://www.norcowe.no/index.cfm?id=289888

NORSEX-Group. "Norwegian remote sensing experiment in a marginal ice zone." *Science*. Vol. 220, no. 4599. 1983: 781–787.

NOU 1983: 24. Satellittfjernmåling.

Nowotny, Helga; Peter Scott and Michael Gibbons. *Re-Thinking Science: Knowledge and the Public in an Age of Uncertainty*. Polity Press, Cambridge. 2001.

Nyberg, Alt. "Synoptic-Aerological Investigation of Weather Conditions in Europe, 17–24 April 1939." *Geografiska Annaler*. Vol. 26. 1944: 273–394.

Ore, Aadne. "Minnetale over Professor Dr.techn. Bjørn Trumpy holdt i den matematisk-naturvidenskapelige klasses møte den 21. februar 1975." *Det norske Vitenskaps-akademi årbok*. Oslo. 1976. [Offprint]

Oreskes, Naomi. "The Rejection of Continental Drift." *Historical Studies in the Physical and Biological Sciences*. Vol. 18, no. 2. 1988: 311–348.

Oreskes, Naomi. "Changing the Mission: From the Cold War to Climate Change." In: Oreskes and Krige 2014: 141–188.

Oreskes, Naomi and John Krige (eds.). *Science and Technology in the Global Cold War*. MIT Press, Cambridge MA. 2014.

Oreskes, Naomi and Ronald Rainger. "Science and Security before the Atomic Bomb: The Loyalty Case of Harald U. Sverdrup." *Studies in the History and Philosophy of Modern Physics*. Vol. 31, no. 3. 2000: 309–369.

Orr, James C. *Global Ocean Storage of Anthropogenic Carbon (GOSAC)*. EC Environment and Climate Program. ENV4-CT97-0495.

Final Report. IPSL/CNRS, France. 2002. http://ocmip5.ipsl.jussieu.fr/OCMIP/reports/GOSAC_finalreport_lores.pdf

Palmén, Erik. "Über die Bewegung der ausstertropischen Zyklonen." *Acta Societatis Scientiarum Fennicae*. Vol. 3, no. 7. 1926.

Palmén, Erik. "Registrierballonaufstiege in einer tiefen Zyklone." *Mittheilungen des Meteorologischen Instituts der Universität Helsingsfors*, no. 26. 1935.

Paulsen, Herfinn Schieldrup. "Meteorologiske elementer og byplanlegging." *Naturen*. 1959.

Paulsen, Herfinn Schieldrup. "Lysbruk ved bilkjøring." *Utvalg for Trafikksikkerhetsforskning*. 1969.

Pedersen, Finn; Harald Johansen and Odd Eide (eds.). *"Dette er Vervarslinga på Vestlandet..."* Nordanger, Bergen. 1968.

Peterson, Ray G.; Lothar Stramma and Gerhard Kortum. "Early concepts and charts of ocean circulation." *Progress in Oceanography*. Vol. 37, no. 1. 1996: 1–115.

Petterssen, Sverre. "Kinematical and Dynamical Properties of the Field of Pressure, with Application to Weather Forecasting." *Geofysiske Publikasjoner*. Vol. X, no. 2. 1933: 1–92.

Petterssen, Sverre. "Practical Rules for Prognosticating the Movement and Development of Pressure Centers." Union géodésique et géophysique internationale. *5e assemblée générale. Lisbonne, septembre 1933. Procès-verbaux des séances de l'Association de météorologie II. Mémoires et discussions*. Dupont, Paris. 1935: 5.

Petterssen, Sverre. *Weather Analysis and Forecasting*. McGraw-Hill Book Company, New York. 1940.

Petterssen, Sverre. *Introduction to Meteorology*. McGraw-Hill Book Company, New York and London. 1941.

Porter, Theodore. "How science became technical." *Isis*. Vol. 100, no. 2. 2009: 292–309.

Pratt, Larry J. (ed.). *The Physical Oceanography of Sea Straits*. NATO ASI Series C: Mathematical and Physical Sciences. Vol. 318.

Kluwer Academic Publishers, Dordrecht/ Boston/London. 1990.

Revelle, Roger. "How I became an oceanographer and other sea stories." *Annual Review of Earth and Planetary Sciences.* Vol. 15, no. 1. 1987: 1–24.

Rice, Anthony L. "Forty years of land-locked oceanography; the Institute of Oceanographic Sciences at Wormley, 1953–1995." *Endeavour.* Vol. 18, no. 4. 1994: 137–146.

Roberts, Peder. *The European Antarctic. Science and Strategy in Scandinavia and the British Empire.* Palgrave Studies in Cultural and Intellectual History, New York. 2011.

Roll-Hansen, Nils et al. *Universitetet i Bergens historie, bind 2.* Universitetet i Bergen/Bergens Museum. 1996.

Roll-Hansen, Nils. 1996. "Biologien ved Bergens Museum og Universitetet i Bergen." In: Roll-Hansen, Goksøyr et al. 1996: 10–125.

Rossby, Carl-Gustaf. "A generalization of the theory of the mixing length with applications to atmospheric and oceanic turbulence." *Massachusetts Institute of Technology Meteorological Papers.* Vol. 1, no. 4. 1932.

Rossby, Carl-Gustaf and Raymond Montgomery. "The layer of frictional influence in wind and ocean currents." *Massachusetts Institute of Technology Meteorological Papers.* Vol. 3, no. 3. 1935.

Rossby, Carl-Gustaf et al. "Relation between variations in the intensity of the zonal circulation of the atmosphere and the displacements of the semi-permanent centers of action." *Journal of Marine Research.* 1939: 38–55.

Rozwadowski, Helen M. *The Sea Knows No Boundaries. A Century of Marine Science under ICES.* International Council for the Exploration of the Sea, Copenhagen/University of Washington Press, Seattle and London. 2002.

Rozwadowski, Helen M. and David K. van Keuren (eds.). *The Machine in Neptune's Garden. Science.* History Publications, Sagamore Beach, USA. 2004.

Rozwadowski, Helen M. *Fathoming the Ocean: The Discovery and Exploration of the Deep Sea.* Harvard University Press. 2008.

Runcorn, Keith. "Palaeomagnetic Comparisons between Europe and North America." *Philosophical Transactions for the Royal Society of London. Series A, Mathematical and Physical Sciences.* Vol. 258, no. 1088. 1965: 1–11.

Sakshaug, Egil and Håkon Mosby. "En oversikt over norsk oseanografis historie fram til den 2. verdenskrig." In: Sakshaug et al. 1976: 218–224.

Sakshaug, Egil; Finn-Erik Dahl and Nils-Petter Wedege (eds.). *Norsk oseanografi – status og perspektiver.* Norsk Oseanografisk Komite, Oslo. 1976.

Sandström, Johan Wilhelm. "Über die Anwendung von Professor V. Bjerknes' Theorie der Wirbelbewegungen in Gasen und Flüssigkeiten auf meteorologische Beobachtungen in den höheren Luftschichten." *Kongl. Svenska Vetenskaps-Akademiens Handlingar.* Vol. 31, no. 4. 1900. [Offprint]

Sandström, Johan Wilhelm. "XI. Theorie Der atmosphärischen Cirkulation." In: Arrhenius 1903: 736–762.

Sandström, Johan and Bjørn Helland-Hansen. "Über die Berechnung von Meeresströmungen." *Report on Norwegian Fishery and Marine Investigations.* Vol. 2, no. 4. 1903: 1–43.

Sawyer, John Stanley. "Dynamic meteorology and weather forecasting: Review." *Quarterly Journal of the Royal Meteorological Society.* Vol. 83, Issue 358. 1957: 560.

Schröder, Wilfried. "Emil Wiechert und seine Bedeutung für die Entwicklung der Geophysik zur exacten Wissenschaft." *Archive for History of Exact Sciences.* Vol. 27, no. 4. 1982: 369–389.

Schwach, Vera. *Havet, fisken og vitenskapen. Fra fiskeriundersøkelser til havforskningsinstitutt. 1860–2000.* 2000.

Schwach, Vera. "Faded Glory: The Norwegian Vøringen-Expedition, 1876–1878." In: Benson and Rozwadowski 2007: 31–70.

Schübeler, Frederik Christian. *Væxtlivet i Norge, med særligt hensyn til plantegeographien*. Festskrift til Kjøbenhavns Universitets 400 aars Jubileum. W.C. Fabritius's bogtrykkeri, Kristiania. 1879.

SCOR/IAPO/UNESCO. "An intercomparison of some current meters, report on an experiment at WHOI Mooring Site 'D', 16–24 July 1967 by the working group on Continous Current Velocity Measurements." *UNESCO technical papers in marine science*. No. 11. 1967: 1–70.

Shapiro, Melyvn A. and Sigbjørn Grønås (eds.). *The Life Cycles of Extratropical Cyclones*. American Meteorological Society, Boston. 1999.

Skaar, Endre. "Distribution of precipitation in the Sognefjord Region." *Geilo Symposium, Norway. WMO/OMM No. 326*. Geneva. 1972.

Smed, Jens. *Jens Smed Archive: An ICES history lesson*. ICES, Copenhagen. 2012.

Solheim, Svale. *Nemningsfordomar ved fiske*. Oslo. 1940.

Spinnangr, Finn. "Nedbørstudier på Vestlandet." *Naturen*. 1925: 221–238.

Spinnangr, Finn. "Meteorologisk observatorium paa Fanaraaken i Jotunheimen." *Naturen*. Vol. 50, no. 11. 1926: 342–349.

Spinnangr, Finn. "Om terrengets virkning på vindene i Vest-Norge og fiskebankene utafor." *Norsk Tidsskrift for Sjøvesen*. Horten. 1939: 434–455; 1940: 104–123, 216–226.

Spinnangr, Finn. "Om varslinga av tåke i Vest-Norge." Bergen Museums Årbok. *Naturvitenskapelig rekke*, no. 5. 1940: 1–39.

Spinnangr, Finn and Harald Hjort. "On Thunderstorm Forecasting in Western Norway." *Bergen Museums Årbok. Naturvitenskapelig rekke*, no. 9. 1941: 1–38.

Spinnangr, Finn. "On the Influence of the Orography on the Winds in Southern Norway." *Bergen Museums Årbok. Naturvitenskapelig rekke*, no. 3. 1942: 1–63.

St.meld. nr. 35 (1975–1976). Om forskningens organisering og finansiering.

St.meld. nr. 39 (1998–1999). Forskning ved et veiskille.

St.meld. nr. 46 (1988–1989). Miljø og utvikling. Norges oppfølging av Verdenskommisjonens rapport.

Storetvedt, Karsten M. "Paleomagnetic Dating of some Younger Dikes in Southern Norway." *Nature*. 205, February 6, 1965: 585–586.

Storetvedt, Karsten M. *Our Evolving Planet. Earth History in New Perspective*. Alma Mater Forlag, Bergen. 1997.

Storetvedt, Karsten M. "Kontinentaldriften i et bergensk perspektiv." In: Utaaker 1999: 51–55.

Storetvedt, Karsten M. *Global Wrench Tectonics. Theory of Earth Evolution*. Fagbokforlaget, Bergen. 2003.

Storetvedt, Karsten M. *Når Grunnlaget Svikter*. Kolofon, Høvik. 2005.

Storetvedt, Karsten M. "'Det skal ikke stå på meg'. Glimt fra dosent Guro Gjellestads virke ved Geofysisk Institutt." Geofysisk Institutt, Bergen. 2008. http://www.uib.no/gfi/55528/guro-gjellestad

Storetvedt, Karsten M. and Guro Gjellestad. "Paleomagnetic Investigtion of an Old Red Sandstone Formation of Southern Norway." *Nature*. 212, October 1, 1966: 59–61.

Sundvor, Eirik and Markvard Sellevoll. "Historia om jordskjelvstasjonen i Bergen og korleis Noreg vart ein oljenasjon." *Naturen*. Vol. 129, no. 3. 2005: 114–131.

Svansson, Artur and Thor Kvinge. "The Ekman Repeating Current Meter." *Historisch-meereskundliches Jahrbuch*. Vol. 19. 2014: 25–36.

Sverdrup, Harald Ulrik. *Der Nordatlantische Passat*. Leipzig. 1917.

Sverdrup, Harald Ulrik and Odd Dahl. *Med Maud over Polhavet: Roald Amundsens ekspedition 1922–1925*. Bio-Film Compagni. 1926.

Sverdrup, Harald Ulrik and Odd Dahl. "Two oceanographic current recorders designed and used on the Maud Expedition." *Journal of the Optical Society of America*. Vol. 12, no. 5. 1926: 537–545.

Sverdrup, Harald Ulrik. *Norwegian North Polar Expedition expedition with the "Maud"*

(1918–1925), Scientific results. Vols. I–V. John Griegs Boktrykkeri, Bergen. 1929–1939.

Sverdrup, Harald Ulrik and Floyd M. Soule. *Scientific results of the "Nautilus" Expedition, 1931: under the command of Capt. Sir Hubert Wilkins. Parts I to III.* Papers in Physical Oceanography and Meteorology. Vol. 2, no. 2. Massachusetts Institute of Technology and Woods Hole Institution, Cambridge MA. 1933.

Sverdrup, Harald Ulrik. "Results of the meteorological observations on Isachsen's plateau: Scientific results of the Norwegian-Swedish Spitsbergen expedition in 1934, Part VI." *Geografiska annaler.* Vol. 18. Stockholm. 1936: 34–47.

Sverdrup, Harald Ulrik. *Hos Tundra-Folket.* Gyldendal Norsk Forlag, Oslo. 1938.

Sverdrup, Harald Ulrik; Martin W. Johnson and Richard H. Fleming. *The Oceans: Their Physics, Chemistry, and General Biology.* Prentice-Hall, New York. 1942.

Sylte, Gudrun. "Den aller kaldaste havstraumen." *Hubro.* No. 2. 2010: 3.

Sylte, Gudrun (ed.). *Bjerknes Centre for Climate Research: 2017 in numbers.* Bodoni, Bergen. 2017.

Sælen, Odd H. "Håkon Mosby. Minnetale i det norske videnskapsakademi." *Det Norske Videnskaps-akademis årbok 1989.* Falch Hurtigtrykk. 1990. [Offprint]

Sætre, Roald and Johan Blindheim. "Jens Eggvin – A Norwegian pioneer in operational oceanography." *ICES CM Documents. Working Paper 2002/W:01.* 2002.

Sætre, Roald. "En bergensk herremiddag anno 1876." *Bergensposten.* No. 2. 2011: 37–40.

Sörlin, Sverker. "The Anxieties of a Science Diplomat. Field Coproduction of Climate Knowledge and the Rise and Fall of Hans Ahlmann's 'Polar Warming.'" *Osiris.* Vol. 26. 2011: 66–89.

Sörlin, Sverker. "Reform and responsiblity – the climate of history in times of transformation." *Historisk tidsskrift.* Vol. 97, no. 1. 2018: 7–23.

Tait, John B. (ed.). *The Iceland Faroe Ridge International (ICES) "Overflow" expedition, May-June 1960. An investigation of cold, deep water overspill into the north-eastern Atlantic Ocean.* ICES Reports et Procès-Verbaux des Reunions vol. 157, Copenhagen. 1967.

Thompson, D'Arcy W. "The Geophysical Institute at Bergen." *Nature.* July 21, 1928: 98–100.

Thompson, D'Arcy W. "Otto Pettersson, 1848–1941." *ICES Journal of Marine Science.* Vol. 15, no. 2. 1948: 121–125.

Thorade, Hermann. "Henrik Mohn und die Entwicklung der Meereskunde." *Annalen der Hydrographie.* Vol. 63, no. 5. 1935: 182–186.

Thornthwaite, Charles W. "A Charter for Climatology. Presidential Address at the First Session of the Commission for Climatology, Washington, March 1953." *WMO Bulletin.* April 1953: 46.

Thorpe, Alan J.; Hans Volkert and Michal J. Ziemanski, "The Bjerknes' Circulation Theorem: A Historical Perspective." *Bulletin of the American Meteorological Society.* Vol. 84. 2002: 471–480.

Tornøe, Hercules. *The Norwegian North-Atlantic Expedition 1876–1878: I. Chemistry.* Grøndahl & Søn, Christiania. 1880.

Tornøe, Hercules. "Om Bestemmelse af Søvandets Saltgehalt ved Hjælp af dets elektriske Ledningsevne." *Nytt magasin for naturvidenskapene.* No. 34. 1895: 232–240.

Tornøe, Hercules. "On the determination of the salinity of sea water by its powers of refraction." *Report on Norwegian Fishery- and Marine Investigations.* Vol. 1, no. 6. 1900: 1–13.

Trumpy, Bjørn. "Ramaneffekt und Konstitution der Moleküle I–VII." *Zeitschrift für Physik.* No. 62, 1930: 806–823; No. 64, 1930: 777–780; No. 66, 1930: 790–806; No. 68, 1931: 675–682; No. 88, 1934: 226–346; No. 90, 1934: 133–137.

Trumpy, Bjørn. "Ramaneffekt und Konstitution der Moleküle V." *Det Kongelige Norske Videnskabers Selskabs Skrifter.* Vol. 4, no. 51. 1931: 194–197.

Trumpy, Bjørn. "Ramaneffekt und Cis-Transisomerie von Wasserstoff und Deuteriumverbindungen." *Det Kongelige Norske Videnskabers Selskabs Skrifter*. No. 29, 1935. [Offprint]

Trumpy, Bjørn. "Raman spectra of some deuterium compounds." *Nature*. 1935: 764.

Trumpy, Bjørn. "Polarisation der Raman-Strahlung und Konstitution der Moleküle. II." *Zeitschrift für Physik*. Vol. 98, no. 11. 1936: 672–683.

Turchetti, Simone. "Swords, Shields and Buoys: A History of the NATO Sub-Committee on Oceanographic Research, 1859–1973". *Centaurus*. Vol. 54, no. 3. 2012: 205–231.

Turchetti, Simone and Peder Roberts. *The Surveillance Imperative: Geosciences during the Cold War and Beyond*. Palgrave Studies in the History of Science and Technology. Palgrave Macmillan US, New York. 2014.

Turner, Roger. *Weathering Heights: The Emergence of Aeronautical Meteorology as an Infrastructural Science*. PhD-thesis. University of Pennsylvania. Publicly accessible Penn Dissertations, Paper 147. 2010.

UN Homepage. http://www.un.org/en/sections/what-we-do/promote-sustainable-development

UN Homepage. http://unfccc.int/bodies/body/6383/php/view/reports.php#c

Union Géodésique et Géophysique Internationale. *5e assemblée générale. Lisbonne, septembre 1933. Procès-verbaux des séances de l'Association de météorologie II. Mémoires et discussions*. Dupont, Paris. 1935.

Union Géodésique et Géophysique Internationale/Association de Meteorologie. *Reunion d'Oslo. Programme et Resume des Memoires*. Los Angeles. 1948.

United Nations. *Report of the United Nations Conference on the Human Environment, Stockholm, 5–16 June 1972*. United Nations Publications, A/CONF.48/14/Rev.1. Sales No. E.73.II.A.14. Switzerland, 1973.

Universitetet i Bergen/Eiendomsavdelingen. *Geofysen – Allégaten 70, Jahnebakken 3, 5. Forvaltningsplan UiB 2017*. Bergen. 2017.

Unsigned. "Obituary: Hans Wilhelmsson Ahlmann." *Geographical Journal*. Vol. 140, no. 3. 1974: 523–524.

Utaaker, Kåre. "Studies in local and micrometeorology at Kleppe 1: Investigations of the air temperature observed in various types of thermometer screens." *Universitetet i Bergens Årbok, Matematisk-naturvitenskapelig serie*. No. 4, 1956. [Offprint]

Utaaker, Kåre. "Studies in local and micrometeorology at Kleppe 2: Thermistors used as free air and soil thermometers." *Universitetet i Bergens Årbok, Matematisk-naturvitenskapelig serie*. No. 1, 1958. [Offprint]

Utaaker, Kåre. "Studies in local and micrometeorology at Kleppe 3: Soil temperature measurements." *Universitetet i Bergens Årbok, Matematisk-naturvitenskapelig serie*. No. 10, 1961. [Offprint]

Utaaker, Kåre. "The Local Climate of Nes, Hedmark." *Universitetet i Bergens Skrifter*. No. 28. 1963.

Utaaker, Kåre. "Meso-Climatic Research in Norway. The Sognefjord Project." *WMO Bulletin*. No. 1. 1965.

Utaaker, Kåre. "A Study of Energy Exchange at The Earth's Surface." *Universitetet i Bergens Årbok, Matematisk-Naturvitenskapelig Serie*. No. 1. 1966.

Utaaker, Kåre. "Klimatologi." *Moderne norsk geofysisk forskning. Foredrag holdt ved Oslo Geofysikeres Forenings 25-årsjubileum, 10. mai 1974*. Oslo. 1974. [Offprint]

Utaaker, Kåre. "Carl Ludvig Schreiner Godske (1906–1970)." *Department of Applied Mathematics, Report No. 117*. Universitetet i Bergen. 1998.

Utaaker, Kåre (ed.). *Bergen Geofysikeres forening, 1949–1999: 50 år*. Bergen Geofysikeres Forening/Allkopi, Bergen. 1999.

Utaaker, Kåre. "C.L. Godske." Biographical sketch. Published online in 2008: http://www.uib.no/gfi/54202/cl-godske

Utaaker, Kåre and Endre Skaar. "Local Climates and Growth Climate of the Sognefjord Region. Part 1: A Survey of the

Project and Climatic Tables." *Acta Agriculturae Scandinavica*. Supplement 18. 1970.

Vaughan, Thomas Wayland et al. *International Aspects of Oceanography. Oceanographic Data and Provisions for Oceanographic Research*. National Academy of Sciences, Washington, D.C. 1937.

Villinger, Bernhard and Henry C. Stetson. *Scientific results of the "Nautilus" Expedition, 1931: under the command of Capt. Sir Hubert Wilkins. Parts IV to V. Papers in Physical Oceanography and Meteorology*. Vol. 2, no. 3. MIT and Woods Hole Institution, Cambridge MA. 1933.

Vinzent, Jutta. *Identity and Image. Refugee artists from Nazi Germany in Britain (1933–1945)*. Schriften der Guernica-Gesellschaft, Vol. 16. Weimar. 2006.

Vollset, Magnus. "Asking too much? Postwar climate research in Norway, 1947–1961." *History of Meteorology*. Vol. 7. 2015: 83–97.

Warren, Bruce Alfred and Carl Wunsch. *Evolution of Physical Oceanography: Scientific Surveys in Honor of Henry Stommel*. MIT Press, Cambridge MA. 1981.

Wasserfall, Karl Falch. "On periodic variations in terrestrial magnetism." Geofysiske Publikasjoner. Vol. 5, no. 3. 1930a: 2–33.

Wasserfall, Karl Falch. "On the relation between the rotation of the sun and variations in atmospheric temperature." *Geofysiske Publikasjoner*. Vol. 5, no. 10. 1930b: 3–18.

Wasserfall, Karl Falch. "Er langsiktige værspådommer mulig?." *Naturen*. 1931: 135–143, 293–302, 360–369.

Wasserfall, Karl Falch. "Praktiske forsøk på langsiktige værspådommer." *Naturen*. 1933: 298–307.

Wasserfall, Karl Falch. "Solaktiviteten og den syn- og målbare virkning på solen selv og på vår klode." *Naturen*. 1941: 229–242.

Weart, Stephen R. *The discovery of global warming. Revised and expanded edition*. Harvard University Press, Cambridge and London. 2008.

Wegener, Alfred. *The Origin of Continents and Oceans*. Translated by George Anthony Skerl. E.P. Dutton, London/New York. 1922.

Weidemann, Hartwig. *A Manual of Current Measuring Instruments and Methods. International Association of Physical Oceanography*. Publication Scientifique No. 21, preliminary report. 1960.

Werenskiold, Werner. "Mohn og oseanografien." *Norsk geografisk tidsskrift*. Vol. 6, no. 6. 1936: 293–302.

West, John B. "Torricelli and the Ocean of Air: The First Measurement of Barometric Pressure." *Physiology*. Vol. 28, no. 2 (March). 2013: 66–73.

Wille, Carl. *The Norwegian North-Atlantic Expedition 1876–1878: IV. Historical Account*. No. 1. Grøndahl & Søn, Kristiania. 1882a.

Wille, Carl. *The Norwegian North-Atlantic Expedition 1876–1878: V. Magnetical Observations*. Grøndahl & Søns Bogtrykkeri, Kristiania. 1882b.

Wolff, Torben. *The Birth and First Years of the Scientific Committee on Oceanic Research (SCOR)*. International Council for Science. Scientific Committee on Oceanic Research. Schor History Report no. 1. 2010.

World Commission on Environment and Development. *Our Common Future*. Oxford University Press, Oxford. 1987.

World Meteorological Organization. "A report of the international conference on the assessment of the role of carbon dioxide and of other greenhouse gases in climate variations and associated impacts." Report from conference in Villach, Austria 9–15 October 1985. WMO. 1986.

Wüst, Georg. "The Major Deep-Sea Expeditions and Research Vessels 1873–1960: A contribution to the history of oceanography." *Progress in Oceanography*. Vol. 2. 1964/Contribution No. 679 of Lamount Geological Observatory, Columbia University, New York. 1964: 1–52.

Østerhus, Svein and Tor Gammelsrød. "The abyss of the Nordic Seas is warming." *Journal of Climate*. Vol. 12, no. 11. 1999: 3297–3304.

Newspapers
Aftenposten
Bergens Aftenblad
Dagbladet
Morgenbladet
NTB
På Høyden
Verdens Gang/VG

Interviews
Lennart Bengtsson, May 20, 2016.
Helge Drange, October 27, 2016.
Arne Foldvik, May 5, 2016.
Gunnar Furnes, October 19, 2016.
Tor Gammelsrød, May 13, 2016.
Yngvar Gjessing, May 9, 2016.
Sigbjørn Grønås, February 5, 2018.
Peter M. Haugan, October 31, 2016.
Eystein Jansen, May 19, 2016; June 1, 2016.
Ola M. Johannessen, March 27, 2017; May 3,
    2017.
Nils Gunnar Kvamstø, March 29, 2017.
Thor Kvinge, May 13, 2016.
Steinar Myking, May 18, 2016.
Karsten Storetvedt, May 20, 2016.
Svein Østerhus, November 17, 2017.

Notes

Chapter 1 – Calculating the world
pp. 11–22

1   Damas 1922: 17 (http://www.vliz.be/imis-docs/publications/277806.pdf).

2   Damas 1922; Annual Report of the Geophysical Institute 1921–22: 70–81. Until 1946, the annual reports were printed in the Bergen Museum yearbooks.

3   Annual Report 1921–22: 76.

4   Annual Report 1921–22: 77–78.

5   "Den internationale veirvarsling," *Tidens Tegn*, July 25, 1921: 6.

6   "Den internationale meteorolog-konferance i Bergen aapnet," *Tidens Tegn*, July 27, 1921: 6.

7   *Report of the proceedings of the seventh meeting of the International Commission for the Investigation of the Upper Air. Held in Bergen 25th-29th July 1921*. A/S. John Grieg, Bergen 1921: 10 (UBBHF-SC/MS. 2080/I.23).

8   In addition to *Tidens Tegn*, the newspapers *Morgenavisen*, *Bergens Tidende*, *Bergens Aftenblad* and *Bergens Annonce-Tidende* all printed reports on the conference, which took place on July 25–29, 1921 (UiBHF-SC /MS.2080/Z.4). The attendees were from Belgium, Denmark, France, Great Britain, Holland, Italy, Japan, Java, Norway, Spain, Sweden and Switzerland.

9   The numbers are based on information from the annual reports from the Geophysical Institute, as well as an undated report summarizing visits to the Weather Forecasting Office found in UIBHF-SC/MS.2080/I.23. Shorter stays, such as those of people attending the five-day conference in 1921, are not counted.

10 For the first presentation of the frontal theory, see: Bjerknes 1918: 1–8. The cold and hot fronts presented in this paper would later be supplemented by the ocular front, and a model of the full life cycles of cyclones.

11 The "Ryder storm" in October 1922 served to prove the point. It was named after the director of the Meteorological Institute in Denmark, Carl Ryder, who unlike the Bergen school meteorologists did *not* predict the storm. See: Friedman 1989: 1–2.

12 Bjerknes and Solberg 1922: 3–18. The complete journal is available online in full text: http://www.ngfweb.no/geophysica_norvegica.html

13 Bjerknes 1921: 1–90.

14 See chapter 3.

15 Schröder 1982: 369–389.

16 The institute in Leipzig was named by the institute in Göttingen, and it was intended to eventually include all branches of geophysics. However, under the first director, Vilhelm Bjerknes, both the seismograph and the equipment for carrying out upper-air observations were put aside in order to focus exclusively on the physics

of the atmosphere. Bjerknes 1938: 49–62; Bjerknes 1997 [1938]: 7; Ehrmann 2013: 14.

17 Sundvor and Sellevoll 2005: 114–131; Helland-Hansen 1925: 1–24.

18 The plans are found in UBBHF/ MS.2080/I.18.

19 Mills 2003: 1–11, quote on p. 2.

20 Hovland 2009.

21 See, for instance, the US National Academy of Sciences Biographical Memoirs (http://www.nasonline.org/publications/biographical-memoirs/) and the "Memorial speeches" (*minnetaler*) published by the Norwegian Academy of Science and Letters (http://www.dnva.no/c42044/seksjon/vis.html?tid=42045).

22 Thompson 1928: 98–100.

23 BCCR. Annual report 2006: 3.

Chapter 2 – The first Bergen school pp. 23–54

1 Letter from H. Mohn and G.O. Sars to the Norwegian Home Department, dated Christiania, March 19, 1874, asking for funding for the expedition. In: Wille 1882a: 6.

2 Wille 1882a: 5.

3 Johansen 2014: 24–30.

4 Schwach 2007: 38.

5 Letter from H. Mohn and G.O. Sars to the Norwegian Home Department, dated Christiania, March 19, 1874, asking for funding for the expedition. In: Wille 1882a: 6.

6 The concept of several "Bergen schools" was first used by Sakshaug and Mosby 1976: 218–224.

7 Tornøe 1880: 66. See also: Mohn 1879: 14–25.

8 Tornøe 1880: 66–73.

9 Nilsen and Vollset 2016: 44–76.

10 Mohn 1870; Kutzbach 1979: 76–80.

11 Mohn 1883.

12 Guldberg 1876/1880: 20 (https://archive.org/details/tudessurlesmouvoounivgoog, https://archive.org/details/tudessurlesmouv03mohngoog).

13 Mohn 1887: 143. See also: Mohn 1880; Mohn 1885.

14 Mohn 1887: 161.

15 Guldberg and Mohn 1876/1880 (https://archive.org/details/tudessurlesmouvoounivgoog; https://archive.org/details/tudessurlesmouv03mohngoog). As pointed out by oceanographer Hermann Thorade in 1935, Guldberg and Mohn's dynamic approach remained a relatively solitary work for about two decades, "an abandoned mine." Thorade 1935: 182, as quoted in: Mills 2009: 95.

16 Kutzbach 1979: 45–58; Nilsen and Vollset 2016: 68–70.

17 Mohn 1883: 149.

18 Tornøe 1880: 75.

19 Tornøe 1895: 232–240. The idea was first presented at a meeting of the Norwegian Academy of Science in October, 1893.

20 Tornøe 1900: 1–13.

21 Ling and Pope 1901: 170–181.

22 Mohn 1887: 192.

23 Werenskiold 1936: 293–302.

24 In his vindication, Werenskiold highlighted the formation of bottom water (rediscovered by Nansen in 1906), the equatorial counter-current (rediscovered by H.U. Sverdrup in 1934), methods for calculating what happens when a current reaches shallower water (presented by Helland-Hansen in 1934), and that his equations – although cumbersome – reached identical results to Bjerknes's circulation theorem (operationalized by Helland-Hansen and Sandström in 1903). The article was, however, characterized by Werenskiold's interpretation of Mohn in light of later findings. This, and the fact that it was published in the Norwegian journal of geography rather than geophysics or oceanography, might explain why Werenskiold's text seems to have been forgotten.

25 Mills 2004: 54–55; Mills 2009: 94–95. The only exception was Professor of Geography at the University of Kiel, Otto Krümmel, who included Mohn's methods in his

curriculum, and had a handful of doctoral students use them in their doctoral work.

26 Schwach 2007: 53–59.

27 Sætre 2011: 37–40.

28 Mills 2009: 92–95. See also: Helland-Hansen and Nansen 1909: 7ff.

29 Wille 1882: 7–9.

30 Vera Schwach has estimated the total cost of the expedition at around 470,000 NOK, which included about 10,000 NOK per year between 1879 and 1900 for analyzing and publishing the findings (Schwach 2007: 42). In comparison, the annual running costs for the country's only university in the late 1870s were about 220,000 NOK (Bureau Central de Statistique. *Annuaire Statistique de la Norvège, Première Année.* T.H. Steen, Kristiania. 1879: 91). Among those who worked on the biological samples from the *Vøringen* expedition was Kristine Bonnevie, Norway's first female professor in 1912. Kristine Bonnevie was also the sister of Vilhelm Bjerknes's wife Sofie Honoria, a gifted mathematician.

31 Sakshaug and Mosby 1976: 215, 217–8; Haugan 2017: 153–4. The expedition is also thoroughly detailed in Nilsen and Vollset 2016: 71–82.

32 The magnetic survey was conducted in collaboration with the captain onboard, Carl Wille, who penned the results: Wille 1882.

33 Mills 2009: 94. The quote is from Peterson, Stramma and Kortum 1996: 105. See also: Kutzbach 1979.

34 Jølle 2011: 15.

35 For more on the zoological station, see: Heuss 1991 [1940].

36 Fridtjof Nansen's doctoral thesis has been presented as a classic in both zoology and neuroscience. For more on his doctoral thesis "The Structure and Combination of the Histological Elements of the Central Nervous System" from 1887, as well as his time in Bergen, see: Bock and Helle 2016; Jølle 2011: 41–45.

37 Nansen 1928 [1890]: 6. An English translation with the title *The First Crossing of Greenland* was published the same year.

38 Ekman 1931: 170.

39 Johansen 1896: 296.

40 Fulsås 2004: 173–209.

41 Nansen 1897. The full title of the English translation published the same year by Harper & Brothers Publishers, is: *Farthest North. Being the Record of a Voyage of Exploration of the Ship "Fram" 1893–96 and of a Fifteen Months' Sleigh Journey by Dr. Nansen and Lieut. Johansen.*

42 Bjerknes. 1943: 1.

43 Devik 1962: 8–9; Friedman 1989: 18; Mills 2009: 99.

44 Bjerknes 1891a: 74–79; Bjerknes 1891b: 92–101; Bjerknes 1891c: 513–526; Bjerknes 1892: 69–76; Bjerknes 1895a: 58–63; Bjerknes 1895b: 121–169.

45 Bjerknes 1898: 1–35 (http://www.biodiversitylibrary.org/item/107113#page/231/mode/1up). For a presentation of the equations and a discussion on the significance of this paper, see: Thorpe, Volkert and Ziemianski 2003: 471–480 (http://journals.ametsoc.org/doi/pdf/10.1175/BAMS-84-4-471). For more on how the equations were developed over time, see: Nebeker 1995; Harper 2008.

46 Bjerknes 1997 [1938]: 7.

47 Bjerknes 1904: 1–7.

48 For more on Otto Pettersson, see: Thompson 1948: 121–125.

49 Rozwadowski 2002: 15–16.

50 Rozwadowski 2008: 18.

51 For the history of ICES, see: Rozwadowski 2002, and Smed 2012 (https://issuu.com/icesdk/docs/jens_smed_archive).

52 Friedman 1989: 39–42; Mills 2009: 103.

53 Bjerknes in Nansen 1906: 1–12.

54 Mills 2012: 83–84.

55 Ekman 1902: 1–27; Ekman 1905: 1–52; Ekman, 1906: 423–430; 472–484; 527–540; 566–583.

56 Ekman 1908: 1–47; Ekman 1926: 1–32; Svansson and Kvinge 2014: 25–36.

57 Bjerknes 1906. (https://archive.org/
stream/fieldsofforce00bjeruoft#page/n4/
mode/1up). See also: Bjerknes 1900/1902;
Bjerknes 1900: 251–276.

58 The same lecture was first given in
Fysiska Sällskapet (the Physical Society)
in Stockholm in 1904, see: Bjerknes 1904:
1–7. An English translation by Yale Mintz
from 1954 was reproduced in: Shapiro 1999
(http://folk.uib.no/ngbnk/Bjerknes_150/
In%20memory%20of%20Vilhelm%20
Bjerknes.pdf).

59 In addition to papers, Bjerknes's Carnegie
research resulted in four monographs. The
first two were published in 1910 and 1911
(*Dynamic Meteorology and Hydrography. 1.
Statistics*, coauthor: J.W. Sandström; *2. Kinematics,* co-authors: Theodor Hesselberg and
Olaf Devik), the third in 1933 (*Physikalische Hydrodynamik mit Anwendung auf
die dynamische Meteorologie*, co-authors:
Jacob Bjerknes, Halvor Solberg and Tor
Bergeron) and the last published by his
assistants posthumously in 1957 (*Dynamical
meteorology and weather forecasting*, written
by Carl Ludvig Godske, Tor Bergeron, Jacob
Bjerknes and R.C. Bundgaard).

60 Bjerknes, Hesselberg, Devik and Sandström 1911. For more on Vilhelm Bjerknes's
students, see: Eliassen 1982: 1–11 (http://
www.annualreviews.org/doi/pdf/10.1146/
annurev.fl.14.010182.000245).

61 Sandström 1900. See also: Sandström 1903:
736–762.

62 Vilhelm Bjerknes, quoted in: Eliassen
1994: 4.

63 As early as 1901, the two had published the
first paper on how the theorem could be
used on oceanographic observations, see:
Bjerknes and Sandström 1901.

64 In 1908, Sandström took a job at the
Swedish Hydrographical Bureau, where
he conducted experiments on thermal circulation, how temperature influences the
dynamics of the ocean. Later, Sandström
studied, among other topics, the influence
of melting ice on ocean currents, and how

the Gulf Stream influences the weather in
Scandinavia.

65 Mosby 1958: 321–323.

66 Kristian Birkeland (1867–1917) was a
Norwegian scientist most famous for his
studies of the northern lights. He organized
several expeditions to northern Norway
and established a network of observation
stations to collect magnetic field data. His
theory that the aurora borealis was caused
by interactions between electrons from the
sun and the Earth's magnetic field were
confirmed by satellite observations only in
the late 1960s. For more on Birkeland, see
chapter 3.

67 Sandström and Helland-Hansen 1903:
1–43; Ellingsen 2013: 81–85 (http://hdl.han-
dle.net/1956/6733). The solenoids would
later be criticized for not accounting for
friction, and for ignoring variations over
time.

68 Helland-Hansen in Murray and Hjort 1912:
159.

69 An almost identical graphical presentation
of solenoids, but based on the atmosphere
rather than the oceans, was published by
Johan Sandström in: Arrhenius 1903: 741
(https://archive.org/stream/lehrbuchder-
kosmi02arrhuoft#page/740/mode/2up).

70 Forland and Haaland 1996: 10–186;
Roll-Hansen 1996: 128–243.

71 In 1906, the institution was reorganized yet
again, and given the name *Fiskeridirektor-
atet* (the Directorate of Fisheries). For the
history of this institution, see: Schwach
2000.

72 The information on the "Havforsk-
ningskursus" is taken from the Bergen
Museum's annual reports [*Bergen Museum
Aarsberetning*] 1902–1914. For an overview
of the course curriculum see: Murray and
Hjort 1912 (http://www.biodiversitylibrary.
org/item/29956#page/9/mode/1up). Several
of the specialized chapters were written by
the teachers of the oceanographic courses,
such as Bjørn Helland-Hansen's chapter
"Physical oceanography." See also: Forland
and Haaland 1996: 96–99.

73 Bergen Museum Aarsberetning 1909: 55.

74 Helland-Hansen and Nansen 1909: 7.

75 Helland-Hansen and Nansen 1909: 30–31.

76 Helland-Hansen and Nansen 1909: 31.

77 It must be added that the measurements of ocean temperatures were of high quality, and have since been used in calibrating models and climate studies. See: Dickson and Østerhus 2007: 56–75.

78 Helland-Hansen and Nansen 1909: IV-V.

79 Helland-Hansen in Hjort and Murray 1912: 306. For more on how this research continued, see chapter 4.

80 Jølle 2011: 49.

81 The plans were published in the local newspaper *Bergensposten* August 19 and 20, 1890. See: Helland-Hansen and Brinkmann 1925: 2.

82 Helland-Hansen and Brinkmann 1925: 7.

83 Sakshaug and Mosby 1976: 229. The authors present physical oceanography as a specialty within marine biology and zoology developed to understand the physical conditions for life in the oceans, and highlight Henrik Mohn as the founding father of both oceanography and meteorology. The *Vøringen* expedition is mentioned as a culmination of grand expeditions for deep-sea marine research, but without further elaboration.

84 Jølle 2009: 611–637.

85 Jølle 2009: 611–637. See also: Schwach 2000: 125.

86 Jølle 2009: 632.

87 Sewer spills in the neighboring Puddefjorden, which served as a water reservoir for the aquarium, caused so much pollution that both the aquarium and the biological station had to keep their doors closed periodically. The lack of income from ticket sales, and fact that the price of herring, the staple food for the seals, skyrocketed due to the World War, forced the aquarium to close for economic reasons in 1916. At the same time, plans were made to move the biological station from the center of town to Herdla, some 30 kilometers north of Bergen, where pollution was not an issue. The new station opened in 1922.

88 The initiative also, specifically, asked for a chemist to aid both physical and biological oceanographic research. The committee comprised the director at Bergen Museum, Jens Holmboe, the chairman of the board at the Museum, Klaus Hanssen, and Helland-Hanssen. Helland-Hansen later stressed that he was not present at the meeting where the establishment of a new institution was suggested (Helland-Hansen and Brinkmann 1925: 17–18).

89 Helland-Hansen and Brinkmann 1925: 19.

90 The first cruise is described in: Auerbach 1914: 3–54; Helland-Hansen 1914: 61–83; Helland-Hansen 1913a; Helland-Hansen 1913b. Offprints are available in UBBHF-SC/MS.2080/A.10. For more on the vessel, its uses and its unorthodox construction, see chapter 5.

91 For more on the public lectures ("Foredrag for hvermansen"), see: Forland and Hovland 2002: 19–21.

92 Initially, Bjerknes was offered equipment to establish an upper-air observatory, but he declined, arguing that it was more important to study how the observations from the increasing number of such observatories could aid in weather forecasting. The name "Geophysical Institute" was adapted from Emil Wiechert's Geophysical Institute in Göttingen, established in 1898, which used seismic methods for studying the Earth's interior. Bjerknes 1997 [1938]: 7.

93 Bjerknes 1997 [1938]: 9.

94 See for instance: Bergeron, Devik and Godske 1962: 1–37 (http://folk.uib.no/ngbnk/Bjerknes_150/In%20memory%20of%20Vilhelm%20Bjerknes.pdf). For an alternative perspective that sees Bjerknes and the later Bergen school as one of several interlinked research programs in meteorology, see: Bergeron 1959: 440–474; Bergeron 1981: 443–473. For presentations by historians that rely heavily on Vilhelm Bjerknes's own perspectives, see: Grønås 2005: 357–366; Jewell 1984: 783–807.

95 Fleming 2016. Fleming is the founder and first president of the International Commission on the History of Meteorology, see: http://www.meteohistory.org

96 Harper 2008. A similar perspective was presented in: Grønås 2005: 357–366.

97 Friedman 1989.

98 Nilsen and Vollset 2016: 219–228.

Chapter 3 – Useful curiosity
pp. 55–92

1 *Annual Report for the Geophysical Institute*, 1917–18: 67.

2 The observatory was established by physicist Kristian Birkeland in the late 1890s, but did not have full-time staffing.

3 Devik and Thrane 1967.

4 Helland-Hansen 1925: 4.

5 Based on documents from the Geophysical Commission, found in UBBHF-SC/MS.2080/I.22–23 and O.8. By the 1930s the institutions included the Geophysical Institute and the Magnetic Bureau in Bergen, the Norwegian Institute for Cosmic Physics in Oslo, the Aurora Borealis Observatory in Tromsø, and the Meteorological Institutes in Oslo, Bergen, and Tromsø. The commission also managed the grants from the Birkeland Fund and the Nansen Fund.

6 Nilsen and Vollset 2016: 163. On the one hand, the opening of the Geophysical Institute in Bergen in 1917 created a second institutional hub for meteorological research. On the other, changes in the leadership at the Meteorological Institute contributed to its reduced importance: in 1913, after 47 years, Henrik Mohn retired as head of the Meteorological Institute (he died in 1916). His successor, Axel Steen, died after only two years in the post, and was replaced by Vilhelm Bjerknes's former assistant, Theodor Hesselberg.

7 Helland-Hansen 1925: 1–3.

8 Helland-Hansen 1925: 8–9.

9 This section draws heavily on Ellingsen 2017 (submitted). For a discussion on the Bergen school actors' own competing historical narratives, see Chapter 10.

10 Nebeker 1995: 47–58. Nebeker's emphasis is on the physics of weather forecasting.

11 Harper 2008. To Harper, the Bergen school is the first chapter in the story of numerical weather forecasting, the weather being calculated by computers.

12 Friedman 1989. To Friedman, the Bergen school was a both a new set of methods that were constructed and disseminated, and a "package" which also included disseminating the narrative of Bergen as a genesis.

13 Petterssen 1941. To Petterssen, the Bergen school represented a change from meteorology as a descriptive science to an exact science that could make predictions based on intimate knowledge of the underlying forces producing weather phenomena.

14 Fleming, Jankovic and Coen 2006. According to the editors of this anthology, the "universality" in the title is a result of the influence of Bergen school: Before the Bergen school, all weather forecasting was based on local experience. The Bergen school made weather forecasting independent of place.

15 Thorpe, Volkert and Ziemanski 2002: 471–480. Here, the main argument is that Bjerknes's circulation theorem was what connected physical theory and empirical practices.

16 Fleming 2016. In this triple biography, the science of the atmosphere started with Vilhelm Bjerknes, although the concept "atmospheric science" was developed later.

17 Between 1913 and 1918, Vilhelm Bjerknes published a total of ten issues in the series *Synoptische Darstellung atmosphärischer Zustände über Europa*, the two last in collaboration with Robert Wenger.

18 Bjerknes 1943: 12.

19 Helland-Hansen 1925: 5.

20 Bjerknes 1919: 8.

21 Bjerknes 1943: 13.

22 Bjerknes 1921: 1–8.

23 Bjerknes 1943: 13; Bjerknes 1919: 5.

24 Bjerknes, 1943: 13.

25 Friedman 1989.

26 Letter from Bjørn Helland-Hansen, Th. Hesselberg, and C. Størmer to the Ministry of Church Affairs and Education, dated Kristiania, December 17, 1919. UBBHF-SC/MS.2080/S.2.1

27 Bergeron 1962: 19.

28 Bjerknes 1920: 321–346.

29 Letter from B.J. Birkeland to the Ministry of Church Affairs and Education, dated Bergen, December 1919. UBBHF-SC/MS.2080/S.2.1. See also: Nilsen and Vollset 2016: 163ff.

30 For four years, Calwagen was central to the Bergen forecasting enterprise, and the main lecturer for many of the courses. Tragically, his career ended abruptly on August 10, 1925, when he died in an airplane accident at Kjeller while conducting upper-air observations. Hesselberg 1927: 15–17 (http://gallica.bnf.fr/ark:/12148/bpt6k96335490/f31.image); Friedman 1989: 232–234.

31 Summary found in UBBHF-SC/MS.2080/I.23.

32 Letter from V. Bjerknes to the Ministry of Church Affairs and Education, dated Kristiania, May 10, 1920. UBBHF-SC/MS.2080/S.2.1; Bjerknes, V. "Betænkning om den norske veirvarsling og dens utvikling" [Report on the Norwegian weather forecasting, and its development]. Attachment to the budget proposal for 1920–21, dated Bergen, September 1919. UBBHF-SC/MS.2080/S.2.1.

33 The eight were: J.W. Sandström, head of meteorology at the Meteorological-Hydrological Bureau in Stockholm; Theodor Hesselberg, head of the Norwegian Meteorological Office; Olaf Devik, in charge of establishing a weather service for northern Norway; H.U. Sverdrup, head of the scientific work for Amundsen's polar expedition on the *Maud*; Robert Wenger, who had taken over Bjerknes's position as professor of geophysics and head of the Geophysical Institute at the University of Leipzig; his son Jacob Bjerknes, meteorologist in Bergen; Halvor Solberg, meteorologist in Kristiania (now Oslo); and himself. Attachment to the budget proposal for 1920–21, dated Bergen, September 1919: 10. UBBHF-SC/MS.2080/S.2.1

34 Bjerknes, V. Attachment to Budget application for 1920–21, addressed to Kirke og undervisningsdepartementet. UBBHF-SC/MS.2080/S.2.1.

35 Bjerknes and Solberg 1923: 3–61.

36 Bjerknes and Solberg 1926: 3–18.

37 Fujiwhara 1949. English translation found in UBBHF-SC/MS.2080/J.5.2. After returning to Japan, Fujiwhara became general director at the Central Institution for the Training of Meteorologists, professor at Tokyo Imperial University, and starting in 1941, the director of the Japan Meteorological Agency.

38 Fujiwhara 1949.

39 Quoted in: Pedersen, Johansen and Eide 1968: 19.

40 Bjerknes 1919: 9–10.

41 Pedersen et al. 1968: 24–26; Hovland 2007: 34–36; Nilsen and Vollset 2016: 205–207.

42 *Annual Report for the Geophysical Institute*, 1917–18: 68.

43 Hartvedt 1999 [1994]: 40.

44 UBBHF-SC/MS.2080/I.18.

45 Hovland 2007: 25–27.

46 Store Norske Leksikon: "Norges historie fra 1905 til 1939." ["Norwegian history from 1905 to 1939"]. Aschehoug and Gyldendal, Online: https://snl.no/Norges_historie_fra_1905_til_1939

47 Hovland 2007: 27–38.

48 Unless otherwise stated, this section is based on: Manne 2009: 103–113.

49 "Havets Rigdomme, og hvilken betydning de kan faa for norsk industri" ["The ocean's riches, and what they mean for Norwegian industry"]. *Aftenposten*, October 26, 1920; "Opsigtvækkende Meddelelser om de norske Saltfabriker. Vældige Muligheter for Magnesium." ["Remarkable report on the Norwegian salt factories. Great opportunities for magnesium."] *Verdens*

Gang, October 27, 1920; "Sjøvann som raastoff for norsk industri. Proforessor Helland-Hansens foredrag idag" ["Seawater as raw material for Norwegian industry. Professor Helland-Hansen's lecture today"]. *Dagbladet*, October 26, 1920; "Sjøvandet som raastof for norsk industri. Professor Helland-Hansens interessante foredrag i raastofkomiteen" ["Seawater as raw material for Norwegian Industry. Professor Helland-Hansen's interesting lecture in the raw materials committee"]. *Morgenbladet*, October 26, 1920.

50 Letter from Bjørn Helland-Hansen to Nansen, dated Bergen, August 12, 1921. UBBHF-SC/MS.2080/I.30.1.c.

51 Letters from Vilhelm Bjerknes in: UBBHF-SC/MS.2080/S.2.1.

52 Helland-Hansen and Nansen 1917: 1–341; Helland-Hansen and Nansen 1920a: 1–456; Helland-Hansen and Nansen 1920b: 12–28, 101–116, 347–361.

53 Helland-Hansen and Nansen 1912: 415.

54 Helland-Hansen and Nansen 1920b: 17.

55 Helland-Hansen and Nansen 1920b: 361.

56 Krogness 1917: 63–81, 104–120, 185–192, 236–246, quote p. 78.

57 Egeland and Burke 2013: 39.

58 Nansen presented the findings at the January 31, 1918, meeting of the Washington Academy of Sciences, see: *Journal of the Washington Academy of Sciences*. Vol. VIII, 1918: 135–138. See also: Huntington 1918: 483–491.

59 Helland-Hansen and Nansen 1920: 23.

60 Calculating different averages, and trying to identify overlapping and sometimes opposite cyclic patterns, was the method used most frequently in climatology.

61 Bjerknes 1919: 321.

62 The expression "exact science" was first used in Bjerknes 1904: 1–7. An English translation by Yale Mintz from 1954 was reproduced in Shapiro 1999 (http://folk.uib.no/ngbnk/Bjerknes_150/In%20memory%20of%20Vilhelm%20Bjerknes.pdf). In 1905, Bjerknes gave the same lecture to the Carnegie Institution of Wash-

ington, which resulted in the grant that was renewed until the Second World War.

63 Bjerknes 1923: 1–88.

64 Bjerknes 1926: 3–24.

65 https://www.nobelprize.org/nomination/archive/. Bjerknes was consistently blocked by individuals in the Nobel Committee, with a range of inconsistent explanations, see: Friedman 1981: 793–798.

66 Letter from Helland-Hansen to Nansen 1924, in: Jølle 2009: 615.

67 Ahlmann and Helland-Hansen 1918: 783–792.

68 Based on the annual reports from the Geophysical Institute. For more about Ahlmann, see obituaries published in *Geographical Journal*. Vol. 140, no. 3. 1974: 523–524; Hoppe and Schytt 1975: 1–5. The latter has a publication list.

69 Ahlmann and Helland-Hansen 1918: 787.

70 Ahlmann and Helland-Hansen 1918: 789.

71 Ahlmann's later studies are more quantitative. Among other things, he developed a system of glacier classification based on snow and ice temperatures, and studied the mass balance of glaciers and how they relate to climate variations. For an overview and a selected biography, see: Hoppe and Schytt 1975: 1–5 (http://www.geosociety.org/documents/gsa/memorials/v06/Ahlmann-HW.pdf).

72 Damas 1922; Annual Report of the Geophysical Institute 1921–22: 70–81.

73 Annual Report from the Geophysical Institute, 1921–22: 76.

74 Helland-Hansen, B. and G. Gran. "*Armauer Hansen*'s tredie Ekspedition til Atlanterhavet." ["*Armauer Hansen*'s third expedition to the Atlantic"]. *Bergens Aftenblad* and *Morgenbladet*. July 15, July 19, July 24, July 26, and August 5, 1922.

75 Some of the results were used in Helland-Hansen and Nansen 1927.

76 Letter from Bjørn Helland-Hansen to Harald Ulrik Sverdrup, dated Bergen, February 24, 1926. (UBBHF-SC/MS.2080/I.30.1.t).

77 Annual Report 1921–22: 77–78.

78 Nierenberg 1996: 339. The memoir also highlights Sverdrup's many students and colleagues, in particular Robert S. Arthur, John Crowell, Dale Leipper, Richard Fleming, Walter Munk, and Roger Revelle. According to Revelle, "Sverdrup was at that time (and probably still should be considered) the greatest oceanographer of our century." (Revelle 1987: 7).

79 According to historian Robert Marc Friedman, Sverdrup is "forgotten by most." Friedman 2004: 146.

80 Sverdrup 1917.

81 Sverdrup 1938.

82 Friedman 2004: 152.

83 Sverdrup 1933: 3–21; Friedman 2003 [2001]: 158–172.

84 *Roald Amundsens "Maud"-ekspedisjon 1922–1925* (Film documentary); Sverdrup 1933: 13.

85 Sverdrup 1938: 10.

86 Sverdrup 1938: 6.

87 Fjeldstad 1929; Mosby 1933.

88 Fjeldstad 1923a; Fjeldstad1923b; Harris 1904: 255–61; Harris 1911.

89 The ridge was first directly observed by Soviet oceanographers in 1948 using sonar equipment. They gave it the name Lomonosov Ridge after the 18th-century Russian polymath Martin Losomov. For more on Fjeldstad, see: Mork 1986. Offprint in: UBBHF-SC/MS.2080/I.30.1.w.

90 Fjeldstad 1935: 3–35.

91 For an updated summary of Fjeldstad's work on internal waves, see: Fjeldstad 1964.

92 Based on the annual reports from the Geophysical Institute.

93 Letter from Helland-Hansen to H.U. Sverdrup, dated November 21, 1925. UBBHF-SC/MS.2080/I.30.1.t.

94 The protocol, including list of participants, abstracts, and reports on the discussions, is found in UBBHF-SC/MS.2080/Å.4.

95 Jølle and Myklebust 2014: 445–469.

96 Sverdrup 1929–1939 (https://catalog.hathitrust.org/Record/007180300). Six hundred copies of the monographs were printed: 250 for the host institution to distribute, 250 for scientific libraries and institutions, and 100 copies for the author.

97 Mosby 1958: 321–323.

98 Friedman 2004: 170. A year prior, Sverdrup had been offered the post as the first director of Woods Hole Oceanographic Institution in Massachusetts. The position was eventually given to Henry Bigelow after Sverdrup's refusal. (Han 2010: 206–8).

99 Letter from Bjørn Helland-Hansen to Harald Ulrik Sverdrup, dated Bergen, May 21, 1930:3. UBBHF-SC/MS.2080/I.30.1.t.

100 For more on the *Norvegia* expedition's uneasy relationship balancing scientific exploration, national politics and patron Lars Christensen's commercial whaling interests, see: Roberts 2011: 63–75.

101 Sverdrup and Soule 1933; Villinger and Stetson 1933.

102 Friedman 1994: 23–26; Sverdrup 1936: 34–47.

103 Letter from Bjørn Helland-Hansen to Adolf Hoel, dated Bergen, June 18, 1934. UBBHF/MS.2080/H.8.

104 Vinzent 2006: 28–29.

105 Müller-Blensdorf, Ernst. "Information on my project of erecting a Polar Monument in Norway." Undated, p. 3. UBBHF-SC/MS.2080/H.8.

106 Undated letter from the "Committee for the erection of an International Nansen monument" to all subscribers of the Norwegian News Agency. UBBHF-SC/MS.2080/H.8.

107 Letter from B. Helland-Hansen and H.U. Sverdrup to Bergen Formannskap, dated Bergen, December 5, 1934; letter to Annet Sjöforvarsdistrikt, Bergen, from Bjørn Helland-Hansen and H.U. Sverdrup, dated Bergen, October 12, 1934. UBBHF-SC/MS.2080/H.8.

108 Among the documents in UBBHF-SC/MS.2080/H.8 is a ten-page typed list of quotes of support, as well as letters from states asking for their "own" national polar exploration hero to be included, dating from 1936.

109 The promotional material is found in UBBHF-SC/MS.2080/H.8; quote from a letter dated Oslo, July 1936. UBBHF-SC/MS.2080/H.8.

110 Letter from Sem Sæland to Bjørn Helland-Hansen, dated Oslo, October 5, 1936: 1.

111 Undated letter from the "Committee for the erection of an International Nansen monument" to all subscribers to the Norwegian News Agency. UBBHF-SC/MS.2080/H.8.

112 Letter from Bjørn Helland-Hansen to Adolf Hoel, dated Bergen, November 26, 1937 (UBBHF-SC/MS.2080/H.8). Nansen International Office for Refugees won the Nobel Peace Prize in 1938. Fridtjof Nansen received his Nobel Peace Prize in 1922 for his work on behalf of displaced victims of the First World War, including the "Nansen passport" for stateless refugees.

113 Letter from Olaf Devik to Ernst Müller-Blensdorf, dated November 24, 1937. UBBHF-SC/MS.2080/H.8.

114 Vinzent 2006: 105–6.

115 Helland-Hansen also ran his own oceanographic program, which will be further examined in the next chapter.

Chapter 4 – Moving in and moving on
pp. 93–124

1 Thompson 1928: 98.

2 Hovland 2007: 65; Helland-Hansen 1929. Apart from 75,000 NOK from a lottery fund, the building was financed wholly by local donations. The largest contributions came from Prime Minister and local shipping magnate Johan Ludwig Mowinckel, shipowners Jacob Christensen, Halfdan Kuhnle and Haakon Wallem, merchant Wilhelm Giertsen, and Consul Thorvald Halvorsen.

3 For a description of the architecture and use of "Geofysen," see: Universitetet i Bergen/Eiendomsavdelingen 2017: 44–51.

4 For a thorough contemporary account of the building and the work leading up to it,

see: Helland-Hansen 1929: 3–24. Offprint found in UBBHF-SC/MS.2080/A.10.

5 The research institute was named after Christian Michelsen, who was Norway's prime minister when Norway gained its independence from Sweden in 1905. He established a shipping company, and was for many years a government-appointed board member of Bergen Museum. Michelsen died in 1925, and the institute was founded in 1930 with a legacy from his estate. The main purpose of the institute was to keep research talent in the country, thereby avoiding a brain drain by offering free research positions with attractive conditions: at least the same salary as university professors, no teaching, and no administrative duties. (Hovland 2007: 70.).

6 Thompson 1928: 98.

7 Helland-Hansen 1918: 357–9.

8 Sverdrup, Johnson and Fleming 1942: 143 (http://ark.cdlib.org/ark:/13030/kt167nb66r/).

9 Murray and Hjort 1912: 125.

10 Op. cit.: 278–279.

11 Op. cit.: 122–125.

12 Op. cit.: 280.

13 Op. cit.: 262–263.

14 Ellingsen 2013: 112–117.

15 Ekman 1926; Ekman and Helland-Hansen 1931. The Ekman current meter was in use for a long time, but as early as 1926 it was supplemented by an electrical current meter developed by H.U. Sverdrup and Odd Dahl during the *Maud* expedition (1917–1925).

16 Letter from Bjørn Helland-Hansen to Dr. Johs Schmidt, dated Bergen, January 16, 1924. UBBHF-SC/MS.2080/I.23; Letter from Bjørn Helland-Hansen to Captain C.F. Drechsel, dated Bergen, June 27, 1924. UBBHF-SC/MS.2080/I.23.

17 Letter from Bjørn Helland-Hansen to Captain C.F. Drechsel, dated Bergen, June 27, 1924. UBBHF-SC/MS.2080/I.23.

18 Rozwadowski 2002: 74–76.

19 Op. cit. Helland-Hansen suggested that Drechsel asked Hjort to prepare an over-

view of research activities in Norway in order to force him to actually reach out to his colleagues.

20 The section for meteorology consisted of Jacob Bjerknes, Vilhelm Bjerknes, Olaf Devik, Theodor Hesselberg, Halvor Solberg, and Harald Ulrik Sverdrup – all Bergen school affiliates. The section for oceanography consisted of Fridtjof Nansen and Bjørn Helland-Hansen. The section for geomagnetism and air electricity consisted of Ole A. Krogness, Carl Størmer, Sem Sæland, and Lars Vegard. Nansen was appointed president for the national committee, with Hesselberg, director of the Norwegian Meteorological Institute, as secretary. *Annual Report*, Geofysisk Institutt 1924/25.

21 Helland-Hansen and Nansen 1927: 2. Also published as *The Cruises of the 'Armauer Hansen'*. Vol. I, no. 1. 1927.

22 Helland-Hansen 1937: 101–103. The reference to theoretical investigations referred to research such as Jonas Ekman Fjeldstad's work on tides and internal waves, and Sverdrup's use of curl to calculate the speed and direction of rotating vector fields.

23 Helland-Hansen 1940: 52–67. UBBHF-SC/MS.2080/A10. In addition to Helland-Hansen (leader), the members of the committee appointed to organize the project were: Martin Knudsen, Denmark; D. Cot, France; Albert Defant, Germany; Sir John Edgell, Britain; and Columbus Iselin, USA.

24 Defant, Neumann, Schröder and Wüst 1940–1941. See also: Wüst 1964. The other research vessels participating were the *Dana* (Denmark), the *Explorer* (Scotland), the *Atlantis* (USA), the *General Greene* (USA), and the *Thor* (Iceland).

25 Helland-Hansen 1940: 10.

26 Rossby, C.G. "The Gulf Stream in the light of experimental fluid mechanics." Undated manuscript. Found with accompanying correspondence in: UBBHF-SC/MS.2080/I.30.1.d.

27 Rossby 1932; Rossby and Montgomery 1935; Montgomery 1940. On how Rossby

became a Bergen school missionary, and what this entailed, see: Nilsen and Vollset 2017: 219–230.

28 Eliassen 1995.

29 Bjerknes 1917: 345–349. The paper built on research by the German doctoral student Herbert Perzold, who was killed at Verdun in 1916.

30 Bjerknes 1919: 1–8.

31 Bjerknes 1926: 3–38; Eliassen 1995: 10.

32 Bjerknes 1924: 3–38. See also: Eliassen 1995: 10–11.

33 Spinnangr 1926: 342–349.

34 Mieghem 1941: 97–117.

35 Palmén 1926.

36 Bjerknes 1930: 107.

37 Bjerknes and Palmén 1933; Bjerknes 1933:3–54; Bjerknes and Palmén 1935: 3–15; Bjerknes and Palmén 1937: 5–62. For an overview and analysis of these and later ascents, see for instance: Nyberg 1944: 273–394.

38 Palmén 1935.

39 Bjerknes and Palmén 1937: 5–62.

40 Bjerknes 1937: 460–466. For a discussion, see: Brunt 2011 [1939]: 365–369.

41 Rossby 1939: 38–55.

42 Untitled report from J. Bjerknes, dated Bergen, September 15, 1938. UBBHF-SC/MS.2080/I.23.

43 Dubois, Multhauf and Ziegler 2002: 26, 33.

44 Annual Report, Geophysical Institute, 1937–1938. The Bergen radiosonde was presented at the IUGG meeting in Washington in 1939 by Håkon Mosby; more on Mosby in Chapter 6.

45 Petterssen 1933: 1–92.

46 Petterssen 1935: 5.

47 Godske 1944: 50.

48 Fjørtoft 1952; Kristiansen 2017: 199–202, 225–229, 236–237. In 1950, Fjørtoft took part in making the first computer-generated weather forecast at the Institute for Advanced Study in Princeton, New Jersey, and five years later he was appointed director of the Norwegian Meteorological Institute.

49 See correspondence in UBBHF-SC/ MS.2080/I.30.1.z.

50 Krogness worked as a consultant to the Bergen municipality in investigating rusting water pipes, identifying telluric currents as the culprit (Krogness 1930a; Krogness 1930b). He also served as an expert witness in court, both after the line ship *Haakon VII* sank south of Florø in 1929, a tragedy in which 18 persons perished, and in a patent case in Oslo. Lastly, he established a magnetic station at Paradis south of Bergen, in addition to the one at Dombås that he had established in 1916 with astronomer Sigurd Einbu, who was the on-site manager.

51 Enebakk 2014: 587–608.

52 Wasserfall 1931:135–143, 293–302, 360–369, quote p. 136; Wasserfall 1930a: 2–33; Wasserfall 1930b: 3–18; Krogness 1933: 266–277; 308–317; 339–347; 361–366.

53 Wasserfall's forecasts were published in the newspaper *Aftenposten* on April 19, 1930; December 29, 1930; April 15, 1931; November 3, 1931; April 23, 1932 and October 11, 1932. For a discussion on the methodology, see: Wasserfall 1933: 298–307; Wasserfall 1941: 229–242.

54 "I should perhaps be allowed to add that a contributing factor for my application, which I initially had not thought of submitting, was that Krogness asked me to continue his work; I have not mentioned this before, but now that my application has been withdrawn, I might as well let you know." Letter from Olaf Devik to Sem Sæland, dated November 15, 1934. UBBHF-SC/MS.2080/I.30.1.0.

55 For more about the Haldde observatory, see: Larsen 2000 (http://munin.uit.no/ bitstream/handle/10037/4392/book.pdf?sequence=1).

56 Letter from Olaf Devik to Bjørn Helland-Hansen, dated Bergen November 15, 1934; Letter from Olaf Devik to the Bergen Museum board, dated November 15, 1934. UBBHF-SC/MS.2080/I.30.1.0.

57 Ore 1976: 94.

58 A full list of Trumpy's publications is printed in Ore 1976: 99–104.

59 The statistics are based on the authors' analysis of the 1264 publications listed in the annual reports from the Geophysical Institute in the period 1928–1988. During this time, the oceanographers published 100 papers in international journals and the meteorologists 67 papers, while Section C had 153 publications in international journals. As a percentage of all publications, the international journals accounted for 24.1 percent of the total publications among the oceanographers; 19.3 percent for the meteorologists, and 53.4 percent of the publications from Section C. These figures include publications in the *Geophysica Norwegica* (*Geophysical Publications*), which was produced by the Geophysical Commission in Norway, but aimed at international audiences. To a large extent, the discrepancies were made up by reports, for which the distribution was as follows: oceanographers 153 reports (36.9 percent); meteorologists 108 reports (31.1 percent); Section C 24 reports (8.4 percent). See more on publication practices in chapter 7.

60 The Norwegian Hydrographic Service [Statens Sjøkartverk] was an independent institution established in 1932 to produce official nautical charts. Prior to this, the service had been part of the Norwegian Geographical Survey, established for military needs in 1773. That the collaborator represented the oceans, and not the land, reflects how magnetic maps had been understood as significant for navigation at sea.

61 Annual Report, 1945–46: 59. Intensity describes the strength of the magnetic field at a given point, inclination its vertical angle, and declination the difference in angle between the geographic and the magnetic North Pole.

62 Annual Report, 1944–5: 53.

63 Annual Report, 1936–37: 73.

64 Trumpy, B. "Ramaneffekt und Konstitution der Moleküle I–VII." *Zeitschrift für*

Physik. B. 62, 1930: 806–823; B. 64, 1930: 777–780; B. 66, 1930: 790–806; B. 68, 1931: 675–682; B. 88, 1934: 226–346; B. 90, 1934: 133–137; Trumpy, B. "Ramaneffekt und Konstitution der Moleküle V." *Det Kongelige Norske Videnskabers Selskabs Skrifter*, Bd. 4, no. 51, 1931: 194–197; Trumpy, B. "Raman spectra of some deuterium compounds." *Nature.* 1935: 764. Trumpy, B. "Ramaneffekt und Cis-Transisomerie von Wasserstoff und Deuteriumverbindungen." *Det Kongelige Norske Videnskabers Selskabs Skrifter.* No. 29, 1935. Trumpy, B. "Polarisation der Raman-Strahlung und Konstitution der Moleküle. II" *Zeitschrift für Physik.* Vol. 98, no. 11. 1936: 672–683.

65 Dahl 1981: 153
66 Annual Report, 1942–43: 60.
67 Dahl 1981: 36.
68 Dahl 1926.
69 Sverdrup and Dahl 1926: 537–545.
70 Kvinge, Sælen and Cleveland 2005: 7.
71 Dahl 1981: 91; Dahl and Ramberg 1927. The inspiration for the postcard stunt came from one of Amundsen's staple methods for financing his polar expeditions: selling postcards and stamping them at the highest latitude achieved by the expedition, before posting them on the way home.
72 Dahl 2002: 43–47.
73 In his will, Michelsen specified four areas of priority for the institute: humanities, natural sciences, technology and medicine, and "cultural and scientific work to foster tolerance between nations and races – religious, social, economic and political." In 1992, the division for science and technology was separated into a new institute, Christian Michelsen Research, while the social science division retained the original name.
74 Dahl 1981: 146.
75 Dahl 1981: 166–167.
76 University of California. *Catalogue of Officers and Students for 1938–39: Section 1 (supplement).* February 1939: 4.
77 Annual report, 1946–47: 64–65.

78 Dahl 2002: 217–218. For more on Dahl, see chapter 6.
79 Munk and Day 2000: 309–369.
80 Letter from T. Wayland Vaughan to B. Helland-Hansen, dated La Jolla, December 11, 1935. UBBHF-SC/MS.2080/I.30.1.t.
81 Letter from H.U. Sverdrup to Bjørn Helland-Hansen, dated La Jolla, November 17, 1936. UBBHF-SC/MS.2080/I.30.1.t. Also Vaughan had corresponded with Helland-Hansen on a monthly basis, starting in 1933. See: VBBHF-SC/MS.2080/I.33.1.aa.
82 Letter from Bjørn Helland-Hansen to Bureau Chief Einar Brun at the Ministry of Church Affairs, dated Bergen, November 6, 1929. UBBHF-SC/MS.2080/I.8.
83 Letter from T. Wayland Vaughan to Harald U. Sverdrup, dated March 26, 1936. UBBHF/MS.2080/I.30.1.t.
84 Revelle 1987: 8. See also: Morgan 1999: 34–46.
85 Fleming 2004: 75–83; Fleming 2001; Nilsen and Vollset 2016: 233–242.
86 Holmboe had been an assistant to Vilhelm Bjerknes in 1925, and worked as a weather forecaster in Tromsø for two years before being recruited as a weather forecaster in Bergen in 1932. He also spent Antarctic summers as a meteorologist for Lincoln Ellsworth's Wyatt Earp expeditions, flying over Antarctica in 1933–35.
87 Nilsen and Vollset 2016: 219–226; Fleming 2016: 77ff; Byers 1960 (http://www.nasonline.org/publications/biographical-memoirs/memoir-pdfs/rossby-carl-gustaf.pdf).
88 Vaughan et al. 1937: 80–88.
89 Based on the list of publications in the Annual Reports from the Geophysical Institute.
90 Bjerknes and Solberg 1923. At the time of publication, Solberg had already left Bergen.
91 See for instance: Annual Report, 1943–44.
92 The protocol, including list of participants, abstracts, and reports on the discussions, is stored in MS.2080/Å.4.

Chapter 5 – The ocean and atmosphere as field
pp. 125–142

1   Diary of Aagot Borge, September 18, 1947. Gunnar Ellingsen's archive. Mareel is phosphorescence on the ocean caused by certain plankton species that produce light upon entering stirred water. It is typically seen in the evening or night in the wake of a boat or around the oars during rowing.

2   Ibid.

3   Thompson 1928: 98–100.

4   Roll-Hansen 1996: 22ff.

5   Auerbach 1914: 33.

6   The *Helland-Hansen* was built in 1957 and took over as the Geophysical Institute's research vessel after the *Armauer Hansen*. In 1976, it was wrecked by a series of surges at Svinøy off the northwestern coast of Norway. See chapter 6.

7   Interview with Tor Gammelsrød, May 19, 2016.

8   Livingstone 2003: 40–48.

9   Fridtjof Nansen quoted in Jølle 2011: 49. On Hjort's and Nansen's line of argument for practical field research as opposed to museum work, see Roll-Hansen 1996: 30–39.

10   Friedman 1995: 28–37. Also Swedish geophysicists had a special focus on the role of the field in geophysical sciences, such as glaciologist Hans W:son Ahlmann, see Sörlin 2011: 66–89.

11   Kohler 2002.

12   Mosby, Håkon. "Utredning om nytt forskningsfartøy" ("Considerations on new research vessel"), probably 1952–3. UBBHF-SC/MS.2080/L.

13   "Litt om geofysikk" ("Short note about geophysics"). Note by Bjørn Helland-Hansen, probably 1930s. UBBHF-SC/MS.2080/A8.

14   Jacob Bjerknes commented to his wife Honoria that Bergen offered rich opportunities to observe lines of convergence. Jewell 2007: 43–60, quote p. 43.

15   Letter from Bjørn Helland-Hansen to C.F. Drechsel, June 27, 1924. UBBHF-SC/MS.2080/I.8.

16   "Utredning om nytt forskningsfartøy" ("Report on new research vessel") by Håkon Mosby, probably 1954. UBBHF-SC/MS.2080/L.

17   Retrospective note by Odd Sælen, undated. UBBHF-SC/MS.2080/I.16.

18   In 1928 he spoke in Berlin on the topic "Ocean research with small vessels" ("Havforskning med smaa fartøier"). Annual report, 1927–28.

19   Annual report, 1936–1937.

20   On "Bergen values" in oceanography, see Hamblin 2014. Hamblin argues that the influence of "Bergen values" on international postwar oceanography included both potentials and limitations.

21   Thompson 1928.

22   Auerbach 1914: 12.

23   Diary of Olav Mosby, July 26 ,1927. Peter M. Haugan's Archive.

24   Diary of Olav Mosby, July 23, 1927. Peter M. Haugan's Archive.

25   Diary of Olav Mosby, August 4, 1927. Peter M. Haugan's Archive.

26   Diary of Olav Mosby, July 24, 1927. Peter M. Haugan's Archive.

27   Diary of Olav Mosby, July 16, 1927. Peter M. Haugan's Archive.

28   Diary of Aagot Borge, September 18, 1947. Author's Archive.

29   Letter from Bjørn Helland-Hansen to his mother Nikoline Mathilde Helland, May 20, 1922. UBBHF-SC/MS.2080/I.30, "Personalsaker."

30   Ibid.

31   Letter from Håkon Mosby to an unknown recipient, probably a family member, on board the *Norvegia*, November 12, 1927. UBBHF-SC/MS.2080/"Mosby."

32   Letter from Bjørn Helland-Hansen to his wife Annemis, July 29, 1930. UBBHF-SC/MS.2080/I.30. "Personsaker. Personalia."

33   Diary of Olav Mosby, July 26, 1927. Peter M. Haugan's archive.

34 Diary of Olav Mosby, July 1927. Peter M. Haugan's archive.

35 Diary of Olav Mosby, July 23, 1927. Peter M. Haugan's archive.

36 "Forskningsfartøyet Armauer Hansen." Note by Odd Sælen. Undated. UBBHF-SC/MS.2080/I.16.

37 Diary of Olav Mosby, July 23, 1927. Peter M. Haugan's archive.

38 There were several traditions of belief concerning dreaming about or mentioning horses at sea. Solheim 1940: 61–62. Most of them connected dreaming about or naming horses to the coming of bad weather, Fulsås 2003. On traditional superstitions about mentioning different kinds of animals by their proper names at sea, see Døssand, Løseth and Elstad 2014: 533–535. Harald Beyer Broch also points to a superstition connected to horses in his anthropological study of superstition among fishermen at sea today. Broch 2016: 73–90.

39 Diary of Olav Mosby, August 15, 1927. Peter M. Haugan's archive.

40 Schwach 2000: 131.

41 Liljequist 1980/81: 409–415.

42 Nilsen and Vollset 2016: 157–162.

43 Jewell 1980/81: 484 (quoting Erik Bjørkdal in 1943).

44 Nielsen and Vollset 2016: 169–170.

45 "Report of the Proceedings of the seventh meeting of the International Commission for the Investigation of the Upper Air Held in Bergen 25th-29th July 1921." UBBHF-SC/MS.2080/I.30 (p. 15).

46 "Report of the Proceedings of the seventh meeting of the International Commission for the Investigation of the Upper Air Held in Bergen 25th-29th July 1921." UBBHF-SC/MS.2080/I.30 (p. 24).

47 Letter from Jacob Bjerknes to the Bergen Museum Research Foundation, November 26, 1932. UBBHF-SC/MS.2080/I.30. Ballooning had a special place in the Bjerknes family. His uncle Ernst tells a story from his childhood about how he and his father and brothers discovered remains of the French balloon "La Ville D'Orleans," which unintentionally left Paris, crossed the North Sea and ended up in Norway in November, 1870. Bjerknes 1944: 35.

48 Letter from Jacob Bjerknes to the Bergen Museum Research Foundation, November 29, 1934. UBBHF-SC/MS.2080/I.30.

49 Annual Report, 1934–35: 71.

50 Letter from Jacob Bjerknes to Nils Russeltvedt, March 16, 1935. UBBHF-SC/MS.2080/I.30.

51 Letter from Nils Russeltvedt to Jacob Bjerknes, March 6, 1935. UBBHF-SC/MS.2080/I.30.

52 Letter from Nils Russeltvedt to Jacob Bjerknes, March 6, 1935. UBBHF-SC/MS.2080/I.30.

53 Letter from Jacob Bjerknes to Nils Russeltvedt, August 31, 1935. UBBHF-SC/MS.2080/I.30.

54 Letter from Jacob Bjerknes to the Bergen Museum Research Foundation, November 29, 1934. UBBHF-SC/MS.2080/I.30.

55 Annual report, 1937–38.

Chapter 6 – The oceans under surveillance
pp. 143–176, 241–246

1 The exception is 1958–1962, when meteorologist Carl Ludvig Godske was head of the institute. For more on Godske and his farewell to the Bergen school of meteorology, see the next chapter.

2 Forland and Haaland 1996: 178–182.

3 Nilsen and Vollset 2016: 249–262.

4 Mosby 1954: 378–384.

5 Haugan 2008 (http://www.uib.no/gfi/57708/bygningens-historie).

6 Mohr 1969: 193; Forland and Haaland 1996: 224.

7 The section for oceanography had seven positions, the meteorologists five, and the Section C had only one position. In addition, four positions were shared among the three sections. Hovland 2007: 100.

8 Roll-Hansen 1996: 214–215.

9   The research councils created in the aftermath of the Second World War were: the Norwegian Council for Scientific and Industrial Research [Norges teknisk-naturvitenskapelige forskningsråd, NTNF, 1946], the Norwegian General Science Council [Norges almennvitenskapelige forskningsråd, NAVF, 1949], and the Norwegian Agricultural Research Council [Norges landbruksvitenskapelige forskningsråd, NLVF, 1949].

10 Njølstad and Wicken 1997: 504.

11 Sælen 1990: 3.

12 Roberts 2011: 63–75.

13 Ocean circulation was increasingly recognized as interesting for physical oceanographers, see: Mills 2009.

14 Mosby 1933; Mosby 1934. The latter dissertation focused on the surface and deep waters of the Atlantic Ocean and western part of the Indian Ocean south of the 55th parallel.

15 Mosby 1934. A similar view had been put forward by German oceanographer Wilhelm Brennecke in 1921 and formed one of the main debates in German oceanography in the 1920s, see: Mills 2009: 144–150; Mills 2005: 246–264.

16 Mosby 1936: 281–286. In a comparison between satellite techniques and five empirical formulas published in 1988, Mosby's equations yielded the best predictions. Frouin, Gautier, Katsaros and Lind 1988: 1016–1023.

17 Deacon 1935: 326–7.

18 Mosby 1951: 377. The review was probably written in 1937, see Ellingsen 2013: 97–98.

19 Mosby 1951: 215.

20 Mosby 1976: 209–323. The overview comprised both physical and biological research.

21 Mosby 1976: 221–222. The same argument is made in Mosby 1986: vii.

22 Mosby 1986: vi–ix.

23 Defant 1941: 191–260.

24 Mosby 1962: 289–313; Mosby 1972: 1–60.

25 Annual report, 1952–53: 83–4.

26 Forland and Haaland 1996: 286.

27 Bjørn Helland-Hansen was onboard while it was being built, but died on September 7, 1957, at the age of 79.

28 The *Armauer Hansen* took part in the first cruise with the *Helland Hansen*, a six-week expedition to the southern Norwegian Sea starting on June 4, 1957. The vessel was sold on January 6, 1959.

29 Mosby 1958: 402–412, quote p. 412.

30 Interview with Steinar Myking, May 18, 2016.

31 For more on this and other expeditions, see: Wüst 1964: 1–52.

32 Tait 1967 (http://ocean.ices.dk/Project/OV60/RPV157.pdf).

33 Nilsen and Vollset 2016: 274–276.

34 Hogstad 1998.

35 Hogstad 1998: 17.

36 Hogstad 1998; Nilsen and Vollset 2016: 274–276.

37 See for instance: Mosby 1959: 1–62; Bøyum 1965: 1–19; Mosby 1970: 1–60; Gammelsrød, Østerhus and Godøy 1992: 68–75; Østerhus and Gammelsrød 1999: 3297–3304.

38 Committee on Oceanography. "12. Marine sciences in the United States 1958." *Oceanography 1960 to 1970*. National Academy of Sciences/National Research Council., Washington D.C. 1959: 4.

39 Rice 1994: 139.

40 Hovland 2007: 100.

41 The International union of Geodesy and Geophysics lists the following number of attendees at their General Assemblies: Stockholm (1930): 331, Lisbon (1933): 200, Edinburgh (1936): 344, Oslo (1948): 368, Brussels (1951): 918, Rome (1954): 923, Toronto (1957): 923, Helsinki (1960): 1375, Berkeley (1963): 1938, Zurich (1967): 2200, Moscow (1971): 2577, Grenoble (1975): 2564, Canberra (1979): 1944, Hamburg (1983): 3204, Vancouver (1987): 3939, Vienna (1991): 4331, Boulder (1995): 4481, Birmingham (1999): 4052, Sapporo (2003): 4151, Perugia (2007): 4375, Melbourne (2011): 3392, Prague (2015): 4231. The number of participants at the first three General Assemblies (Rome 1922, Madrid 1924, and

Prague 1927) are not listed. http://www. iugg.org/assemblies/

42 Wolff 2010.

43 Ellingsen 2013: 159–166; Hamblin 2005: 99–139.

44 Hamblin 2005: xxvii, 219.

45 Krige 2000: 105–106. In addition to education, a dynamic industry, economic growth, and security, science and technology were seen as playing an important cultural role in strengthening the transatlantic community as well as NATO's reputation.

46 Doel 2003: 645–666, quote p. 639.

47 Mosby 1959: 179; Ellingsen 2005: 163–164.

48 Mosby 1959: 181.

49 Mosby 1959: 181. In 1959 SCOR started a series of International Oceanographic Conferences, which continued until 1988 under the name "Joint Oceanographic Assemblies." In addition, the organization set up working groups with narrower scopes. The first working groups were on radioactivity in the oceans, carbon dioxide in the oceans, and measuring the biological production of the sea. For the fourth, on the physical properties of the sea, SCOR and IACOMS appointed the same members. By 2017, SCOR had established 155 such groups. SCOR also arranged large-scale ocean research projects focusing in particular on the Indian Ocean.

50 Mosby was the third Norwegian who served as president of IAPO, after Bjørn Helland-Hansen (1934–1946) and Harald Ulrik Sverdrup (1946–51).

51 In 1954, Robert S. Dietz from the US Coast and Geodetic Survey pointed out: "As there are only a few workers, Norway has lost the preeminence she once enjoyed in oceanography." (Dietz Papers, Office of Naval Research London Technical Report ONRL-85-54, quoted in Ellingsen 2013: 172). The understanding of a "fall" or "stagnation" in physical oceanography was shared by the Norwegian oceanographers themselves. (Ellingsen 2013: 189).

52 Mosby, Håkon. "Letter from Norway in Response to Circular Letter Inviting Proposals, Oslo, September 11, 1956." UNESCOR, Unesco/NS/OCEAN/CIRC/3, quoted in: Hamblin 2005: 104. For more details, see: Hamblin 2005: 104–111.

53 Weidemann, Hartwig. *A Manual of Current Measuring Instruments and Methods.* International Association of Physical Oceanography. Publication Scientifique No. 21, preliminary issue submitted to the Editorial Committee on Current Measurement for discussion during the UIGG Congress, Helsinki 1960. In: UBBHF-SC/MS.2080/B.8. For a more in-depth description of most of the instruments indexed, see: Johnson and Wiegler 1959; Barnes 1959.

54 Barnes 1959: 93. Mosby also developed other instruments. One of them, an instrument to measure currents at twelve different elevations off the seafloor, was mockingly given the name "Mosby's Christmas tree" by his colleagues. He also tried to develop a tool to measure turbulence using two pendulums of different weights, but this never succeeded.

55 For a brief historical overview of the definition of salinity, and the instruments used to measure it, see: Golmen and Østerhu 1983: 135–137.

56 In 1955, Chr. Michelsen Institute moved into its own offices across the street from the Geophysical Institute, making way for the later staff expansion. Until then they had been located in one of the wings on the first floor of the Geophysical Institute. See: Enge 1998: 3 (http://docplayer.me/16725914-Fysisk-institutt-50-ar.html).

57 Speech by Håkon Mosby to NATO's Parliamentary Congress, November 6, 1963, quoted in Ellingsen 2013: 176, 179.

58 Ellingsen 2013: 178.

59 Dahl 1969; Ellingsen 2013.

60 Annual report, 1965–66: 130. For more on earth surveillance in the geosciences, see: Turchetti and Roberts 2014.

61 SCOR/IAPO/UNESCO. "An intercomparison of some current meters, report on an experiment at WHOI Mooring Site 'D,'

16–24 July 1967 by the working group on Continuous Current Velocity Measurements." *UNESCO technical papers in marine science.* No. 11. 1967: 1–70; Annual report, 1967–68: 122.

62 Hufford and Seabrooke 1970: 1–5.

63 Ellingsen 2013: 181–182.

64 Bø, Foldvik and Kvinge 1973: 147–148.

65 Gade 1979: 189–198; Gade 1993: 139–165; Miles, Travis et al. 2016: 16–29.

66 Mork 1966: 1–22. An exception was Mork's follow-up to Jonas Fjeldstad's work on internal waves, using observations from the weather ship station M. Mork developed a theory for how a passing low pressure area could influence temperatures several hundred meters below the surface, including new mathematical methods for solving the equations. Mork 1972: 184–190.

67 Gammelsrød 2018: 116–126.

68 Mork 1971; Orvik and Mork 1993: 114–126. Martin Mork also collaborated with colleagues at the Marine Research Institute, Chr. Michelsen Institute and the University of Rhode Island to develop the dropsonde "Yvette," used to measure small-scale vertical shear, differences in speed and direction of currents, and in situ density. Evans, Rossby, Mork and Gytre 1979: 703–718. He later organized a cruise to Bermuda to study the dynamics of the Gulf Stream, using underwater gliders.

69 Baker in Warren and Wunsch 1981: 405. The instrument was also called the "NATO-buoy" and the "Bergen current meter." See Ellingsen 2013 for a history of the instrument.

70 Baker 1981: 406.

71 More on how the geophysicists in Bergen created the University's IT-department can be read in the next chapter.

72 For a description of the instrument, and how the data were recorded and exported, see: Dahl 1969: 103–106; Ellingsen 2013.

73 The company is now called Aanderaa Data Instruments, and has been a subsidiary of Xylem, Inc. since 2011. Ellingsen in Hovland 2007: 102–115.

74 Instead of being an actual bottle, the Niskin bottles consisted of a tube open to the water at both ends, made of plastic to make them more maneuverable, and designed to be used in multi-bottle arrays. Part of the development was financed by the US Office of Naval Research. Niskin 1964: 591–594.

75 The *Helland Hansen* was upgraded with new CTD facilities in 1976, as well as a new sonar, two echo sounders, and new equipment for hydrographic measurements. Three years later, Halvor Røyset and A. Evjen developed a software suite to work with the CTD data directly on a computer. In addition to converting from the instrument tape to a digital computer tape, the software conducted automatic quality control looking for outliers and established average values according to different parameters. It also produced graphical TS diagrams and station curves, and could draw contours of vertical sections using a Calcomp plotter. (Annual Report, 1978) For more on the early development of the CTDs, see: Warren and Wunsch 1981: 216–219.

76 Interview with Steinar Myking, May 18, 2016.

77 Turchetti 2012: 205–231.

78 Ellingsen 2013: 162–3.

79 Turchetti 2012: 211–212.

80 Turchetti 2012: 210.

81 The other Scandinavian NATO-members, Denmark and Iceland, also ban the storage of nuclear weapons in peacetime.

82 Ellingsen 2013: 173–174.

83 Mosby became head of the Faroe-Shetland project, while Henri Lacombe from France, who had already headed several expeditions with similar goals, became head of the Gibraltar project. The latter began in May-June 1961, when the *Helland Hansen* and five research vessels from France, Italy, and Belgium took part. Pratt 1990: ix.

84 Mosby 1967: 6.

85 Ellingsen 2013: 174–5. The subcommittee also organized non-field experiments, such

as the Hindcasting project, which started in 1960. This was aimed at analyzing previous observations from the North Atlantic and the Norwegian Sea in order to develop predictive models for currents and sea levels. The subcommittee also gave economic support to the construction of a rotating model of the Gibraltar Strait to analyze mechanisms involved in a laboratory setting.

86 Ellingsen 2013: 148, 182–187.
87 Hamblin 2005: 231.
88 Krige 2000: 101–102.
89 Krige 2006: 257.
90 Thor Kvinge was a visiting professor in 1964–65, followed by Ola M. Johannessen.
91 For an overview of current meter moorings and the paths for cold shelf water entering the deep Weddell Sea circulation, see: Foldvik 2004. For an analysis of the long-term variability of the oceanographic circulation in the area, see: Nicholls and Østerhus 2004.
92 Sylte 2010: 3.
93 Nilsen and Vollset 2016: 336–337.
94 For an accessible overview, see: Gade in Hurdle 1986: 183–189.
95 Gade 1963: 1–62; Johannessen 1965; Gade 1967: 1–104; Johannessen 1968: 3–38.
96 Sverdrup, Johnson and Fleming 1942. For more about the genesis of *The Oceans*, see: Day 2003 (http://scilib.ucsd.edu/sio/hist/day_the_oceans_history.pdf). For Sverdrup's efforts at Scripps during the war, see: Oreskes and Rainger 2000: 309–369.
97 Hognestad 1999. Prior to this, in 1952–53, Mosby was head of the Norwegian Geophysical Association, established in 1917.
98 For more on the Norwegian Coastal Current Project (1975–80), the first interdisciplinary collaboration which included marine biologists, see chapter 9.
99 Eggvin 1940: 1–153 (http://hdl.handle.net/11250/114836). For more on Eggvin, who remained head of the Marine Research Institute's division for physical oceanography until 1969, see Schwach 2000; Sætre and Blindheim 2002/W:01.

(http://hdl.handle.net/11250/100525). A brief obituary, written by his successor Lars Midttun, was printed in *Fiskets Gang*. No. 9, 1989: 24.
100 Schwach 2000: 226.
101 Schwach 2000: 246.
102 Geophysical Institute, Annual reports. More on the meteorologists in the next chapter.
103 Interview with Svein Østerhus, November 17, 2017. See also: Mork in Utaaker 1999: 77.
104 Mosby 1971: 35; 80–85.
105 Mosby 1971: 71.
106 Interviews with Ola M. Johannessen, March 27, 2017; Interview with Thor Kvinge, May 13, 2016. Mosby was Johannessen's supervisor for his graduate thesis, and then hired him as an assistant using NATO funds.
107 Oreskes 2014: 170.
108 Oreskes 2014: 170. See also: Nowotny, Scott and Gibbons 2001.
109 Oreskes 2014: 171.
110 The popularizations were limited to a longer article with the title "Physical oceanography, a natural science" ["Fysisk oseanografi, en naturvitenskap"], written by Herman Gade in the local daily *Bergens Tidende*, and three papers in the University's outreach journal *Naturen*: a popularization on fjord dynamics and one on the ocean circulation system, both written by Gade in 1976, and a report on the hunt for the world's coldest ocean current, written by Gade and Arne Foldvik in 1978.
111 Porter 2009: 292–309.
112 Porter 2009: 292.
113 After independence in 1905, the Norwegian state exam (cand. real.) required passing four out of seven minors, followed by a major thesis in one of the subjects. Each minor consisted of lecture series that could last up to two years, reading, and a final exam at the end; the thesis consisted of supervised research, often on a subject decided by the professor. Over time, the minors grew from specialties to full-fledged disciplines with specialties of

their own, and by the Second World War the required number of minors had been reduced to two. Still, the time from enrollment to handing in the thesis remained seven to eight years. In Sverdrup's new model, inspired by the system he had encountered in the United States, the minors were replaced by shorter courses. This gave students more flexibility to compose their undergraduate degrees, and abandoned the system whereby it could take many years of studying before one could take the examinations. The reorganization made it possible to expand the number of students, gradually opening the Norwegian universities for mass education. Finally, the Sverdrup plan was more teaching intensive, and the need for lecturers became a rationale for new university positions. A draft of the plan from 1950, for discussion among Norwegian geophysicists, is found in UBBHF-SC/MS.2080/T.5. In 1959 the Faculty of Mathematics and Science was the first at the University of Bergen to introduce the Sverdrup plan, and the following year, the Geophysical Institute had its first staff expansion, a teaching position. The position was filled by Arne Foldvik, who, since his days as a student, had been attached to Einar Høiland's "Oslo school" of meteorology based at the Institute of Astrophysics at the University of Oslo. More on Foldvik's research and the shift from meteorology to oceanography will appear in the following chapter.

114 Mosby 1971: 101–103.
115 Annual report, 1955–56.
116 Interview with Steinar Myking, May 18, 2016.
117 Dahle and Kjærland 1980: 2–13.
118 Interview with Steinar Myking, May 18, 2016.

Chapter 7 – Matters of scale in meteorology and earth physics
pp. 247–278

1 Unsigned. "The International Union of Geodesy and Geophysics." Note, dated August 11, 1948. UBBHF-SC/MS.2080/I.30.1.jj. Correspondence and planning documents are found in this folder and in UBBHF-SC/MS.2080/I.30.1.v.
2 Letter from Carl Størmer to Bjørn Helland-Hansen, dated Oslo, June 19, 1928. UBBHF-SC/MS.2080/I.30.1.g.
3 Instead, fellow student Anton Jakhelln, who planned to specialize as an oceanographer, became Nansen's last assistant. Nansen died in 1930.
4 Published by Springer Verlag, Berlin. The following year, a French translation with the title *Hydrodynamique physique, avec applications à la météorologie dynamique* was published by Les Presses Universitaires de France, Paris.
5 Godske 1934 (http://adsabs.harvard.edu/full/1933POslO...1K...1G); Bjerknes and Godske 1936.
6 "No one has more reason than me to be thankful for Godske's efforts, and I could not decide to use my legal right to displace him from the professorship in Bergen." Letter from Jacob Bjerknes to Bjørn Helland-Hansen, dated Santa Monica, California, December 3, 1945:1 UBBHF-SC/MS.2080/I.30.1.s.
7 Letter from C.L. Godske and Finn Spinnangr to Herr Kansleren for St. Olavs Orden, dated Bergen, May 9, 1947. UBBHF-SC/MS.2080/T.5.
8 Letter from Godske to Jacob Bjerknes, dated Bergen April 28, 1949. (UBBHF-SC/MS.2080/T.5). By then, Godske's frustration had been growing for more than a year. As Godske put it in a letter to Bergeron in January 1948, found in the same box: "[The paragraph] is so unreadable that I almost feel like sending it to you as a representative sample of a Bob-ian original manuscript – but my heart bleeds thinking of the sorrows this would cause

and I will take on the task of the first D.B.P. (De-Bobification Process)." Letter from Carl Ludvig Godske to Tor Bergeron, dated Bergen, January 15, 1948.

9 Letter from Carl Ludvig Godske to Robert "Bob" Bundgaard, dated October 7, 1949. UBBHF-SC/MS.2080/T.5.

10 Utaaker 2008 (http://www.uib.no/gfi/54202/cl-godske).

11 Letter from Carl Ludvig Godske to Erik Palmén, dated Bergen, March 6, 1951. UBBHF-SC/MS.2080/T.5.

12 Sawyer 1957: 560; Fleming 2016: 72–73. The reference to rapid development in other branches of meteorology referred in particular to numerical forecasting and the use of simpler quasi-geostrophic equations to make the calculations more viable.

13 Vollset 2015: 83–97.

14 Utaaker 1998: 6. For more on Høiland's research group in Oslo, and his Institute for Weather and Climate Research, see: Vollset 2015: 83–97 (http://meteohistory.org/wp-content/uploads/2015/09/07-Vollset-Asking-too-much.pdf).

15 Godske 1948: 21. UBBHF-SC/MS.2080/I.30.1.jj. One worry reflected in the correspondence was that the growing aviation industry was turning weather forecasters into "valets of the aerial taxi drivers." Godske, as quoted in a letter from Tor Bergeron to W. Blecker, dated Jämtland, Sweden, August 9 1950. UBBHF-SC/MS.2080/T.5. See also: Kristiansen 2016: 76. The expression paraphrased the English mathematician Lewis Fry Richardson, who in 1922 had published the first attempt at calculating the weather using Vilhelm Bjerknes's equations.

16 Nilsen and Vollset 2016: 278.

17 Nilsen and Vollset 2016: 276.

18 Fleming 2004: 75–83; Nilsen and Vollset 2016: 233–242.

19 After the first meeting of the Norwegian Geographical Society after the war, with King Haakon VII among those in the audience, Petterssen's lecture on how Norwegian meteorologists had been involved in the war effort was referred to as a highlight. (Unsigned. "Nye metoder i værvarslingen" ["New methods of weather forecasting"], *VG*. September 20, 1945.)

20 Nilsen and Vollset 2016: 272–280. In 1948, Petterssen accepted the position as head of meteorological research for the US Air Force and returned to the United States.

21 Between 1945 and 1954, 12 of 22 new members of the Norwegian Geophysical Association were weather forecasters. Rather than using the criterion of publications, a recurring argument for membership was that a forecast made using Bergen school methods was a scientific product. Dannevig 1991: 20.

22 Nilsen and Vollset 2016: 274–276.

23 Nilsen and Vollset 2016: 272–280. In 1948 Petterssen moved to the United States, where he served as the head of meteorological research for the US Air Force for four years, after which he became professor of meteorology at the University of Chicago.

24 Harper 2008: 75–76, 87; Turner 2010: 140.

25 Nilsen and Vollset 2016: 222–226.

26 Letter from Carl-Gustaf Rossby to Col. H.H. Bassett, dated March 13, 1943. Institute Archives and Special Collections, MIT Libraries, Cambridge, Massachusetts, University Meteorology Committee Papers (MC 511), Box 1, as quoted in Turner 2010: 154. Emphasis in the original.

27 Harper 2007: 80; Liljequist 1981: 421–423; Eliassen, Blanchard and Bergeron 1978: 387–392. During his visit to Moscow, he was assisted by his Russian student Vera Romanovskaja, who later became his wife.

28 Nebeker 1995: 191. Friedman 1989: 246.

29 Nilsen and Vollset 2016: 343–345.

30 Nilsen and Vollset 2016: 331.

31 Godske continued to make the same argument throughout his career, see for instance: Godske 1969: 52–63.

32 Godske 1944.

33 Nilsen and Vollset 2016: 253.

34 Spinnangr 1939: 434–455; 1940: 104–123, 216–226; Spinnangr 1940: 1–39; Spinnangr and Hjort 1941: 1–38; Spinnangr 1942: 1–63.

35 Johansen 1964: 37–41.

36 Andersen 1969: 1–20; Andersen 1972: 1–20; Andersen 1973a: 1–20; Andersen 1973b: 1–22; Andersen 1975: 377–399.

37 Nilsen and Vollset 2016: 265–268.

38 Godske 1944: 3. See also: Utaaker 1998: 5. For more about the origins of micrometeorology, see: Coen in Fleming et al. 2006: 115–140.

39 Letter from Olav Skard to Dr. Godske, dated Ås, July 15, 1945. UBBHF-SC/MS.2080/T.5.

40 See for instance: Utaaker 1963; Utaaker 1966; Gjessing 1969; Skaar 1972; Utaaker 1965; Utaaker and Skaar 1970.

41 Kåre Utaaker, Endre Skaar and Yngvar Gjessing produced in-house reports on the consequences of hydropower development in a number of locations: Tokke-Vinjevassdraget (1976), Tovdalsvassdraget (1977), Skjåk (1977), Alvdal (1978), Sarvsfossen-Nomelandsmo (1979), Glomma-Renavassdraget (1980), Sundsbarmvatn (1983), Røldal (1983), Surnadal (1984), Etne (1985), Orkdalen (1986), and Nerskogen (1986). These and other publications from the Geophysical Institute's internal report series are available online: http://www.uib.no/gfi/55145/rapportserie. See also: Faugli, Erlandsen and Eikenæs 1993 (http://publikasjoner.nve.no/publikasjon/1993/publikasjon1993_13a.pdf).

42 In a letter to Olav Skard at the Agricultural College in 1945, Godske explained: "Ordinary climatology, where one works schematically, requires quite long time series, at least 5–10 years. However, if one takes into account the weather characteristics for each year, by analyzing the material statistically according to weather types, it will be possible to reduce the length of the time series considerably." Letter from C.L. Godske to Olav Skard, dated Bergen, July 6, 1945: 3. UBBHF-SC/MS.2080/T.5.

43 Letter from Godske to Lindebrekke, dated August 26, 1946. UBBHF-SC/MS.2080/T.5.

44 Godske 1971; Bergen Byarkiv. "Godskegården." *Oppslagsverket oVe*. 2010 (http://www.bergenbyarkiv.no/oppslagsverket/2010/10/05/godskegarden/).

45 Utaaker 1998: 11.

46 For more on Godske's philosophy and world view, see: Godske 1946: 9–26; Godske 1971.

47 Utaaker 1956; Utaaker 1958; Utaaker 1961.

48 Paulsen 1959.

49 Annual Report from the Geophysical Institute 1954–55: 103. A year later, he concluded that that the difference in light conditions inside and outside the smoggy area did not seem sufficient to influence the growth conditions of the forest.

50 Paulsen 1969.

51 Annual Report, Geophysical Institute. 1973: 133. The reference was probably alluding to the United Nations Conference on the Human Environment, held in Stockholm in the summer of 1972, where man's capacity to impact his environment and the need to safeguard the environment for future generations was the major topic: http://www.un-documents.net/aconf48-14r1.pdf

52 From the early 1950s to the end of the Cold War in 1989, more than 30 geophysicists from the Soviet Union visited Bergen. Meteorologists and oceanographers from Bergen also made frequent visits to the other side of the Iron Curtain, especially from the 1970s onward, under the auspices of a cultural exchange program between Norway and the Soviet Union. For more information on the Bergen Radiation Observatory, see: http://www.uib.no/en/rg/meten/57124/radiation-observatory

53 World Meteorological Organization. "Resolution 12: Collection and publication of data in physical meteorology" and "Annex XIII." *Fourteenth session of the executive committee. Geneva, May 20–20 June 1962.* WMO – No. 121.RC.21. Geneva. 1962: 69–70, 202–204 (https://library.wmo.int/pmb_ged/wmo_121_en.pdf). Incidentally, the same meeting decided the standard nomenclature for the outer atmosphere, layers based on its temperatures. In the troposphere (0–10 km altitude), tempera-

tures decrease by altitude. In the strato-sphere (10–50 km altitude) the temperature increases with height. In the mesosphere (50–90 km altitude), it decreases yet again, before finally increasing in the upper ther-mosphere that reaches into space.

54 Volumes 22–46 (1986–2010) of the *Radia-tion Yearbook* are available online: http://www.uib.no/en/gfi/56743/radiation-year-books. Until 1988, the Geophysical Insti-tute also published yearbooks containing the annual observations from the Magnetic Bureau's station at Dombås. The first volume of *Results from the Magnetic Station at Dombås* was published in 1935 with observations going back to 1916.

55 Nilsen and Vollset 2016: 357.

56 The first EMMA computer soon became outdated, and in the first decade it was supplemented, and eventually replaced, by an IBM 1620 I (1962/63), an IBM 1620 II (1964) and an IBM 360/50H (1967). For an overview of the first scientific computers in Norway, see: Berntsen in Bubenko, Impagliazzo and Sølvberg 2003: 23–32.

57 Charney, Fjørtoft and von Neumann 1950: 237–254; Kristiansen 2017; Harper 2004: 84–91; Harper 2008: 91–150; Nebeker, 1995: 135–181; Nilsen and Vollset 2016: 310–313.

58 Unsigned. "Nye metoder for værvarsling" ["New methods for weather forecasting"]. *VG*. June 20, 1958: 5.

59 Peter Andersen worked on weather types, while Jon Helge Knudsen studied how the stochastic (random) processes could be used to construct more realistic models for practical prediction. The latter would, in 1971, go on to defend his PhD at Princeton.

60 Godske 1965: 1.

61 Godske, Andersen, Jakobsson and Johan-sen 1969: 2.

62 Godske 1952.

63 Godske 1962: 5–7. Bjørgum changed depart-ments, opening the permanent position for Godske's field assistant, Kåre Utaaker. In the early 1960s, Bjørgum received funding from the American military to study turbulence, including a year doing research

at the Mathematics Research Center at the University of Wisconsin in Madison, operated by the US Army.

64 Utaaker 1974: 36.

65 Thornthwaite 1953: 46.

66 Vollset 2016: 93–94.

67 Monomer 1999: 15–17. For an early example of weather typing, see: Bigelow 1897.

68 Fleming 2004: 80

69 Krick and Fleming 1954: 180–181, in: Flem-ing 2004.

70 Lorenz 1963: 130–141.

71 Nielsen and Vollset 2016: 336.

72 Interview with Arne Foldvik, May 5, 2016.

73 Roll-Hansen 1996: 167–177.

74 Graue 1998 (http://docplayer.me/16725914-Fysisk-institutt-50-ar.html).

75 Gjellestad was the first female scientific leader in Bergen, and her students have different opinions on how this affected her activities. According to Karsten Store-tvedt, male chauvinism was rampant, and she faced an uphill struggle for respect among the other geophysicists. According to Jacqueline Naze Tjøtta, one of Gjelle-stad's female students, gender was never a concern. Kalvaag 2008: 34–35.

76 Wegener 1922; Storetvedt in: Utaaker 1999: 51–55.

77 Oreskes 1988: 311–348.

78 Runcorn 1965: 1–11.

79 Storetvedt 2008 (http://www.uib.no/gfi/55528/guro-gjellestad).

80 After Runcorn's lecture, Gjellestad received funding from L. Meltzers Høyskole-fond, Nansenfondet and the Norwegian Research Council for Science and the Humanities.

81 Storetvedt 1965: 585–586; Storetvedt and Gjellestad 1966: 59–61.

82 For a thorough chronicle of the clash between "movists" and "fixists," focusing in particular on the birth of plate tectonics in the 1960s, see: Frankel 2012.

83 In addition to Newcastle, the annual reports from the 1960s and 1970s list guests from Liverpool, London, Cambridge, Brno, Lund, Copenhagen, Aarhus, Den

Burg, Munich, Ann Arbor, Utah, Ottawa, Toronto, Moscow, Tokyo, and Yamaguchi.

84 In comparison, 14–18 percent of the meteorologists' publications in the 1960s, 1970s and 1980s were in international journals, while for the oceanographers the proportion increased from 7 to 31 percent in the same period.

85 Annual report, 1969–70.

86 Storetvedt 1999: 52.

87 Storetvedt 2003: 18; Storetvedt 1997: 22. By then Storetvedt had held several leading positions internationally: from 1976 to 1986, he was a board member of the European Geophysical Society (EGS), EGS treasurer (1982–88), and foreign member of the budget and finance committee of the American Geophysical Union (1988–90), and he has traveled extensively for field work and research stays.

88 Storetvedt 1997: 403. In an interview in May 2016, Storetvedt likened himself to the Bergen school of meteorology: "It is not often two great theories originate at one institute, as is the case with the Bergen School and me. But I tackle bigger questions, like the origins of life."

89 Storetvedt 2003: 21.

90 Storetvedt 1997; Storetvedt 2003; Storetvedt 2005. Global Wrench Tectonics is also presented at his website http://www.storetvedt.com

91 Storetvedt 2003: 13.

92 Storetvedt's perspective is explicitly influenced by Thomas Kuhn's *Structures of Scientific Revolutions* (1962), and uses many of Kuhn's concepts to position himself as the forerunner of a scientific revolution long overdue: Paleomagnetism is riddled with anomalies, he contends, but today's scientific leaders, in a host of different scientific disciplines, have invested interests in continuing to promote a paradigm that takes "hot earth" as its starting point.

93 SQUID is an acronym for Superconducting QUantum Interference Device.

94 Goksøyr 1996: 241–242.

Chapter 8 – New ways of organizing geophysical research
pp. 279–308

1 The University of Bergen also took part in the establishment of the University Centre in Svalbard (UNIS) in 1993, together with the universities of Oslo, Trondheim and Tromsø. In this chapter, however, we focus on the centers actually set up in Bergen. For a description and discussion of UNIS, see Misund and colleagues 2017: *A Norwegian pillar in Svalbard: The development of the University Centre in Svalbard (UNIS)*.

2 Gibbons and colleagues 1994, Forman 2012, Forland 1996, Helsvig 2011, Benum 2007.

3 The complete list of participants also includes the Geological Institute in Bergen, the River and Port Laboratory in Trondheim, the Norwegian Sea Mapping Authority in Stavanger, the Norwegian Oceanographic Data Centre in Bergen, and two Biological Stations located in Trondheim and Flødeviken, respectively. Den norske kyststrøm 1976: 2.

4 Mork was employed at GFI in 1965, and served at head of the Department of Oceanography from 1973. He became head of the Geophysical Institute in 1974. Geophysical Institute. Annual Report 1974: 135.

5 JONSDAP is an acronym for Joint North Sea Data Acquisition Project. The other participants were Belgium, the Netherlands, Sweden, Denmark, and France. Norwegian oceanographers contributed by taking hydrographic measurements along the northern boundary of the area investigated, a section that stretched from the Norwegian coast to the Orkney Islands. They also anchored buoys in the same area and collected chemical as well as biological data. Den norske kyststrøm 1977. See also Rozwadowski 2002: 326.

6 Weart 2008: 104.

7 Geophysical Institute. Annual Report 1971. Wiin-Nielsen only spent two years at GFI. In 1973, he became the first director of the newly established, intergovernmental

European Centre for Medium-Range Weather Forecasts (ECMWF), located in Reading, UK. This institute has maintained its world leadership in weather forecasting ever since. In 1979, Wiin-Nielsen was appointed Secretary General of the World Meteorological Organization (WMO).

8   Geophysical Institute. Annual Reports 1976 and 1977.

9   Geophysical Institute. Annual Report 1976: 151. See also Hovland 2007: 120.

10   Interview with Arne Foldvik May 5, 2016.

11   The project conducted for Statoil lasted for about 16 months. Geophysical Institute. Annual Reports 1976 and 1978.

12   Gibbons et al. 1994: vii, 1–8.

13   Gibbons et al. 1994: 13.

14   The massification of education was the outcome of a variety of factors, including the democratization of politics and society after WWII, the growth of a public sector requiring more white-collar workers and university graduates, and a growing industrial economy employing more highly skilled and educated workers. Gibbons et al. 1994: 73.

15   Gibbons et al. 1994: 49.

16   Forman 2012: 76–79. The other values and beliefs identified by Forman were disciplinarity, proceduralism, autonomy, and solidarity. In a strict sense, *proceduralism* referred to trust in the efficacy of scientific method. In a broader sense, it referred to a faith in the axiom that the means sanctify the ends. In postmodernity, he argued, even lying is generally considered acceptable if it is a means toward a legitimate end. (72–76) The modern valuation of *autonomy* for human individuals as well as for corporations could be traced back to the Protestant movement's demands for the individual's right to an unmediated relationship to God. The valuation of corporate autonomy had been a particularly crucial precondition for the establishment of autonomous scientific institutions during modernity. The Romantic revolution in the late 18th century contributed another kind

of legitimation for the autonomy of scientific institutions, as it provided a new and deeper foundation for "pure, disinterested, for-its-own-sake, autonomous science." Autonomy for individual scientists in their work life, as well the autonomy of the scientific institutions, was rejected in postmodernity. Due to a widespread suspicion of institutions claiming to serve "the public interest" in a disinterested way, it is now more common to support and defend the autonomy of private and interested corporations. The final value abandoned in postmodernity was *solidarity*. The willingness of individuals to undertake "substantially self-denying, even self-sacrificing, behaviors, with the intent of benefiting either [a] collective [or] individual persons" was a sentiment of crucial importance to the maintenance of scientific disciplines during modernity. Without solidarity, trust in science and regard for disciplinarity would have been impossible (84–90).

17   Forman 2012: 59. The word "fall" is inserted in brackets because Forman in fact used the term "rise" here. But for the phrase to make any sense he must have meant fall, and not rise.

18   Forman 2012: 91.

19   Forland 1996: 451–452, 500.

20   Forland also indicated that a third factor made a difference. Norwegian politicians at the time tended to be interested in the fate of the institutions of higher learning to the extent that policies could benefit the region that they represented. For this reason, they prioritized setting up new colleges in rural areas at the expense of the established universities in Bergen and Oslo.

21   Forland 1996: 450–451, 460, 493. The concept of "legitimation crisis" seems to be borrowed from the sociologist Jürgen Habermas. Habermas published a book entitled *Legitimationsprobleme im Spätkapitalismus* in 1973, translated to English with the shorter title *Legitimation crisis* in 1976. Habermas 1976.

22 Helsvig 2011: 17.

23 Helsvig 2011: 18. See also Gibbons et al. 1994: 158.

24 In spite of the clear differences between these groups, and the critiques they directed against the universities, Helsvig interprets both of them as testimonies of an emerging crisis of trust and legitimacy for the universities in this period. Helsvig 2011: 17–18.

25 One example of a report discussed and circulated by OECD was sociologist Joseph Ben-David's *Fundamental research and the universities. Some comments on international differences.* Benum 2007: 555.

26 Benum 2007: 553.

27 Benum 2007: 555.

28 Benum 2007: 560.

29 St.meld. nr. 35. (1975–1976): Om forskningens organisering og finansiering (Report to the Storting no. 35 [1975–1976] on the organization and funding of research).

30 Benum 2007: 554–555.

31 Benum 2007: 551.

32 Our emphasis. Forland 1996: 452. In a similar way, Helsvig declares that "the most recent history of UiO is not least a history of the efforts to clarify identity, to create internal and external trust and *regain legitimacy* in Norwegian society. This history is the topic of this book." Our emphasis. Helsvig 2011: 9 and 20.

33 Forman 2012: 58.

34 Latour 2005.

35 St.meld. nr. 35 (1975–1976): 1.

36 St.meld. nr. 35 (1975–1976): 5.

37 St.meld. nr. 35 (1975–1976): 57–59.

38 Our translation. St.meld. nr. 35 (1975–1976): 56.

39 St.meld. nr. 35 (1975–1976): 25, 52.

40 Benum does reflect briefly on this question. He suggests that "[t]he new OECD discourse could support the government's and other research organs' demands of research, but research of 'societal relevance' could obviously be a result of the scientists' own engagement in societal questions. Here we may assume that subtle interactions occurred, of which we know little." Benum 2007: 572.This momentary open-mindedness and uncertainty regarding sources of influence does not conform well to the conclusion that the OECD discourse had "considerable consequences for science-policy thinking and for the framework conditions of science" in Norway.

41 Den norske kyststrøm 1976: 1.

42 Our translation. Den norske kyststrøm 1976: 4. This quote is an early instance among geophysicists in Bergen of what Forman has called a "discipline-disregarding discourse." Forman 2012: 57.

43 Den norske kyststrøm 1976: 5.

44 Den norske kyststrøm 1976: 7.

45 Den norske kyststrøm 1976: 5.

46 See for instance Helsvig 2011: 17–19, Forman 2012: 72–76, and Forland 1996: 450–451.

47 Interview with Ola M. Johannessen 27.03.2017.

48 Satellite data would also help develop numerical models to describe the new phenomena, and they would assist in the testing and improvement of such models. Finally, satellite measurements could bring forth new insights about the earth's internal structures. Johannessen 1985: 8, 9.

49 Johannessen 1985: 3, 8, 9.

50 Johannessen 1985: 5–6.

51 Johannessen 1985: 6.

52 Interview with Ola M. Johannessen, March 27, 2017.

53 Landmark became head of the space division of NTNF in 1977. In this position, he had contributed some of the funding required to realize the NORSEX project initiated in 1978. In 1993, he became head of the board of the Nansen Center. Interview with Ola M. Johannessen, March 27, 2017.

54 Forland 1996: 496, 505.

55 At a workshop convened on the topic of remote sensing technology in geophysics in 1969, the "basic concept" of satellite-based oceanography was generally accepted. For this reason, discussions

revolved around specific questions like what would be the appropriate choice of orbit for the satellites. Conway 2016: 132.

56 Conway 2016: 135.

57 Johannessen 1985: 6.

58 Our translation. NOU 1983: 60.

59 During the eight months of work, the committee conferred with various departments and research communities, made two study-trips to Sweden and the UK, and convened 11 committee meetings. NOU 1983: 3.

60 NOU 1983: 63.

61 Interview with Ola M. Johannessen, March 27, 2017.

62 NORSEX started up on the initiative of the Geophysical Institute, the Christian Michelsen Institute and the Norwegian Technical Science Research Council (NTNF). The other Norwegian institutions not mentioned above were the Tromsø Satellite Telemetry Station, the Norwegian Institute of Technology, the Norwegian Hydrographic Office and the IBM center in Oslo. Den norske kyststrøm 1979: 1–2.

63 In the NORSEX-project, technical questions about how signals from satellites could be translated into reliable data to the oceanographers appears to have been more crucial than really gaining new insights about the sea. Den norske kyststrøm 1979: I. See also Johannessen 1985: 4.

64 See NORSEX group 1983.

65 See Johannessen et al. 1992.

66 Forland 1996: 501.

67 As the events and activities that we have added to the list actually left behind visible traces in Johannessen's application to the Dean of the Faculty of Mathematics and Natural Sciences, we may conclude that the connections between these events and the subsequent emergence of the new institution are better documented. We have not speculated about people's "mental states."

68 The Nansen Center has received some basic funding from the government since 2012. Earlier, the center depended solely on external funding. Nansen Environmen-

tal and Remote Sensing Center. Annual Report 2012: 4.

69 Interview with Ola M. Johannessen, March 27, 2017.

70 Nansen Environmental and Remote Sensing Center. Annual Report 1991: 9–23.

71 Nansen Environmental and Remote Sensing Center. Annual Report 1991.

72 Nansen Environmental and Remote Sensing Center. Annual Report 1992.

73 Nansen Environmental and Remote Sensing Center. Annual Report 1993.

74 Nansen Environmental and Remote Sensing Center. Annual Reports 1996 and 1997.

75 Nansen Environmental and Remote Sensing Center. Homepage. https://www.nersc.no/about/nansen-group. Accessed February 12, 2018.

76 Interview with Ola M. Johannessen, March 27, 2017.

77 Interview with Helge Drange, October 27, 2016.

78 Interview with Helge Drange, October 27, 2016.

79 Bjerknes Centre for Climate Research. CoE-application to the Research Council of Norway.

80 Bjerknes Centre for Climate Research. Annual Report 2003: 2.

81 Bjerknes Centre for Climate Research. CoE application to the Research Council of Norway: Front page.

82 Bjerknes Centre for Climate Research. CoE application to the Research Council of Norway: 1.

83 Bjerknes Centre for Climate Research. CoE application to the Research Council of Norway: 2.

84 Bjerknes Centre for Climate Research. CoE application to the Research Council of Norway: 3–5.

85 St.meld. nr. 39 (1998–1999): Forskning ved et veiskille.

86 St.meld. nr. 39 (1998–1999): 133.

87 The meeting was organized by the World Meteorological Organization (WMO), the UN Environment Programme (UNEP), and the International Council of Scientific

Unions (ICSU). Weart 2008: 146. See also Edwards 2013: 358 and 390.

88 World Commission on Environment and Development 1987: 168.

89 Edwards 2013: 396.

90 St.meld. nr. 46 (1988–1989).

91 Den interdepartementale klimagruppen 1991: 23.

92 In 1990, scientists affiliated with the Nansen Remote Sensing Center wrote a report to the Interdepartmental Climate Group. The report explained in detail why oceanography was indispensable to climate studies. We will investigate the content of this report in chapter 9.

93 Norges forskningsråd 2008. See also Nilsen and Vollset 2016: 402.

94 Interviews with Helge Drange, October 27 2016 and Nils Gunnar Kvamstø, March 29, 2017.

95 Bjerknes Centre for Climate Research. Annual Report 2007: 3.

96 Forman 2012: 57.

97 Gibbons et al. 1994: 18, 152.

98 Interview with Ola M. Johannessen, March 27, 2017.

99 På Høyden, March 6, 2017: Han gjorde Bergen til hovudstad for klimaforsking (He made Bergen a capital of climate research).

Chapter 9 – Geophysics in the age of climate change
pp. 309–334

1   In 2006, an international evaluation committee established that the Bjerknes Centre for Climate Research was about to become "one of the leading centres worldwide" in climate research. Bjerknes Centre for Climate Research. Annual Report 2006: 3.

2   Three feature stories in Norwegian newspapers can serve as examples: Bergens Tidende, August 19, 1997: Klima og alvor (Climate and concerns), by Sigbjørn Grønås; Bergens Tidende, July 22, 1997: Klimausikkerhet og klimadebatt (Climate uncertainty and climate debate), by Eystein Jansen; and Aftenposten, February 17,

1994: Kan drivhuseffekten druknes? (Can the greenhouse effect be drowned?), by Peter M. Haugan and Ola M. Johannessen.

3   Nilsen and Vollset 2016: 400.

4   Aftenposten, May 30, 1984: Norges rolle i klimaforskningen (Norway's role in climate research).

5   The title of the lecture was "Ett studium av kolcykeln och hur lufthavets koldioxidhalt kan komma att öka på grund av framtida användning av fossila bränslen" (A study of the Carbon Cycle and how the CO_2-level of the atmosphere may come to increase due to future burning of fossil fuels). Geophysical Institute. Annual Report 1983: 137. Bert Bolin would later become the first leader of the UN Intergovernmental Panel on Climate Change (IPCC).

6   Nilsen and Vollset 2016: 399.

7   Geophysical Institute. Annual Report 1988: 144.

8   Weart 2008: 146. See also Edwards 2013: 358 and 390.

9   World Commission on Environment and Development 1987: 174–177. Our Common Future contains some references to scientific work on the greenhouse effect and global warming. Among these are Bolin (ed.) 1986: *The greenhouse effect, climatic change and ecosystems*, and World Meteorological Organization: *A report of the international conference on the assessment of the role of carbon dioxide and of other greenhouse gases in climate variations and associated impacts*, Villach, Australia, October 9–15 1985. See World Commission on Environment and Development 1987: 202–203.

10 World Commission on Environment and Development 1987: 175.

11 World Commission on Environment and Development 1987: 176.

12 Borowy 2014: 165–172.

13 "The initial task for the IPCC as outlined in UN General Assembly Resolution 43/53 of 6 December 1988 was to prepare a comprehensive review and recommendations with respect to the state of knowledge of the

science of climate change; the social and economic impact of climate change, and possible response strategies and elements for inclusion in a possible future international convention on climate." IPCC. Homepage: https://www.ipcc.ch/organization/organization_history.shtml

14 St.meld. nr. 46 (1988–1989): 159.

15 The government's White Paper "elevated" global warming to an issue of somewhat greater importance and concern than the other issues addressed in *Our Common Future*. It was one of the first topics addressed in several chapters of the White Paper. St.meld. nr. 46 (1988–1989).

16 Den interdepartementale klimagruppen 1991: 23.

17 Geophysical Institute. Annual Report 1988: 144. We know that the report was written after March 1989, because the author discusses research activities conducted in February-March, 1989.

18 Geophysical Institute. Annual Report 1988: 144.

19 Midttun 1985: 1234. Friedman 2004: 136.

20 Historian Edgar Hovland has suggested that findings published by Foldvik on this topic around 1977 are one of the "great novelties" in the annals of Norwegian oceanography. Hovland (ed.): 130–131. In 1985, oceanographer Lars Midttun suggested that recent studies confirmed that the conclusions drawn by Nansen in 1906 were correct. Midttun 1985: 1234.

21 Friedman 2004: 134.

22 The 1988 Annual Report briefly refers to experiments conducted in February and March 1989, indicating that the report was completed after March 1989. Geophysical Institute. Annual Report 1988: 144.

23 NTB tekst, February 7, 1989: Drivhuseffekten utfordrer arktisforskere (The greenhouse effect challenges Arctic scientists).

24 Aftenposten, February 24, 1989: Vil vite hvorfor polisen smelter (Wants to know why the polar ice is melting).

25 Original title in Norwegian: Drivhuseffekten og klimautviklingen.

26 Braathen et al. 1990: 245.

27 Original title, in Norwegian: Havets innvirkning på det atmosfæriske CO_2-budsjettet og den globale klimautviklingen. Haugan et al. 1990: 3.

28 Haugan et al. 1990: 4. Fourteen years later, Johannessen and colleagues published a paper in which they proposed that the Arctic sea ice would disappear entirely during the summer if atmospheric CO_2 doubled. See Johannessen and colleagues 2004.

29 Haugan et al. 1990: 5, 20.

30 Haugan et al. 1990: 5, 62.

31 Haugan et al. 1990: 5, 90, 98.

32 Haugan et al. 1990: 5.

33 Haugan et al. 1990: 97.

34 Braathen et al. 1990: 245.

35 Haugan et al. 1990: 97.

36 Interview with Nils Gunnar Kvamstø, March 29, 2017.

37 Grønås and Lystad 2017: 137. Interview with Sigbjørn Grønås, February 5, 2018.

38 Interview with Nils Gunnar Kvamstø, March 29, 2017.

39 Interview with Nils Gunnar Kvamstø, March 29, 2017. MICOM (Miami Isopycnic Coordinate Ocean Model) was developed by a team at the Rosenstiel School of Marine and Atmospheric Science, University of Miami, Florida.

40 Interview with Nils Gunnar Kvamstø, March 29, 2017.

41 The institutions from Oslo were the Meteorological Institute, NILU and the Institute of Geosciences at the University of Oslo. Bjerknes Centre for Climate Research became a participant in 2002. Norges Forskningsråd 2008: 34.

42 Norges Forskningsråd 2008: 35. RegClim continued the work of a previous program by the Norwegian Research Council known as Climate and Ozone. Due to recommendations from an international evaluation of the former program, RegClim was designed as a large, nationally coordinated project to strengthen climate modeling. Nilsen and Vollset 2016: 402.

43 Among the members of this group were Mads Bentsen, Asgeir Sorteberg, Tore Furevik, and Frode Flatøy.

44 Furevik et al. 2003: 27. The title of the paper is "Description and evaluation of the Bergen Climate Model: ARPEGE coupled with MICOM."

45 Furevik et al. 2003: 48.

46 Bjerknes Centre for Climate Research. Annual Report 2003: 12.

47 Bjerknes Centre for Climate Research. Annual Report 2006: 6 and 16.

48 Bjerknes Centre for Climate Research. Annual Report 2007: 3.

49 Bjerknes Centre for Climate Research. Annual Report 2006: 3.

50 Bergens tidende. February 2 2007: Mye symbolpolitikk om klima (A lot of symbolic politics on climate).

51 Bjerknes Centre for Climate Research. Annual Report 2004: 16.

52 Bjerknes Centre for Climate Research. Annual Report 2006: 17.

53 Nobel foundation: https://www.nobelprize.org/nobel_prizes/peace/laureates/2007/press.html. Accessed August 30 2017.

54 Bjerknes Centre for Climate Research. Annual Report 2007: 3. The fact that the Nobel Peace Prize Committee chose to award the Peace Prize to IPCC and Al Gore indicates that Jansen's analysis was accurate.

55 NTB, June 25, 1991: 400 millioner i miljø-investeringer på Sleipner Vest (400 million NOK in environmental investments on Sleipner Vest).

56 Marchetti 1977. Subsequent studies are listed in Haugan and Drange 1992: 318 and 320.

57 Haugan and Drange 1992: 318.

58 Nansen Environmental and Remote Sensing Center. Annual Report 1991: 4.

59 Drange 1994.

60 Drange emphasized that their method was not an "artificial" storage of CO_2. It would merely accelerate a natural process. *Bergens Tidende*, June 4, 1992: Bergensk CO-sensasjon. The newspaper *Aftenposten*

covered the story the subsequent day. *Aftenposten*, June 5, 1992: Store oppslag ute: Forsker-suksess med CO_2 (Great attention abroad: Researcher success with CO_2).

61 In addition to Statoil, the oil company Norsk Hydro and the Norwegian Pollution Control Authority sponsored the investigations. Aftenposten, June 5, 1992: Store oppslag ute: Forsker-suksess med CO_2.

62 Drange discussed the topic thoroughly in his PhD thesis in 1994, and publication overviews for these years contain several titles relating to this research activity. These publication overviews can be found in the Nansen Center's Annual Reports.

63 Haugan and Drange 1996: 1022.

64 Bergens Tidende, June 6, 1992: Bergensk CO_2-løsning diskuteres av IEA (Bergen CO_2 solution discussed by IEA).

65 EC Environment and Climate Programme 2018.

66 For a presentation of the project and an analysis of the process leading to the dismissal of application for permissions, see de Figueiredo 2000.

67 "Postponed indefinitely" is our translation of "utsatt på ubestemt tid." Bergens tidende, September 26, 2001: Lar drivhuseffekten seg drukne? (Can the greenhouse effect be drowned?) Feature article by Guttorm Alendal and Helge Drange.

68 Bergens tidende, June 26, 2002. Griper inn mot klimaforsøk. Aftenposten, July 8, 2002. Debate paper by Bjørn Skaar, political advisor at the Ministry of Environment.

69 Aftenposten, July 3, 2002: Miljøvernminister Børge Brendes utspill forundrer. Peter Mosby Haugen elaborated more thoroughly on the topic in a paper presented at an international conference on greenhouse gas control technologies in Kyoto in October 2002. See Haugan 2003.

70 Aftenposten, July 8, 2002: Forhastet om CO_2-prosjekt (Hasty conclusions about CO_2 project). Debate paper by Bjørn Skaar.

71 NTB, July 9, 2002 SFT tillater CO_2-deponering i Norskehavet. See also Haugan 2003.

72 Bergens tidende, August 23, 2002. Brende stanser CO_2-dumping (Brende halts CO_2 dumping).

73 Aftenposten, September 16, 2002: Hav-forskningen rammes av Brendes politikk (Marine research hampered by Brende's policies).

74 Aftenposten, March 3, 2003: Brende må støtte CO_2-forsøk i havet (Brende has to support CO_2 experiments in the ocean).

75 Personal correspondence with Peter M. Haugan, October 31, 2016. Results from the experiment are discussed inter alia in Hove and Haugan 2005.

76 IPCC (2005): IPCC Special Report on Carbon Dioxide Capture and Storage.

77 FME is an abbreviation for Forsknings-sentre for miljøvennlig energi. Among the industrial partners in the consortium were Agder Energi AS, Aker MH AS, Lyse Produksjon AS, National Oilwell Norway AS, NorWind AS, Origo Engineering AS, Statkraft Development AS, and Vestavind Offshore. NORCOWE. Annual Report 2009:5.

78 NORCOWE. Final Report. May, 2017: 14–15.

79 NORCOWE. Final Report. May, 2017: 14–15.

80 NORCOWE. Annual Report 2014: 6.

81 NORCOWE. Final Report. May, 2017: 22. See also NORCOWE. Annual Report 2014: 6–11.

82 NORCOWE. Annual Report 2011: 35.

83 See for instance NORCOWE. Annual Report 2011: 35, and Annual Report 2012: 16–17.

84 NORCOWE. Annual Report 2015: 4, and Final Report. May, 2017: 44–46.

85 NORCOWE. Final Report. May, 2017: 36.

86 Bjerknes centre for climate research. Annual Report 2003: 2.

87 Bjerknes centre for climate research. Annual Report 2010: 3.

88 Bjerknes Centre for Climate Research. Annual Report 2016: 3.

89 Weart (2008): 200–201.

90 "During the 1960s and the 1970s, when [science and technology studies] was emerging as an interdisciplinary social science, we attacked a technocratic elitism that elevated science above other ways of knowing and seemed to place scientists beyond the reach of moral values and democratic ideals." Edwards 2010/2013: 436.

91 Edwards 2010/2013: 437–439.

92 UN Homepage: http://www.un.org/en/sections/what-we-do/promote-sustainable-development. Accessed September 14, 2017.

93 Asdal 2011: 179–180, 192.

94 Asdal 2011: 210.

95 UN Homepage http://unfccc.int/bodies/body/6383/php/view/reports.php#c

Chapter 10 – Reflections
pp. 335–356

1 Bjerknes 1938: 49–62; Bjerknes 1997 [1938]: 1–17 (http://edepositireland.ie/handle/2262/70473); Bjerknes, Hesselberg, Godske and Spinnangr 1944.

2 Annual Report 1957–58; Pedersen et al. 1968: 36.

3 Bergeron, Devik and Godske. 1962. Only weeks after returning to the United States, Lorenz submitted the manuscript that laid the foundation for chaos theory. Lorenz 1963: 130–141.

4 Pedersen, Johansen and Eide 1968.

5 Shapiro 1999.

6 Hovland 2007.

7 Bjerknes et al. 1944: 7–19. An English translation by Ralph Jewell entitled "How the Bergen school came into existence" was published in Shapiro 1999.

8 Bjerknes 1997 [1938]: 4. The sentiment was echoed by meteorologist Lennart Bengtson in his opening address at a symposium celebrating the 75th anniversary of the Bergen school's cyclone model in 1994: "For all of us still reasonably young meteorologists, it is difficult to imagine what meteorology was like before the Bergen school. Weather prediction and meteorological research

before the time of World War I was at a low mark. The quality of weather prediction was inferior, the observing system miserable, the methods of weather forecasting were unimaginative, uninspiring, mechanistic and unphysical. The advent of the Bergen school changed this state of affairs in a fundamental way." Bengtson in Grønås and Shapiro 1994: 1.

9   Bjerknes 1997 [1938]: 8.

10 Bjerknes 1944: 7.

11 Friedman 1989: 18; Mills 2009: 99.

12 Bjerknes 1898: 1–35 (http://www.biodiversitylibrary.org/item/107113#page/231/mode/1up). For a presentation of the equations and a discussion on the significance of this paper, see: Thorpe, Volkert and Ziemianski 2003: 471–480 (http://journals.ametsoc.org/doi/pdf/10.1175/BAMS-84-4-471). For more on how the equations were developed over time, see: Nebeker 1995; Harper 2008.

13 Bjerknes 1997 [1938]: 4.

14 Bjerknes in Jewell 1994 [1943]: 2; See also: Grønås in Drange (ed.) 2005: 357–365.

15 Eliassen 1982: 1–11. (http://www.annualreviews.org/doi/pdf/10.1146/annurev.fl.14.010182.000245).

16 Bjerknes 1944, as translated by Ralph Jewell in "The integrity at work in the Bergen School of meteorology," Grønås and Shapiro 1994: 32.

17 Petterssen 1941: 217.

18 Petterssen also notes that in India, precise rain measurements were written down starting around 400 BC. Later, the India Meteorological Department has pushed its history even further, and now traces its roots back to philosophical writings on the nature of clouds and seasonal weather cycles made around 3000 BC. http://www.imd.gov.in/pages/about_history.php

19 Petterssen 1941: 218.

20 West 2013: 66–73.

21 Godske 1956: 20–22.

22 Petterssen 1941: 217.

23 Godske 1956: 32.

24 Godske 1956: 40–41.

25 Bergeron Bolin 1959: 440–474, quote p. 441.

26 Op. cit.

27 Most of the characteristics of a successful "research school" fit the Bergen school of meteorology: it had a charismatic leader with research reputation and institutional power, an informal setting and leadership style, and loyalty to a focused research program; it developed simple and rapidly exploitable techniques, and had access to publication outlets; its students published early under their own names, and they produced and "placed" a significant number of students. Shortcomings were dealt with directly, such as not having access to pools of potential recruits, or not initially having adequate financial support. For a discussion on research schools as an analytical tool in the history and sociology of science, see: Geison 1981: 20–40; Geison and Holmes 1993.

28 Bergeron 1957: 455.

29 This approach was given the term "quasi-geostrophic approximation," and Rossby's novelty was simplification, stabilizing the equations, and filtering out "noise." Notably, in the process, Bjerknes's equations were defined as "primitive equations." Despite being more advanced, they were simply too demanding when it came to computing power. Since science was understood as progress, having been developed prior to the simplified equations, it was only logical that the former must be "primitive." See: Charney 1948.

30 The last issue of "Geophysical publications" was published in 1985. The Norwegian Geophysical Association is still active, and is organized in branches for geodesy, seismology, glaciology, hydrology, meteorology, oceanography, and astrophysics. This more or less reflects the subdivisions established by the International Union of Geodesy and Geophysics in 1919, which also includes volcanology and chemistry of the earth's interior.

31 Between 2012 and 2016, the Geophysical Institute received 40–52 million NOK per

year in external funding, and the budget
for 2018 is 65.5 million NOK. Between
60 and 70 percent of this comes from the
Norwegian Research Council; the rest of
the external funding is split almost equally
between the European Research Council
(where the Institute has had a success rate
above 40 percent) and other sources.

32   Historian Sverker Sörlin is among those
who argue that nature and history are
increasingly becoming intertwined, and
that historians need to go beyond the history of humans and integrate the history of
nature into our work. Sörlin 2018: 7–23.

33   Dickson and Østerhus 2007: 56–75.

34   Johannessen et al. 2004.

35   Stroeve et al. 2012; Maslowski et al. 2012.
For an overview, see: Meleshko and Pavlova 2018.

36   Meleshko and Pavlova 2018.

37   Johannessen et al. 2016.

38   Helland-Hansen and Nansen 1909: 324.

39   Foldvik, Gammelsrød and Tørresen in
Jacobs 1985: 5–20; Foldvik, Gammelsrød
and Tørresen 1985a: 177–193; Foldvik,
Gammelsrød and Tørresen 1985b: 195–207;
Foldvik 2004.

40   Johannessen 2005: 251–272.

41   Section C (1928–1988) is an exception,
consistently having a larger proportion of
coauthorship and a higher percentage of
publications in international journals than
the oceanographers and meteorologists.

42   Arrhenius 1896: 237–276.

43   Sylte 2017: 3, 18–19. In comparison, the
Geophysical Institute has 74 researchers
and a technical and administrative staff of
18, from a total of 20 countries. The Nansen
Environmental and Remote Sensing
Center has a total staff of 76 people of 26
nationalities. The latter has also spawned
centers in Russia, India, South Africa, and
Bangladesh, as well as an office on Svalbard.

Index

Graphic production: John Grieg, Bergen

ISBN: 978–82–450–2197–4

Cover design by Modest [Rune Døli]
Cover Photo: UiB, Special Collections.
Typeset by Modest [Rune Døli]
Typeset with GT Super

Enquiries about this text can be directed to:
Fagbokforlaget
Kanalveien 51
5068 Bergen
Tel.: 55 38 88 00
Fax: 55 38 88 01
email: fagbokforlaget@fagbokforlaget.no
www.fagbokforlaget.no